FOREST FIRES AND REGIONAL HAZE IN SOUTHEAST ASIA

FOREST FIRES AND REGIONAL HAZE IN SOUTHEAST ASIA

PETER EATON
AND
MIROSLAV RADOJEVIĆ
(EDITORS)

Nova Science Publishers, Inc.
Huntington, New York

Senior Editors: Susan Boriotti and Donna Dennis
Office Manager: Annette Hellinger
Graphics: Wanda Serrano
Information Editor: Tatiana Shohov
Book Production: Cathy DeGregory, Helen Lein, Jennifer Vogt and Lynette Van Helden
Circulation: Ave Maria Gonzalez, Ron Hedges, Andre Tillman

Library of Congress Cataloging-in-Publication Data
Available upon request

ISBN 1-56072-892-2

227 Main Street, Suite 100
Huntington, New York 11743
Tele. 631-424-6682 Fax 631-424-4666
e-mail: Novascience@earthlink.net
Web Site: http://www.nexusworld.com/nova

Printed in the United States of America

CONTENTS

Foreword

THE EMERGENCE OF CONCERN OVER FOREST FIRES

Peter Brimblecombe

School of Environmental Sciences, University of East Anglia, Norwich NR4 7TJ, UK

It is easy to forget with the current intensity of interest in forest fires and air pollution, that they have been studied for many decades. Paul Crutzen had long been convinced of the importance of forest fires for global atmospheric chemistry, and it was soon clear that they could perturb regional atmospheric ozone concentrations (Delaney *et al.*, 1985). Some of the earliest detailed studies were made in the Amazon in the 1980's. These added new information about the chemistry revealing the importance of some hitherto rather exotic oxidized organic compounds such as pyruvic and methacrylic acid (Andreae *et al.*, 1987, Jacob and Wofsy, 1988). A fascinating chemistry began to emerge.

In the mid 1980's I had the chance to work briefly at NASA Langley, and remember Joel Levine taking a special interest in forest fires in chapparal and wetlands (Lebel *et al.*, 1988, Cofer *et al.*, 1989). He was always trying to raise awareness of the magnitude of biomass burning as a global source. Much of this early work was published in interesting papers within the scientific literature, but even the most extensive fires seemed to provoke little attention. The great China forest fire in the Heilongjiang Province in 1987 was one of the largest to occur (Cahoon *et al.*, 1994), but failed to generate widespread public interest.

These studies took place at a time of heightened cold war fears. There was a view that nuclear war would produce large clouds of smoke that would cool the earth. The survivors of any war would face an awful fate in a *nuclear winter* where the Earth froze. Well-known figures such as Paul Crutzen and Carl Sagan drew public attention to this issue (e.g. Crutzen and Birks, 1982), although inevitably moral concerns complicated the scientific argument. Studies of forest fires seemed a way of studying the complex mechanisms involved in triggering a nuclear winter (Seitz, 1986). Robcock (1991) showed that smoke from biomass burning could yield a few degrees of daytime cooling on seasonal time scales.

Over time there was a reduction in public interest and a sense, with evaporating cold war tensions, that the early models may have been oversimplified (Madox, 1988). Nevertheless Turco and Sagan's fuller studies were not available until 1990 (Turco *et al.*, 1990). A re-interest in Nuclear Winter was triggered by the Gulf War (Sagan, 1991) of the 1990's which again stressed the environmental impact of large fires. The fog of war allowed for some exaggerated accounts of the magnitude of impact of the fires and likely numbers of deaths. However, scientific studies were valuable in addressing issues concerning large fires that were poorly understood. In particular, the lofting of plumes and buoyancy changes under solar irradiation seemed an important influence on plume behaviour (Herring and Hobbs, 1994), along with reduced light for photosynthesis and cooling beneath plumes. The effects on health were much more difficult to ascertain given the movement of populations and shifts in health care across the period. The enduring images from Kuwait of this time became visual ones. They ranged from the pictures of birds coated in oil to fire fighters working in scenes of fire lit blackness closer to Dante's inferno than an earthly desert.

In terms of environmental impacts it was apparent that cloud scavenging will reduce the atmospheric lifetimes of the particles (Hudson and Clarke, 1992). This is an important consideration in the plumes from intense fires. Observations of smoke particles from a large wildfire in Oregon suggested that ageing of particles had important effects on particle coagulation (Radke *et al.*, 1995). If coagulation was effective then the impact of smoke on climate change could be overestimated.

By the early 1990's the potential impacts of forest smoke in Southeast Asia were becoming more apparent (Lai and Tang, 1993). From 1991 my students drew my attention to newspaper articles that were increasingly common in Malaysia and Singapore. There had been earlier fires, with those of 1982-1983 (Nichol, 1999) especially significant in Indonesia, but little international attention had been paid to them. The fires of 1994 triggered some scientific work (e.g. Nichol, 1997) and increasing levels of haze in Southeast Asia raised concerns about the growth and survival of understory plants when Tang *et al.* (1996) showed a decrease of leaf carbon gain under haze conditions.

However, once again this seemed to be remote from world attention. As with all prominent environmental issues there seemed the need for a trigger to gain widespread public attention. Perhaps it is what journalists would call an angle. The fires of 1997 provided us with a range of interpretations that made excellent copy. Ships had collided and now a plane crashed. The looming crisis of the millennium bug gave a *fin-de-siecle* spectre to many issues. *Time Magazine* personified *El Niño* on its cover raising concerns that we were in for the worst weather of the century. The economies of Southeast Asia, after periods of unprecedented growth, suddenly took a turn for the worse with sharp loss of exchange rates. In the face of these transitions, the fires took on new meaning and could be linked to the uncertainty of an approaching millennium. The more cynical saw them as the outcome of industrial greed and the devastation of the rain forests for rapid monetary gain.

The development of the World Wide Web and its live weather-cameras gave an unprecedented freedom to view the effects of forest smoke from an increasingly wider range of sources. Not all sites were equally reliable. While the UK Department of Health carefully considered health impacts on soldiers stationed in Brunei, I remember friends being warned against travel to Asia, by well meaning, but misguided general practitioners. A saturated media had broadened our mental vision of the problem far beyond the geographical spread of smoke.

In the face of a huge demand for information the underlying science became engulfed by media attention. In Malaysia, the government required scientists not to comment on the smog, asking that they should defer to higher authorities before making public statements (Masood, 1997). While on the one hand wanting to avoid alarming their citizens, governments suddenly needed to provide information on matters that had previously attracted only modest attention. Across Southeast Asia reporting agencies used to offering only limited information on air quality were suddenly required to release regular reports.

Almost over night Air Pollution Indices (API) sprung into existence, with only the most meagre of background information (Swinbanks, 1997). Few of us were able to discover quite how these were determined or what relation they bore to health. It is not clear what effect this enhanced reporting had, but the API's seemed to be assembled from standards for urban particles established in other countries. Little attention was paid to the likely compositional or morphological differences between forest fire particles and those found in urban smogs, so it was difficult to know of the reliability of health forecasts. Given the clamour for action and public information, one has to sympathise with the pressures on government departments. However, the gaps in knowledge served to remind us that there had not been sufficient research to support policy decisions. In the long term it is important to modify the indices and approaches set up so hastily. It would seem that Southeast Asia requires its own air pollution standard that reflects the special situation faced by the region during times of forest smoke.

Although the atmospheric impact of forest fires and their chemistry had been studied for more than a decade, the situation in Southeast Asia had unique features that could not be neglected. The region is a heterogeneous archipelago, quite different from the continental nature of Africa or South America. Nevertheless, the spread of pollutants from fires is transnational and can even overwhelm more localised effects at great distances (eg. Wotawa and Trainer, 2000).

The source of the smoke and its impact occurs within a complex political landscape. Sustainable development, which balances short term national needs against the longer term protection of the environment, remains difficult to convert from textbooks into action. This is particularly true in Southeast Asia, which is seeing a period of great social change. The forest fires cannot be separated from the problems of a growing population, resettlement, and a desire for more democratic participation in government. There is relatively little experience in this region in creating environmental legislation, robust enough to withstand external political and commercial pressures. Existing legislation could help, but there are continued claims, hardly restricted to developing countries, that it is inadequately enforced. Indeed, governments within the region are clearly worried by this, e.g. the Indonesian government suggested to the UNFCCC, that: "One major source of CO_2 emission in the forestry sector comes from forest fire, caused by careless application of fire in land clearing. This careless application of fire, combined with the weak enforcement of forestry and land management policies, has caused many forest fires of a catastrophic size, notably in 1991, 1994, and 1997....".

The transboundary nature of the problem requires transnational co-operation for effective solutions. Acid rain and the ozone hole have shown us intergovernmental resolutions on pollutants that are transported over long distances. Models for increasing international co-operation in fire-fighting, environmental policy development and research already exist, but they require an inspired translation into the ASEAN context. Success here might mean that the forest fires could ultimately be seen as a catalyst for increased environmental awareness in the region.

While stressing the need for environmentally sound policies and adequate enforcement to reduce the incidence of the fires, there also remains a need for a better scientific and academic understanding. This book is an important contribution to our vision. The introduction places the studies in their broadest context, recognising the social and climatic drivers behind them. It is immediately clear that the reader will be invited to go well beyond mere science. Modelling, measurement, and inventory are key tools to investigate the problem and these are covered through the middle sections of the book. The later chapters take this academic understanding to the human and political interface by looking at the interactions with health, economics, policy and the media. This book demands that we view the forest fires from an interdisciplinary context. Of course we all know that this is necessary, but it will help us to broaden our perspectives.

Within environmental science the fires provide the opportunity to study questions of broad interest. Taking a somewhat narrower focus as an atmospheric scientist, I note that the chemistry of the aerosol dicarboxylic acids (Narukawa *et al.*, 1999) and PAHs (Cecinato *et al.*, 1997) has special significance in the atmosphere. Scavenging processes (Tabazadeh *et al.*, 1998) and the pH balance (Balasubramanian *et al.*, 1999) in forest smoke offer valuable insights into our understanding of aerosol chemistry, while on a global scale the importance of emissions from biomass burning (Galanter *et al.*, 2000) and changes in trace gas concentrations (Rinsland, 1999) can be studied in this area of research. This creates the impetus for the development of strong regional research programs with relevance to both policy and wider questions within environmental science. We already see the emergence of groups in Southeast Asia with a continuing interest in this subject. If satisfactory funding is available to foster this growing expertise, the future promises that these groups within the region will make significant and fundamental contributions to atmospheric science and more broadly to environmental science.

References

1. Andreae M.O., R.W. Talbot, and S.M. Li, 1987. Atmospheric measurements of pyruvic and formic-acid. *Journal of Geophysical Research* **92** (D6): 6635-6641.
2. Balasubramanian R., T.Victor, and R. Begum, 1999. Impact of biomass burning on rainwater acidity and composition in Singapore. *Journal of Geophysical Research* **104** (D21): 26881-26890.
3. Cahoon D.R., B.J. Stocks, J.S. Levine, W.R. Cofer, and J.M. Pierson, 1987. Satellite analysis of the severe 1987 forest-fires in northern China and southeastern Siberia. *Journal of Geophysical Research* **99** (D9):18627-18638.
4. Cecinato A., P. Ciccioli, E. Brancaleoni, A. Brachetti, M. Zagari, and P.D. Vasconcellos, 1997. PAH as candidate markers for biomass combustion in the Amazonian forest area. *Annali di Chimica* **87** (9-10): 555-569.
5. Cofer W.R., J.S. Levine, D.I. Sebacher, E.L. Winstead, P.J. Riggan, B.J. Stocks, J.A. Brass, V.G. Ambrosia, and P.J. Boston, 1989. Trace gas emissions from chaparral and boreal forest fires. *Journal of Geophysical Research* **94** (D2): 2255-2259.
6. Crutzen P.J. and J.W. Birks, 1982. The atmosphere after a nuclear-war - twilight at noon. *AMBIO* **11**, (2-3): 114-125.
7. Delaney A.C., P. Haagensen, S. Walters, A.F. Wartburg and P.J. Crutzen, 1985. Photochemically produced ozone in the emission from large-scale tropical vegetation fires. *Journal of Geophysical Research* **90** (D1): 2425-2429.
8. Galanter M, H. Levy and G.R. Carmichael, 2000. Impacts of biomass burning on tropospheric CO, NO_x, and O_3. *Journal of Geophysical Research* **105** (D5): 6633-6653.
9. Herring J.A. and P.V. Hobbs, 1994. Radiatively driven dynamics of the plume from 1991 Kuwait oil fires. *Journal of Geophysical Research* **99** (D9): 18809-18826.
10. Hudson J.G. and A.D. Clarke, 1992. Aerosol and cloud condensation nuclei measurements in the Kuwait plume. *Journal of Geophysical Research* **97** (D13): 14533-14536.
11. Jacob D.J. and S.C. Wofsy, 1988. Photochemistry of biogenic emissions over the Amazon forest. *Journal of Geophysical Research* **93** (D2): 1477-1486.
12. Lai Y.K. and S.M. Tang, 1993. A survey of air-quality in Singapore. *Nuclear Instruments and Methods in Physics Research, Section B- Beam Interactions with Materials and Atoms* **75** (1-4): 300- 303.
13. Lebel P.J., W.R. Cofer, J.S. Levine, S.A. Vay and P.D. Roberts, 1988. Nitric-acid and ammonia emissions from a mid-latitude prescribed wetlands fire. *Geophysical Research Letters* **15** (8): 792-795.
14. Maddox J. 1988. What happened to nuclear winter? *Nature* **333** (6170): p.203.
15. Masood E. 1997. Malaysia backs 'gag' on haze scientists. *Nature* **390** (6656) p.107.
16. Narukawa M., K. Kawamura, N. Takeuchi and T. Nakajima, 1999. Distribution of dicarboxylic acids and carbon isotopic compositions in aerosols from 1997 Indonesian forest fires. *Geophysical Research Letters* **26** (20): 3101-3104.

17. Nichol J. 1997. Bioclimatic impacts of the 1994 smoke haze event in Southeast Asia. *Atmospheric Environment* **31** (8): 1209-1219.
18. Nichol J.E. 1999. The global forest fire issue: savannization of tropical humid forests. *Progress in Environmental Science* **1** (1): 97-103.
19. Radke L.F., A.S. A.S. Hegg, P.V. Hobbs and J.E. Penner, 1995. Effects of ageing on the smoke from a large forest-fire. *Atmospheric Research* **38** (1-4): 315-332.
20. Rinsland C.P., A. Goldman, F.J. Murcray, T.M. Stephen, N.S. Pougatchev, J. Fishman, S.J. David, R.D. Blatherwick, P.C. Novelli, N.B. Jones and B.J. Connor, 1999. Infrared solar spectroscopic measurements of free tropospheric CO, C_2H_6, and HCN above Mauna Loa, Hawaii: Seasonal variations and evidence for enhanced emissions from the Southeast Asian tropical fires of 1997-1998. *Journal of Geophysical Research* **104** (D15): 18667-18680.
21. Robock A. 1991. Surface cooling due to forest-fire smoke. *Journal of Geophysical Research* **96** (D11): 20869-20878.
22. Sagan C. 1991. Kuwaiti fires and nuclear winter. *Science* **254** (5037): p.1434.
23. Seitz R. 1986. Siberian fire as nuclear winter guide. *Nature* **323** (6084): 116-117.
24. Swinbanks D. 1997. Forest fires cause pollution crisis in Asia. *Nature* **389** (6649) p.321.
25. Tabazadeh A., M.Z. Jacobson, H.B. Singh, O.B. Toon, J.S. Lin, R.B. Chatfield, A.N. Thakur, R.W. Talbot and J.E. Dibb, 1998. Nitric acid scavenging by mineral and biomass burning aerosols. *Geophysical Research Letters* **25** (22): 4185-4188.
26. Talbot R.W., M.O. Andreae, T.W. Andreae and R.C. Harriss, 1988. Regional aerosol chemistry of the Amazon basin during the dry season. *Journal of Geophysical Research* **93** (D2): 1499-1508
27. Tang Y.H., K. Naoki, F. Akio and M. Awang, 1996. Light reduction by regional haze and its effect on simulated leaf photosynthesis in a tropical forest of Malaysia. *Forest Ecology and Management* **89** (1-3): 205-211.
28. Turco R.P., O.B. Toon, T.P. Ackerman, J.B. Pollack and C. Sagan, 1984. Nuclear winter to be taken seriously. *Nature* **311** (5984): 307-308.
29. Turco R.P., O.B. Toon, T.P. Ackerman, J.B. Pollack and C. Sagan, 1990. Climate and Smoke - An Appraisal of Nuclear Winter. *Science* **247** (4939): 166-176.
30. Wotawa G. and M. Trainer, 2000. The influence of Canadian forest fires on pollutant concentrations in the United States. *Science* **288**: 324-328.

Chapter 1

INTRODUCTION

Miroslav Radojević
Department of Chemistry, University of Brunei Darussalam, Tungku Link BE 1410, Brunei Darussalam

> "The bay itself is surrounded by vast forests, and not long ago a steamer was prevented from entering the river for three days, in consequence of a fierce forest fire, the dense volumes of smoke from which completely obscured the entrance."
> ***Lady Annie Brassey, "The Last Voyage (1886- 87)"***

The above quotation from a 19th century travel book (Brassey, 1889) is a direct reference to Brunei Bay and Brunei River, and it demonstrates that forest fires and the associated smoke pollution are not new to Borneo island. Nevertheless, the haze resulting from biomass burning in Borneo, Sumatra, and other parts of Southeast Asia has been occurring with increasing frequency, severity, extent, and duration over the last 20 years. The regional haze in Southeast Asia can extend over an area of several million km^2 and create a major air pollution disaster on an unprecedented scale (Radojevic, 1997). The haze can affect several countries (Indonesia, Malaysia, Singapore, Brunei and the southern parts of the Philippines and Thailand) for weeks or months at a time and it can impact on the lives of millions.

Although forest fires have been studied extensively in other parts of the world none of the reported studies has considered the health implications of emitted pollutants (Radojevic, 1998). Relatively comprehensive studies of tropical rainforest fires have been carried out in Africa and the Amazon region of Brazil (Crutzen and Goldammer, 1993) but these have generally been confined to sparsely populated areas. Also, most of the chemical components determined in these studies are of little significance for assessing many of the impacts of haze, especially the health effects. The main objective of past forest fire studies was to quantify the flux to the atmosphere of various trace gases which could serve as input parameters in models of biogeochemical cycles and global warming. In view of the potential effects of the haze on the health of large populations and other deleterious environmental impacts of forest fires and haze, there is a need to review the available information regarding the haze in Southeast Asia.

PREHISTORIC FIRES

Although forest fires in Southeast Asia have been extensively recorded since the 19th century (see Chapter 3), there is evidence of fires dating back to prehistoric times, even before the appearance of the first humans. Fossil records suggest the occurrence of biomass fires at many locations around the world throughout the Cenozoic (last 60 million years), although the pre-Quaternary record is incomplete (Clark and Robinson, 1993). The presence of charcoal in coal and rocks rich in land-plant fossils has been taken as evidence of ancient fires, and it is claimed that early fires may have influenced atmospheric composition, climate, and terrestrial ecosystems ever since forests first appeared on the earth in the Devonian, some 400 million years ago (Clark and Robinson, 1993). Climatic conditions are also believed to have played a role in these early fires. Records of past biomass fires have also been derived from the presence of charcoal in swamp, lake, and marine sediments, and in soils.

Ancient forest fires in the tropics are poorly understood in comparison to biomass fires in other ecosystems, however, the presence of charcoal in soils indicates that fires also burned in Borneo, Amazonia, and Africa. Sites of biomass fires as old as 20-50,000 years have been found in Southeast Asia (Kershaw, 1995). Goldammer and Seibert (1990) report evidence of fires occurring between 17,510 and *ca.* 350 years ago in eastern Borneo.

There has been considerable debate over the extent to which natural causes contributed to past forest fires. Although there is evidence of natural causes for fires elsewhere in the world, an anthropogenic cause has been inferred for past fires in Southeast Asia (Kershaw, 1995). Lightning is considered to be an important natural cause of fires in some regions (e.g. savannah) prior to human settlements, however, it is uncertain whether fires occurred in tropical rainforests prior to the presence of humans. Nevertheless, the contribution of natural causes to recent tropical forest fires has become negligible (Goldammer, 1993). Most forest fires in the tropics are set intentionally by human settlers, and past forest fires in Southeast Asia have been attributed almost exclusively to anthropogenic causes (Kershaw, 1995). Evidence of human occupation in east Borneo extends back 1.3 million years, and there is indication of biomass fires during the late Pleistocene (Kershaw, 1995). These early fires in Borneo appear to have been anthropogenic in origin and facilitated by dry weather conditions. According to Kershaw (1995), the increase in tropical forest fires over the last 2,000 years is probably due to an expansion and intensification of agriculture in the region. Traditional anthropogenic causes of past forest fires include: deforestation (conversion of forest to agricultural land), slash-and-burn and other agricultural practices, accidental fires (from fires used for cooking, lighting, etc.), and other traditional uses (communication, religious ceremonies, warfare, etc.).

INCIDENCE OF FOREST FIRES AND HAZE SINCE 1980

Episodes of forest fires and haze in Southeast Asia from the 19th century up to, and including, the 1972 event are reviewed in Chapter 3. Since 1980, regional haze episodes in Southeast Asia have been recorded during the following periods: August-September 1982, September 1983, September 1987, August 1990, August-September 1991, August-October 1994, August-October 1997 and February-May 1998. Industrial pollution and volcanic eruptions, such as Mt. Pinatubo in 1991, have contributed to some of the haze episodes, but forest fires started deliberately by developers clearing land for plantations and settlements, and farmers engaged in traditional "slash and burn" practices have been responsible for most of the forest fires in Southeast Asia. The August 1990 haze affected mainly the Klang Valley in Malaysia and it was due primarily to industrial and vehicular sources of air pollution. Other haze episodes have been mainly the result of forest fires on the islands of Sumatra and Borneo. Many of the fires are in areas of peat swamps and underground coal deposits which may smoulder throughout the year, re-igniting vegetation during the dry season. The episodes of 1994, 1997 and 1998 have been the most severe recent events and the extent of the haze during these episodes is illustrated in Figure 1. Also shown is the location of major forest fires responsible for the haze.

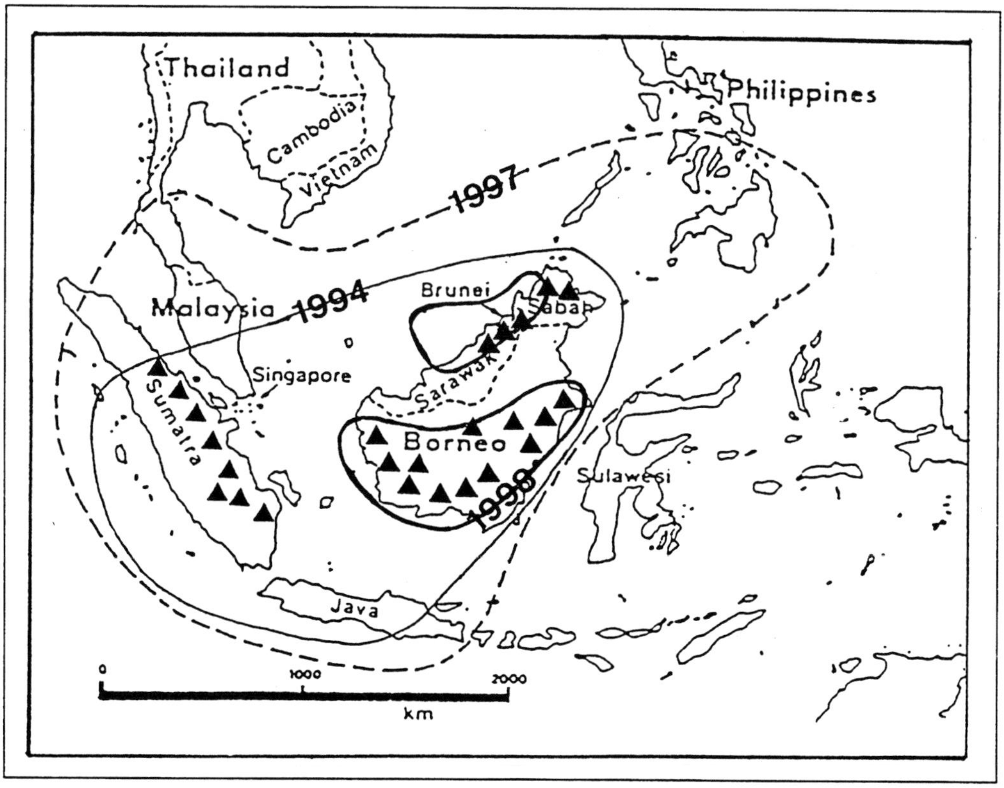

Figure 1 Approximate location of forest fire hot-spots and area affected by regional haze in Southeast Asia during August-October 1994, July-October 1997, and February-May 1998.

The haze is compounded by meteorological and climatic factors. Most of the haze episodes (with the exception of the February-April 1998 episode) occurred during the prevailing southwesterly monsoon winds (May-September) when the atmosphere is relatively stable and the weather is generally dry. Furthermore, the warming effect of the El Niño Southern Oscillation (ENSO) in the Pacific Ocean causes a longer dry season and draught in the region during certain years and severe haze episodes coincide with the El Niño years. The recent draught in Indonesia during 1997 was the worst in 50 years. Local meteorological factors such as the presence of a low-lying inversion and low wind speeds restrict the dispersion and dilution of pollution trapping the smoke close to the ground. Land and sea breezes in coastal areas can also affect the transport of pollutants as well as the extent and severity of the haze. Meteorological factors are also important in clearing the haze. The haze normally clears in October when easterly winds prevail over the region and rainfall increases. The onset of heavy showers in October clears the haze from most of the region.

EFFECTS OF FIRES AND HAZE

The harmful effects of regional haze episodes have caused considerable concern to the government authorities, experts, and the public throughout Southeast Asia. Effects of the haze include the following:

- *Heath Impacts*. The effects of pollutants presents in the haze on public health are a major concern throughout the affected region. Observed symptoms and illnesses include upper respiratory tract infections, asthma, bronchitis, eye and throat irritation, coughing, breathlessness, blocked and runny noses, skin rashes, cardiovascular disorders, and psychological effects. Hospital admissions increased with greater particle concentrations throughout the affected region. Several deaths have been reported due to smoke inhalation in Indonesia, however, impacts on mortality statistics in affected areas have yet to be investigated.
- *Visibility reduction*. The haze greatly reduces visibility, sometimes to less than 100 metres. This can increase the incidence of transportation accidents; the 1997 haze has been implicated in several shipping accidents and one passenger airline accident in the region. Many airports were closed during severe haze and there were numerous flight cancellations throughout the region as a result of low visibility.
- *Loss of biodiversity*. The physical destruction of the forests by fire can threaten many plant and animal species. For example, many orangutans have died in the Borneo fires and their ecosystems have been destroyed by the fires. Many valuable plant species have also been threatened by the fires.
- *Lower productivity*. The haze drastically reduces productivity, both industrial and agricultural, in affected areas. Many offices were closed during the most severe haze condition and industrial activity was severely restricted. Many employees whose health was affected stayed at home, and travel and transport of goods were limited. Destruction

of crops by fires and the difficulty of working outdoors during severe haze lowered agricultural productivity.

• *Downturn in tourism.* The region affected by the haze is a favourite tourist destination, and tourism, both local and from outside the region (e.g. from Europe, America, China, Japan and Australia) is an important source of revenue for many of the countries in Southeast Asia. The extensive coverage of the haze in the international media turned many potential visitors away from the region, and many travel agents and governments advised tourists to avoid visiting areas affected by the haze. This led to a major reduction in tourism to some countries, notably Malaysia, Singapore and Indonesia during late 1997 and early 1998.

• *Economic costs.* All of the aforementioned impacts have economic costs associated with them and these can have major impacts on the economies of affected countries. It has been estimated that forest fires and haze in 1997 had a total cost of US$ 4.5 billion.

• *Weather and climate.* Many of the pollutants emitted by forest fires could have significant impacts on weather and climate. Pollutant gases such as CO_2 and CH_4 contribute to global warming, while smoke particles could affect the earth's radiation balance by scattering and absorbing solar radiation. Furthermore, smoke particles could act as cloud condensation nuclei (CCN) and affect cloud formation and precipitation patterns.

• *Impacts on atmospheric chemistry.* Primary gaseous pollutants emitted by forest fires (e.g. CO, NO, organic compounds) can react in the atmosphere to produce O_3 and other harmful secondary pollutants. Smoke particles can act as a medium for surface catalysed reactions, and they can also reduce the intensity of solar radiation and photochemical activity over areas affected by haze.

• *Rainwater acidity and chemistry.* Forest fires emit acidifying substances (e.g. SO_2, NO_x, organic acids) which contribute to rainfall acidity, however, alkaline substances are also emitted with the smoke particles (e.g. Ca, K) and these could act to neutralise the acidity.

• *Global biogeochemical cycles.* Forest fires can be an important source of many atmospheric constituents (CO, NO, H_2, CH_3Cl, CHN, etc.) and they affect global biogeochemical cycles such as the carbon cycle.

• *International relations.* Since the regional haze is caused by the trans-boundary transport of air pollutants from source areas, there are repercussions on diplomatic relations between nations in Southeast Asia. As these states are members of ASEAN, the haze phenomenon has been widely debated at intergovernmental level, and several international initiatives have been taken by ASEAN to combat and monitor forest fires and haze.

• *Lifestyle.* The restrictions imposed on everyday recreational and work-related activities have a significant impact on the lifestyle of affected populations. Closures of schools and offices, long periods spent indoors, restrictions on travel, greatly reduced visibility, the unpleasant odour of the haze, the health effects, soiling of materials (clothing, vehicles, etc.) and other aspects of the haze had a major social impact in affected areas. Many multinational companies and diplomatic missions evacuated family members of employees and non-essential staff.

HAZE MONITORING AND RESEARCH

While forest fires and resultant air pollution in Southeast Asia affect one of the most densely populated regions of the world, until recently, there has been very little concerted effort to understand the extent of air pollution or its impacts on the population. The scientific community in the region has been slow to react to this event, partly because of absence of co-ordination and partly because of the political "sensitivity" of the issue. The consequence was that monitoring efforts of haze were left to government agencies, which managed the severe air pollution episode just as a heightened level of conventional urban air pollution, using the same indicators (i.e. criteria pollutants).

Most of what is known about earlier haze episodes has come from visual and subjective observations of people residing in the region, anecdotal evidence, reports in the media, satellite images, etc. The few reported studies of earlier haze episodes have been of a general and qualitative nature, and more comprehensive studies have emerged only during the 1997 and 1998 episodes. Only measurements of suspended particles and some meteorological parameters are available at some sites in Malaysia and Singapore during the 1987, 1990 and 1991 episodes. Measurements of criteria pollutants (CO, NO_2, O_3, SO_2, PM_{10}) are available for some sites in Malaysia and Singapore during the 1997 and 1998 episodes and for Brunei during the 1998 episode. Also, some studies have measured harmful organic pollutants (e.g. benzene, aldehydes, polynuclear aromatic hydrocarbons) during the 1997 and 1998 haze episodes. Unfortunately, there is a paucity of scientific studies and measurements in Indonesia, the country most affected by forest fires. While there is now a fairly good density of air quality monitoring stations in Singapore, Malaysia and Brunei, Indonesia does not yet have a satisfactory air quality monitoring network. Satellite monitoring of forest fire hot spots and haze has also improved significantly over the years. Information on haze is now available from several satellites (NOAA, TOMS, etc.) and it can readily be accessed via the Internet. Recently, computer models of the long-range transport of pollutants from Indonesian fires have been developed, however, there is a paucity of relevant laboratory (e.g. biomass combustion) studies. Also, during earlier haze episodes, there was little concern expressed about many of the harmful impacts. Haze impacts have only just started to be investigated, and many studies (e.g. epidemiological studies) have yet to be completed.

OUTLINE OF BOOK COVERAGE

This book summarises, for the first time, in a single volume the current state of knowledge regarding the various aspects of forest fires and regional haze in Southeast Asia. Previously published books relating to forest fires have generally focused on forest fires in Africa and South America, and on global issues such as biogeochemical cycles and global warming (Goldammer, 1990; Crutzen and Goldammer, 1993; Levine, 1991). Also, many of the important impacts of forest fires and haze (e.g. health effects) were not considered in these earlier publications. Although there have been several expert meetings on the haze during the last few years, and although much information is available from the Internet, and from papers and articles on the haze in newspapers, magazines, and scientific journals (e.g. Special Section on Indonesian Fires in *Geophysical Research Letters* **26**, 1999, pp. 2409-2424) there has not previously been a comprehensive book on forest fires and haze in Southeast Asia.

Early chapters of the book examine the relationship between climate, the El Niño effect and drought, and discuss the history of forest fires in the region. This is followed by a comprehensive chapter on the cause of the fires, and the use and misuse of burning in land management. Remote sensing played an important role in both identifying the location of fires and the spread of haze and this is discussed in the next chapter.

Other chapters examine the chemistry of forest fires and haze, the long-range transport of the pollutants and their modelling, the effects of the fires on forest ecosystems, the implications for global climate change, the health impacts, and the economic cost of the fires and haze. Final chapters discuss the role of the media in reporting the haze, government responses to the problem, and wider public policy issues. In the conclusion we discuss the state of existing knowledge on the Southeast Asian fire and haze phenomena and suggest areas for further scientific research and action.

The range of expertise needed for such a book means that no single author could be found to cover authoritatively all the material, and an edited volume was therefore produced. Chapter authors are internationally recognised experts in the various disciplines involved in understanding this complex phenomenon and they have produced a most valuable compilation of knowledge on forest fires and regional haze in Southeast Asia. The resulting book is a unique and authoritative treatment of this important environmental problem which is likely to become even more severe and widespread in the future, not only in Southeast Asia, but also elsewhere in the tropics. Although research into regional haze is still in its infancy, the present book makes an important contribution to understanding the multifarious aspects of this unique phenomenon.

REFERENCES

1. Brassey, A. 1889. *The Last Voyage (1886-87)*. Longman & Co., London.
2. Clark, J.S. and J. Robinson. 1993. Paleoecology of fire. In: *Fire in the Environment: the Ecological, Atmospheric and Climatic Importance of Vegetation Fires*. Eds. Crutzen P.J. and J.G. Goldammer, Wiley, Chichester. pp. 193-214.
3. Crutzen, P.J. and J.G. Goldammer (Eds.), 1993. *Fire in the Environment: the Ecological, Atmospheric and Climatic Importance of Vegetation Fires*. Wiley, Chichester.

4. Goldammer, J.G. (Ed.) 1990. *Fire in the Tropical Biota: Ecosystem Processes and Global Challenges*. Springer-Verlag, Berlin.
5. Goldammer, J.G. 1993. Historical biogeography of fire: tropical and subtropical. In: *Fire in the Environment: the Ecological, Atmospheric and Climatic Importance of Vegetation Fires*. Eds. Crutzen, P.J. and J.G. Goldammer, Wiley, Chichester. pp. 297-314.
6. Goldammer, J.G. and B. Seibert, 1990. The impact of droughts and forest fires on tropical lowland rain forest of East Kalimantan. In: *Fire in the Tropical Biota: Ecosystem Processes and Global Challenges*. Ed. Goldammer, J.G., Springer-Verlag, Berlin. pp. 11-31.
7. Kershaw, P.A. 1995. A history of biomass burning in the tropics: relative contributions of climate and people. In: *Proceedings of the International Symposium on Climate and Life in the Asia-Pacific (10-13 April 1995, University of Brunei Darussalam)*. Ed. Sirinanda, K.U., University of Brunei Darussalam. pp. 130-142.
8. Levine, J. S. (Ed.), 1991. *Global Biomass Burning: Atmospheric, Climatic, and Biospheric Implications*. The MIT Press, Cambridge, Massachusetts.
9. Radojevic, M. 1997. The haze in Southeast Asia. *Environmental Scientist* **6** (6): 1-3.
10. Radojevic, M. 1998. Burning issues. *Chemistry in Britain* **34** (12): 38-42.

Chapter 2

CLIMATE, EL NIÑO, DROUGHT AND FIRE

J. R. E. Harger
Highspan International Environmental Consultants, 36 Vanessa Crescent, Glendowie, Auckland, New Zealand

INTRODUCTION

The "El Niño-Southern Oscillation" (ENSO) consists of a movement of the waters along the equatorial region of the Pacific, associated with a so-called "Kelvin Wave" and coupled with atmospheric changes. This combination causes significant areas of the globe to experience extended droughts or alternatively, periods of heavy rainfall. Over the last 20-30 years or so, the frequency and intensity of this massive movement of water along the equator in the Pacific has increased. The development of dry conditions in Southeast Asia, particularly in Indonesia, has been emphasized since the 1982-1983 ENSO Warm Event. At that time, fires took hold of second-growth vegetation and some of these eventually penetrated into underlying coal seams in Kalimantan in 70-80 separate locations. These underground "hot-spots" continue to burn to date (late 1999) and are one source (along with land clearing, lightning-strikes etc.) of wild-fires that have characterized the ENSO Warm Events. Since 1982 ENSO warm events have promoted conditions permitting fires to break out in Kalimantan and Sumatra in 1982,1983, 1987, 1991, 1993, 1994, 1997, 1998. The worst fires occurred in 1982-1983, 1987, 1991, 1994 and 1997. The system appears to be moving in a more agitated manner as the Earth redistributes incoming heat throughout the depths of the oceans. In effect the response can be likened to that of a simple "heat pump". The following account is based on analysis of long-term temperature records from Indonesia (1866-1994).

The major features in development of the "El Niño - Southern Oscillation" (ENSO) involve oscillation of the Pacific Ocean and the related atmosphere in an unpredictable (chaotic) fashion. The system moves between extremes of so called "warm events", lasting one or two years, and "cold events". In the course of the "warm events", warm sea water moves from the western Pacific along the equator to impact on the west coast of the

American continent. Cold-events are associated with easterly trade winds and induced flows of colder water from the eastern Pacific towards the west. Historical data indicate that ENSO years as experienced by the Island of Java (Indonesia) are either much warmer than non-ENSO years or are only slightly, if at all, warmer than normal (non-ENSO) years. Along the equator in Southeast Asia, cooler wet years almost always follow hot-dry years and vice-versa, within the ENSO warm event cycle. This alternating pattern also extends to include the year immediately following the terminal year of an ENSO warm event set. The initial year of an ENSO warm event set may be either hot with a long dry season or relatively cool (nearer to the temperature of a non-ENSO year) and having a short dry season. For Indonesia, in the years between 1950 and 1999, there have been 10 separate ENSO warm events. The initial year for 7 (1951, 1957, 1963, 1972, 1982, 1991, 1997) has been hot and dry or neutral and three years have been cool and wet (1968, 1976, 1986).

Air Temperature and El Niño Activity

Air-temperature records from Jakarta (Pusat) 06 degrees 11 minutes south, 106 degrees 50 minutes east, population 8.8 million in 1991, exist since 1866 with missing years confined to 1943, 1946, 1947 and 1958. These have been combined ("pooled") with records from Semarang 07 degrees 00 minutes south, 110 degrees 25 minutes east, a smaller city (population 1.1 million in 1989) 450 km to the east, from the years 1982-1991 inclusive. The data include the Jakarta temperatures from 1866-1996. Figure 1 shows the secular trend of the mean annual temperature (in °C) together with regression lines representing the mean temperatures for the warmest and coldest months for ENSO and non-ENSO years respectively. The slopes of the regression lines representing the temperature progression of the coldest and the warmest months with time, differ significantly from each other (Figure 2). When the Jakarta-Semarang temperature records for ENSO warm event years (Kiladis and Diaz, 1989; Brookfield and Allen, 1991; Wang, 1991) are separated from "non-ENSO years" (all other years), the former are significantly warmer than the latter on average by 0.16 °C degrees throughout the time covered by the data. However, the trend towards increasing temperature is consistent for both groups so that each appears to increase in temperature at the same rate. The relationship between temperature of warmest months for ENSO and non-ENSO years is shown in Figure 1.

It is pertinent to note that the divergent pattern shown by the temperature increase for the warmest and coldest months does not exist for Manila, a city removed from the equator and the immediate effects of the Indonesian through-flow of warm seawater associated with the warm-pool.

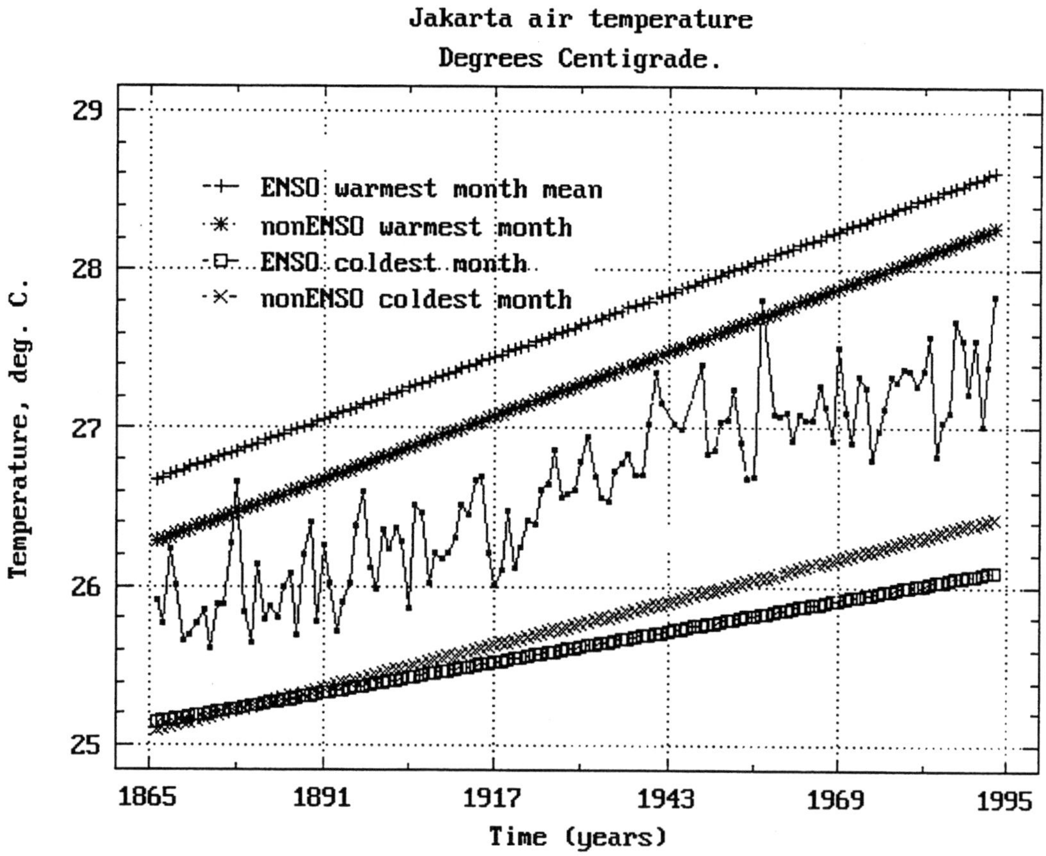

Figure 1 The cities of Jakarta and Semerang on the north coast of Java have warmed progressively over the last 130 years or so at a comparatively constant rate. The temperatures of the warmest and coldest months of El Niño years are now markedly different from "normal years". The El Niño years now swing through a broader range than was the case at the turn of the century.

As previously mentioned, there is an *increasing annual trend in air-temperature* exhibited by the mean monthly values over the period 1866-1993, for Java (Jakarta and Semarang data taken together). This totals 1.64 °C (0.0132 °C per year as the temperature has moved from 25.771 °C to 27.409 °C). The major industrial development in infrastructure for Jakarta has been significant only since 1980 or so and was not apparent before 1970 when the city had the aspect of an extended village with few large buildings (greater than 3-4 stories) and no extensive highways. The 1.65 degree difference between 1866-1991 can presumably be partitioned into: (1) urban heat-island effect, (2) effect of deforestation, (3) effect of secular micro-climate shift, (4) influence of general global warming with particular reference to the tropics.

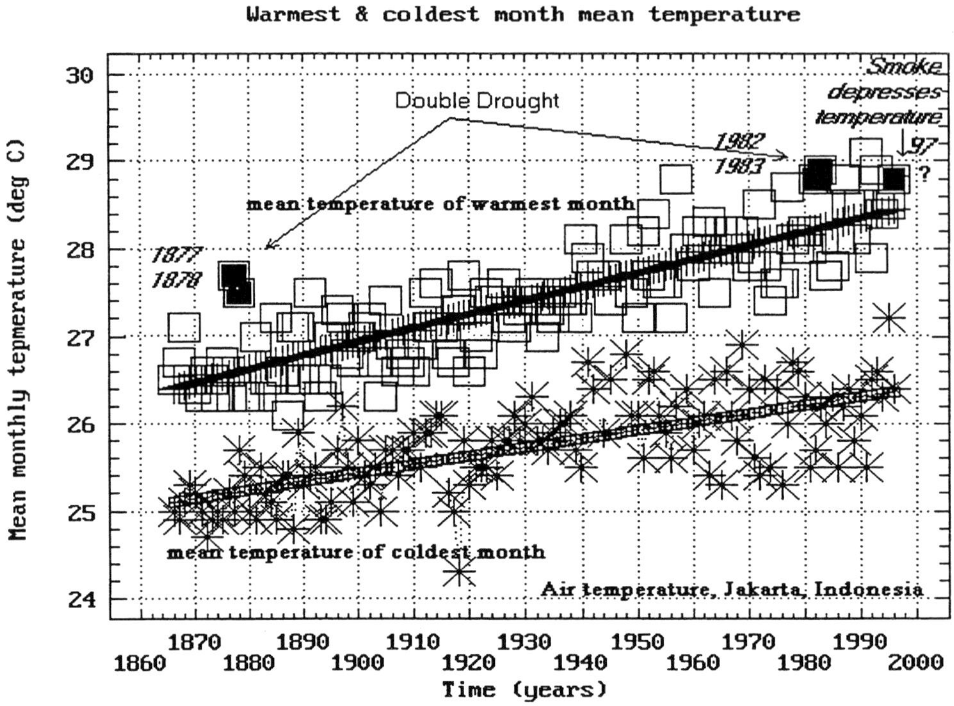

Figure 2 Overall, the mean temperature for the warmest and coldest month from 1866 to 1996.

When the blocks of non-ENSO years in themselves are examined, the deviations from the secular trend for warmest month mean temperatures in successive years are correlated with those of the next immediate year deviation. It thus appears that either continual warming or cooling is taking place from the termination of one ENSO to the initiation of the next. When the deviations around the time-based trend shown by the warmest month average temperatures are summed for the inter-ENSO intervals (the separate non-ENSO years) the resulting index is positively correlated with the following (initial) ENSO warmest month deviation from the overall ENSO warmest month secular trend. This provides an immediate predictive mechanism for the likely strength of an ENSO, in terms of the dry season impact to the Island of Java, should one occur in the next year to break a non-ENSO sequence. The length of the build-up and the build-up achieved seem not to be related (Harger, 1994a). The relationship does not in itself however, predict the occurrence of the "next" ENSO.

The data show that a consistent structure underlies ENSO events for the last century and a quarter. However, as a process monitored by mean monthly air-temperature measurements at Jakarta-Semarang, the system is changing in character with time in association with an overall atmospheric temperature increase in a way that involves increased intra-annual temperature fluctuations (Figure 2). In general, ENSO years are associated with higher temperatures than non-ENSO years, with a significant negative correlation between subsequent years which are thereafter systematically cooler. This is presumably because the ENSO event actively mixes excess heat energy into the ocean-sink to an extent that is in direct proportion to the outstanding positive temperature deviation (Figure 3). A weak ENSO, preceded by a relatively modest temperature build-up in the lead-up non-ENSO years, then results in limited mixing which leads to a relatively warm subsequent year while a strong event leads to extensive mixing and so generally results in a following very much cooler year. Atmospheric temperature build-up possibly associated with the greenhouse effect may be coupled to an increasingly wider temperature swing in west and central Java associated with the warm pool influence but anchored by the ocean-sink.

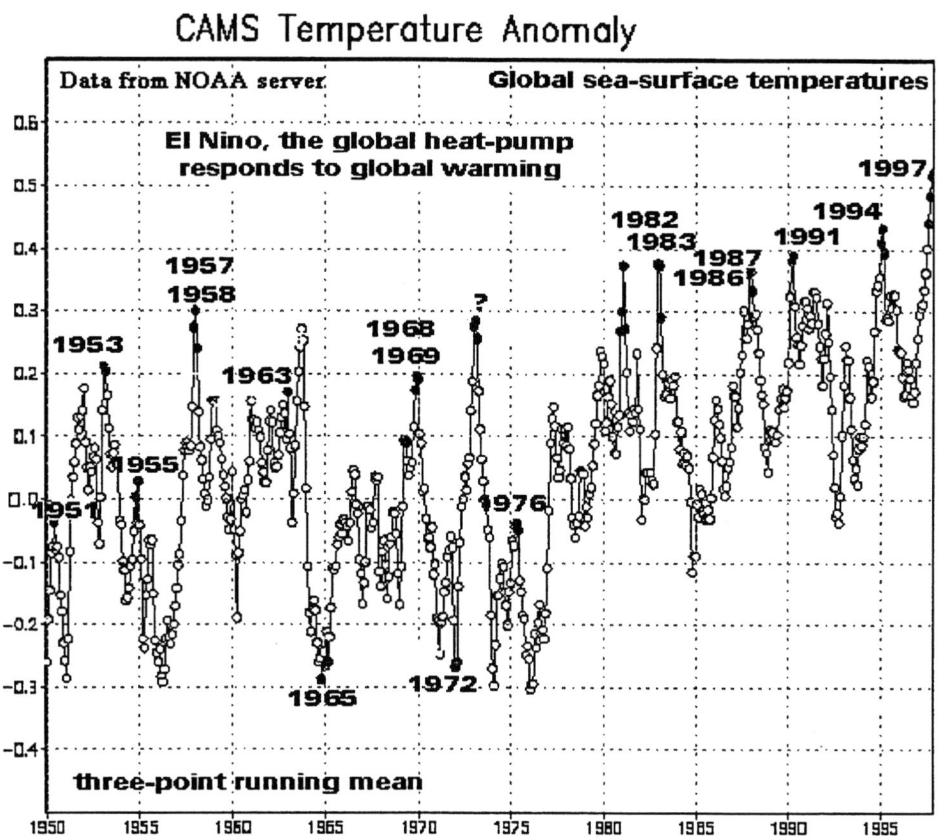

Figure 3 El Niño years are typically (but not always) associated with elevated sea-surface temperatures on a global scale.

TELECONNECTIONS

The influence of the ENSO (warm events) is apparently far reaching. During such events in the Pacific, seasonal rainfall is increased for Kiribati, Tuvalu and the northern

Cooks but decreased in southern Cooks, Fiji, and Tonga. The "great dry event" of 1877-1878 was associated with global impact being the proximal cause of millions of deaths from famine in India and China. Particular effects were noted in association with the 1982-1983 ENSO (warm) event. These are now termed ENSO-climate teleconnections. The effects induced by such teleconnections are termed “climate anomalies” (periods of drought or extensive rainfall) and these tend to occur during most if not all ENSO events. In particular, the effects of ENSO Warm Events are not spread evenly throughout Indonesia and both drought and flooding create substantial disruptions and severe economic dislocation (Indonesian Ministry of the Environment, 1998).

A general scenario of suspected teleconnections associated with ENSO (warm events) may be summarized as warm and dry in southeast Asia extending north and south of the equator with a wet area around the equator in the central Pacific with another dry area in the northeast of south America. Wet areas are located in the southeast and southwest of the U.S.A. as well as northeast of South America, southeast of South America and also east Africa. Warm areas occur in south Asia, East Asia, northwest of North America, southeast Africa, southeast Australia and east of South America.

Fire and Crop Loss in Indonesia

Indonesia lies on the western margin of the ENSO interaction and for the most part enjoys a humid tropical climate except in the eastern most regions. Indonesia presently supports extensive tracts of tropical rain forest, apparently amounting to some 117.9 million ha in 1990, which accounts for some 6.4% of the global total estimated as 1,838 million ha in 1982. For Indonesia, the ENSO-associated warm event drought of 1991 led to the failure of 190,000 ha in paddy with an overall 843,000 ha affected. This event caused unprecedented losses in rice production to Indonesia resulting in 600,000 tons being imported to the previously self-sufficient archipelago. In 1982-1983 the ENSO-associated drought of that time, resulted in 420,000 ha of paddy being affected and failure of 158,000 ha. The event was also accompanied by forest fires which burned 3.7 million ha of generally second-growth timber, mainly in Kalimantan (Borneo). An area of 88,000 ha burned in 1991 (*Jakarta Post*, 30 November 1991) largely in Kalimantan in association with the 1991-1992 ENSO event. An extensive pall of smoke developed over Kalimantan, Singapore and Malaysia during September-October of 1991. Similar fires raged during the height of the 1997 El Niño event (Figure 4) and these burned well into 1998 killing at least 40 people directly and affecting 1.5 million ha. The smoke was responsible for inflicting respiratory problems on 50 million people. Over 3000 flights were cancelled due to haze (Indonesian Ministry of the Environment, 1998).

Figure 4 Smoke from biomass-burning rising from Kalimantan and Sumatra in Indonesia, September 1997 during the El Niño - induced drought. (Image courtesy of the Indonesian Department of the Environment).

The development of dry conditions in Southeast Asia, particularly in Indonesia, has been emphasized since the 1982-1983 ENSO Warm Event. At that time, fires took hold of second-growth vegetation and some of these eventually penetrated into underlying coal seams in Kalimantan (Borneo) in 70-80 separate locations. In many parts of Kalimantan coal seams measuring several meters in thickness occur at shallow depths (a few meters to several meters). Once ignited, the coal burned underground and the overburden subsequently slumped into open pits of smoldering ash. Smoke and coal-distillates vented into the overlying vegetation. Surface fire-pits developed in various places as the underlying coal was exploited by combustion. At a later time, surface fires arose because the vegetation subsequently caught alight due to direct or indirect contact with the burning coal. The combustible gases released from the pits and vents exacerbated the problem. The fires were often located in isolated regions and are extremely hard to stop. The burning area must be completely ringed by wide trenches extending many meters from the surface to the base of the coal seams. These underground "hot-spots" continue to burn to date (late 1999) and are one source (along with land clearing, lightning-strikes etc.) of the wild-fires that have characterized the ENSO Warm Events.

Most of the fires burn through secondary vegetation since primary forest in general tends to resist penetration except under conditions of extreme drought. A recovery period of 8-10 years or so is more than sufficient for the generation of re-growth vegetation capable of supporting further combustion. Forest-use practices involving the establishment of logging roads, clear-cutting and even controlled extraction of trees serve to open up the vegetation and to make it more susceptible to penetration by fire during ENSO Warm Events than would otherwise be the case.

Since 1982 ENSO Warm Events have promoted serious fires in Kalimantan and Sumatra in 1982,1983, 1987, 1991, 1994, 1997, 1998. The worst fires occurred in 1982-1983, 1987, 1991, 1994 and 1997.

Severe droughts in the Island of Java can be defined as periods greater than 6 months in duration where less than 100mm of rainfall are recorded in each month. The frequency of ENSO Warm Events associated with such droughts has increased through the period 1950-1999 from a mean length of 4.5 months to 5.0 months ($p<0.00001$), (Figure 5). The length of the dry season, as defined above, has progressively increased in central Java (Jakarta, Tegal, Yogyakarta, Surabaya, and Banguwangi) from 1909 to the present when all years are considered.

The impact of the 1997 El Niño in Indonesia was greater than any other ENSO event in this century, and hunger appeared in the eastern provinces and a large number of people died from starvation particularly in Irian Jaya (Indonesian Ministry of the Environment, 1998). Others have moved out of traditional areas in search of food. It is not clear that the event was more intense than the "Great Dry" of 1877-1878 but it had a greater impact than the warm-event of 1982-1983. In 1982-1983 a total of 340 deaths were reported due to starvation in Indonesia. The Great Dry promoted famine around the world and prompted the Dutch to put in the extensive meteorological data network existing to this day in expanded form in Indonesia

From the standpoint of a global atmospheric carbon balance, a continuing fire cycle in Indonesia and Southeast Asia associated with ENSO Warm Events every 4-6 years or so makes little difference once the primary forest is oxidized by burning or by other means. Replacement growth generated within 4-10 years after a burn balances the liberation of carbon due to combustion by reclaiming carbon dioxide from the atmosphere. It is the elimination of the climax vegetation in the first instance that causes a net increase in carbon dioxide. However, if productivity falls as the result of nutrient loss caused by such a fire-cycle, aridity and desertification would likely follow and this would further increase the atmospheric load of carbon dioxide.

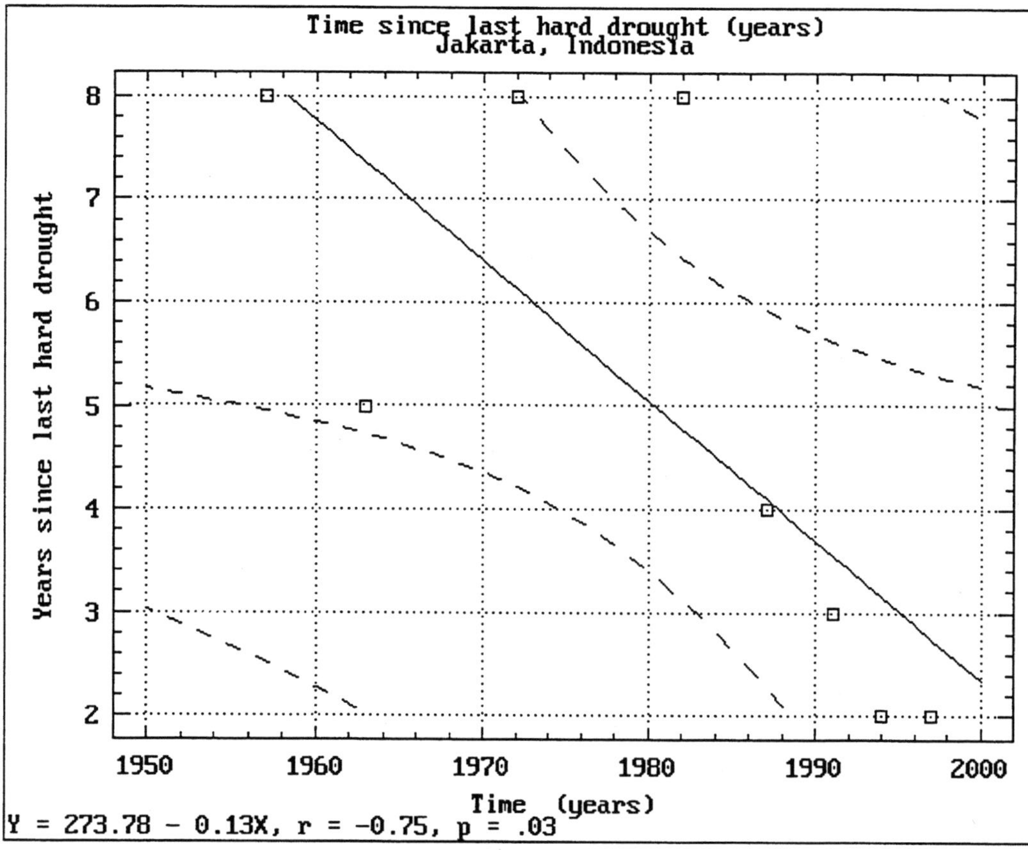

Figure 5 The frequency of severe droughts (greater than 6 consecutive months with <100 mm of rainfall each) in the region of Jakarta, Indonesia has increase markedly since the 1940s. There is no indication of a similar heightened frequency of El Niño events elsewhere in the measurement record between 1866 and 1940.

SUNSPOTS

Rainfall and temperature in west Java are strongly associated with preceding ENSO year variations in the temperatures of the warmest months as measured by deviations from an overall secular trend. Although such ENSO year deviations are presumably related to the southern oscillation as a whole, the present analysis throws no light on the manner in which they arise. However, a significant positive relationship exists between the annual sunspot count from 1886-1991 and the warmest month deviates from the overall secular trend. The ENSO year warmest month residual deviation is significantly correlated with the sunspot activity of the same year ($r = 0.31$, $n = 40$, $p = 0.045$). The significant positive relationship also holds if the sunspot activities in the current (0), the preceding (-1) as well as the year twice removed (-2) are added ($r = 0.33$, $n = 38$, $p = 0.037$) but breaks down if further prior years are added. The residual deviations for both ENSO and non-ENSO years taken together and the sunspot activities are not significantly correlated. There is no relationship between sunspot activity and the secular deviation of non-ENSO years. A stepwise multiple regression procedure identified time as the only significant variable (38%) in accounting for the maximum of the mean monthly temperatures for ENSO and non-ENSO years taken together. This produced an adjusted correlation coefficient of 0.69, n = 120 when time, sunspot activity in both current and previous years are used as independent variables. When ENSO years only are used as the dependent variable together with time, sunspot activity in both current and previous years as independent variables, a model with $r^2 = 0.69$, n = 39 results with time (67%) and also sunspot activity in the current year (3%) as significant predictors.

There is presently no truly satisfactory explanation for the above significant ENSO-sunspot relationships of 126 years duration although the data apparently indicate that sunspot numbers are associated with the ENSO warm event deviations in maximum monthly temperature from the overall positive secular trend. In turn, this may mean that the overall ENSO response is directly modified and perhaps enhanced, but not necessarily triggered by, an energy flux from the sun that is correlated with the appearance of sunspots. Such an interpretation is not widely accepted although Friis-Christensen and Lassen (1991) provide strong circumstantial evidence of a different nature to associate broader trends in a positive relationship between sunspots and temperature-mediated climate changes. Similar broad comparisons between climate and sunspots have been made by Eddy (1977). Reid (1990) has cautiously related globally averaged SST secular variations to sunspot numbers from 1854-1980 using 11 year running means but Kelly and Wigley (1990) tend to contradict the relationship and suggest that greenhouse forcing is the more important consideration.

A significant factor associated with the data reported herein is that the response of ENSO warm events relates to almost every phase of the observed sunspot cycle. The data also indicate however, that sunspots might, at the most, only play a secondary role in the observed secular warming. As a whole, the relationship becomes more coherent after 1900 when the warm month deviation of both ENSO and non-ENSO years is significantly associated with sunspot numbers (r = 0.23, p = 0.038, n = 87). The ENSO years 1877-1888 are major departures from the general pattern with high temperature deviations in a low sunspot period clearly demonstrating that the relationship is not obligate.

ENSO and Global Warming

Atmospheric temperature build-up possibly associated with the greenhouse effect may be coupled to an increasingly wider temperature swing in west and central Java associated with the warm pool influence but anchored by the ocean-sink. In Southeast Asia, longer dry periods coupled with increased temperatures may thus result from an ENSO-driven mechanism, which may force equatorial aridity as global carbon dioxide concentration rises. This may change "ever-wet" conditions currently associated with the humid tropical rainforest of Southeast Asia. Forest fires have become a persistent problem in both Sumatra and Kalimantan (Indonesia) in the ENSO-associated droughts of recent years. The warm event years apparently provide the driving mechanism. The temperature record shown by the Southeast Asian cities studied to date indicates a progressive annual warming for the region of around 0.013 °C over the last century (Harger, 1994a).

ENSO year warm events vary in their effects as estimated by deviation of warmest month mean air temperature from the secular trend exhibited by the Jakarta/Semarang data set and perhaps more widely throughout Southeast Asia. The extent of this temperature deviation from the secular trend for ENSO year provides a key to understanding the teleconnections. These presumably take the form of heat-induced signals passed outwards from Southeast Asia and the associated "warm pool" of surface seawater that accumulates against the western margin of the equatorial Pacific.

In general ENSO years are associated with higher temperatures than non-ENSO years, with a significant negative correlation between subsequent years which are thereafter systematically cooler. This is presumably because the ENSO event acts as a heat-pump and actively mixes excess heat energy into the ocean-sink to an extent that is in direct proportion to the outstanding positive temperature deviation. A weak ENSO, preceded by a relatively modest temperature build-up in the lead-up non-ENSO years, then results in limited mixing which leads to a relatively warm subsequent year while a strong event leads to extensive mixing and so generally results in a following very much cooler year. There is a differentiation between non-ENSO-ENSO blocks showing trends of positive "heat-loading" indices as opposed to those showing decreasing trends and this suggests that heat build-up alone may not be the only critical variable but that perhaps two sub-categories of ENSO initiating mechanisms may be involvedP: those dominated by heat forcing, and those by water-mass forcing (Harger, 1994a). Both however, generate heat redistribution into the Pacific Ocean.

The warming-pulse effect of the ENSO warm-events observed recently are probably not quantitatively different from the "great dry event" of 1877-1878, however, the overall global temperature is now around 0.5 °C higher and perhaps as much as 1.5 - 2.0 °C in the region of the equator. The Jakarta air-temperature data, near the equator, show a clear change in amplitude and degree over the 127-year record (Figure 2). It is of course relatively easy to dismiss these changes in a superficial way by attributing them to a "city effect". Similar overall changes throughout the Indonesian and Philippines archipelagos and from El Salvador are less easily dismissed (Harger, 1994a). However, ecological conditions and vegetation cover have probably also been changed markedly at local levels in the same interval making it difficult to ascribe such observed effects entirely to secular responses of a global nature. It is certain that if the instrumental record of the 1877-1878 ENSO event did not exist, the impact experienced by Indonesia due to the events observed in 1982-1983 and 1991-1995 and 1997-1998 would appear to be quite extra-ordinarily severe in relation to all activity from 1879 forward.

Information drawn from meteorological records in Southeast Asia clearly indicates that each event is unique in terms of the signature, which it imposes on the rainfall and temperature from location to location. Nevertheless, a strong underlying pattern within the context of each event, itself apparently initiated or molded by the character of the proceeding years, can be detected. This pattern permits relatively circumscribed predictions of forward conditions (drought-intensity) for 2 to 3 years, to be made once the event "locks in" for the duration of the warm event and at least one year beyond.

Non-ENSO years leading up to a warm event can be scored in terms of the extent to which they depart from the secular warming trend for the warmest month using data from Jakarta and Semarang on the north coast of Java. The value taken by the cumulative temperature deviations signals the character of the upcoming ENSO event. However, this signal does not allow an exact determination to be made with respect to whether or not an ENSO event will occur in the next year. For the available historical instrumental data, all non-ENSO blocks with markedly positive cumulative tendencies eventually delivered a hot dry season in east Indonesia. This sort of tendency within non-ENSO blocks can thus serve as a caution in the sense that a very hot ENSO event is likely in the offing. The background data can also be used to actually predict the probable intensity of an ENSO in the upcoming year in terms of its drought potential, should such an event take place in reality (Harger, 1994b). In this respect the correlation between the cumulative temperature deviation of the inter-ENSO blocks in relation to the temperature deviation of the first ENSO year is 0.43.

In conjunction with indications generated by models, such relationships could form the basis of an advisory service for forecasting conditions within dry seasons from one year to the next. It is anticipated that a broad description of temperature and rainfall patterns associated with space and time in relation to ENSO events will lead to better forest management and food security for the region as long as sudden changes do not occur.

Global temperature anomalies are also strongly correlated with the warmest month temperatures of the Jakarta/Semarang data set ($r=0.84$, $n=121$, $p<0.00001$), for the period of available data. When the secular trends are removed the residual deviations shown by the Jakarta/Semarang warmest month data are also correlated with the deviations shown by the annual global anomalies ($r=0.53$, $n=121$, $p<0.00001$). This is of some interest since the Jakarta/Semarang warmest month indications are obtained usually by May or June (10/35 ENSO-years) or at least by October (31/35) of any one year. In addition, the Jakarta/Semarang warmest month temperatures are closely related to the warmest month temperature deviations in other regions (Harger, 1994a). As noted above, the temperature of the warmest month of the year in Jakarta Indonesia seems to closely reflect overall global warming and in particular may even be used to predict upcoming droughts and global temperatures. Gaia appears to be running a fever as her heartbeat is increasing both in frequency and intensity.

Western Pacific Dry Events

A consistent structure underlies ENSO events for the last century and a quarter. However, as a process monitored by mean monthly air-temperature measurements at Jakarta-Semarang, the system is changing in character with time in association with an overall atmospheric temperature rise in a way that involves increased intra-annual temperature fluctuations. In general ENSO years are associated with higher temperatures than non-ENSO years, with a significant negative correlation between subsequent years which are thereafter systematically cooler.

The warm month Jakarta/Semarang deviates themselves may be directly related to the size and relative movement of the western Pacific warm pool, particularly in the ENSO warm-event years. The most obvious periods when temperature induced changes in biomass burning of southeast Asia take place correspond with ENSO warm-event years and particularly those in which marked positive deviations from the overall secular trend are readily apparent. In recent years these are: 1997 (depressed by smoke), 1994, 1991, 1987, 1983, 1982, 1972, 1963, 1958 (estimated datum), 1957, 1953 and 1951. In 1997 this relationship (as measured by the ground-level air temperature in Jakarta) was not so obvious because the dense smoke released by the extensive forest fires obscured the sun to such an extent that temperatures were lowered in Jakarta (Figure 2). The heat build-up in Southeast Asia and the western Pacific presumably commences before the Kelvin wave resurgence from the western to the eastern Pacific. Kelvin waves seem to have started in October to mid-December in recent years although in 1993 a Kelvin wave was also detected in August and then in late October. The warmest month of the first ENSO year usually takes place in either September (3 times in the record), October (18) or November (4) but it can also show up in May (7) and June (2) as well. In itself, a significant warm deviation above the secular trend for the average monthly temperature in any of these months signals an upcoming warmer-than-average ENSO with a hit-ratio of 21:6.

With reference to the Jakarta record, the ENSO blocks ending in 1878, 1891, 1902, 1914, 1951, 1953, 1958, 1963, 1969, 1983, 1992 and 1998 were apparently trending warmer than average. The mean duration of the period between warm-trending peak events was around 11 years (for the instrumental record under consideration) and the longest period without such a peak was 37 years (1914-1951). The bulk of the record indicates that of the identified 32 ENSO events, the majority caused one drought-year at a time in Indonesia, preceded and followed by a wet year (normal to an extent). Eleven of these events seem to have presented no particular problem to Indonesia at all. There were a few severe dry events around the turn of the century then the ENSO "went quiet" for around 37 years or so (after 1914) as far as an association with the "Western Pacific Dry Event" is concerned. It has recently become active. Western Pacific Dry Events, as registered in Jakarta are becoming significantly more frequent and perhaps more intense. They may well signal the effects of global warming with the increased numbers of fires in Southeast Asia as the logical consequence.

REFERENCES

1. Brookfield H. and B. Allen. (1991) "Environmental and human responses to climatic variability in the west Pacific Maritime Continent Region". In: *South*

Pacific Environments, Interactions with Weather and Climate. University of Auckland, Auckland New Zealand 2-7 September 1991, proceedings. ISSN-0114-590 Environmental Science Occasional Publication No. 6, Environmental Science, University of Auckland.

2. Eddy J. A. (1977); "Climate and the changing sun". *Climatic Change* 1: 173-190.
3. Friis-Christensen and Lassen K. (1991); "Length of the solar cycle: an indicator of solar activity closely associated with climate." *Science* 254: 698-700.
4. Harger, J.R.E. (1994a) "Air temperature variations and ENSO effects in Indonesia, the Philippines and El Salvador. ENSO Patterns and Changes from 1806-1993." *Atmospheric Environment* **29**: 1919-1942.
5. Harger, J.R.E. (1994b); "ENSO variations and drought occurrence in Indonesia and the Philippines. "*Atmospheric Environment* **29**: 1943-1955.
6. Indonesian Ministry of the Environment (1998); Final Report on the National Workshop on El Niño Impacts Assessment, March 7, 14, 26, 1998. Jakarta Indonesia (unpublished report 18pp plus diagrams).
7. Jones, P. D., Raper S.C.B, Bradley R. S., Diaz H.F., Kelly P.M., and T. M. Wigley (1986); "Northern hemisphere surface temperature variations: 1851-1984." *Journal of Climate and Applied Meteorology* 25: 2.
8. Kiladis, G. N. and H. F. Diaz (1989); "Global Climatic Anomalies Associated with Extremes in the Southern Oscillation." *Journal of Climate* **2** (9): 1069-1090.
9. Reid G.C. (1990); "Influence of solar variability on global sea surface temperatures." *Nature* 329: 142-143.
10. Wang S. (1991); "El Niño events, 1470-1989." Unpublished presentation to the UNEP-Earth-watch Global Environmental Monitoring System workshop on ENSO and Climate Change. Bangkok, Thailand.

Chapter 3

DROUGHT, FIRE AND HAZE IN THE HISTORICAL RECORD OF MALAYSIA

Lesley Potter
Department of Geographical and Environmental Studies, University of Adelaide, Adelaide, South Australia 5005, Australia

INTRODUCTION

It is proposed to confine the analysis in this chapter largely to the equatorial areas of island Southeast Asia, which include the present states of Indonesia, Malaysia, Singapore, Brunei and the southern parts of the Philippines. The natural vegetation of much of this area is tropical moist forest, a vegetation type unlikely to burn readily, unless under exceptional conditions of extended drought or human interference. The monsoonal forests of mainland Southeast Asia are thus excluded, as dry seasons there are longer and fire is a more 'normal' and often annual visitor. For the same reason, the dry teak forests of central and eastern Java are not discussed.

The exceptional climatic conditions inducing fire in the equatorial forest are related to the cyclic droughts accompanying ENSO occurrences, as discussed in Chapter 2. Over the past twenty years, and especially since the major ENSO of 1982-3, much scholarly work has been done in drawing together the literature on earlier drought sequences, both on a world scale (e.g. Kiladis and Diaz, 1986, 1989; Quinn, 1992; Ropelewski and Halpert, 1987, Whetton *et al.*, 1990) and for parts of the region (Allen, 1989; Allen *et al.*, 1989; Brookfield *et al.*, 1995; Harger, 1995a, b; Kripalani and Kulkarni 1997; Nicholls, 1992, 1993; Quinn *et al.*, 1978; Whetton and Rutherford, 1994). Interest in tropical deforestation, global environmental change and fire has also inspired several recent studies of fire ecology, which include chapters on Southeast Asian forests and forest fire (Goldammer, 1990; Levine, 1992; Crutzen and Goldammer, 1993).

Goldammer, who with Seibert has identified fire sequences in East Kalimantan dating back to the Pleistocene (Goldammer and Seibert, 1989, 1990), has suggested that prehistoric fire mapping is one of the upcoming challenges of tropical fire ecology (Goldammer, 1991:84). This author would argue that fire mapping or description from the historical record is no less a challenge. Since an initial attempt to document drought sequences for Borneo (Chapter 8 in Brookfield *et al.*, 1995), further work has uncovered more sources which include forest fire (Potter, 1999) and this chapter presents some additional findings, while extending the study area. There is no doubt that using the known major drought or ENSO events in the past has made it easier to focus on particular years; however, the written record is patchy at best, especially for the period before the extreme event of 1877-78, and even for those two years. Because of time limitations, the decision has been taken to focus on four major drought/fire periods: 1877-78, 1902, 1914, and 1972, while mention will be made in passing of others. While some conclusions may be drawn, there are still gaps and inconsistencies needing further investigation. Nevertheless, the story is slowly being pieced together.

THE NATURE OF THE AVAILABLE SOURCES

Scientific reports of meteorological conditions in Southeast Asia (as elsewhere in the tropical world) were an integral part of the establishment of colonial administration in the areas concerned. Those reporting were invariably European, though not necessarily of the same nationality as the colonising power. Individual traveller-scientists conducting studies or simply making observations in remote districts were a prominent part of the nineteenth century colonial scene, of whom the region had its fair share.

In insular Southeast Asia, detailed collection of meteorological data began in the major centres in the 1860s (Singapore, 1869[1]; Jakarta 1864; Manila, 1865). Kuching in Sarawak has a record dating from 1876, which included the major event of 1877-8, while Sandakan (Sabah) began its record in 1879. The spread of instruments to collect climatic data from other sites across Indonesia began only after 1879, as a direct result of the extreme conditions of the previous two years (Bergsma, 1879).

The dominance of Indonesia's forests in the vegetation of the 'maritime continent' of Malesia has meant a focus in this study on the larger Indonesian islands, especially Sumatra and Borneo, with greatest forest resources and relative proximity to the centres of administration in Java. The ENSO of 1877-8, the earliest of the four historical drought periods under investigation, was also extraordinarily severe, both in its dryness and in the much higher temperatures which occurred (Harger, 1995a; Kiladis and Diaz, 1986). The Jakarta-based newspaper the *Javasche Courant* with its *Binnenlandsche Berigten* ('Inland News') has been very useful in plotting the impact of this event across Indonesia. The government annual reports (*Koloniaal Verslag*) also contain some details, as do the specific reports of the Dutch Residents for particular regions. An excellent source for coverage of the early period in northern Borneo is the *Sarawak Gazette.* Sabah at the time of the 1877-8 drought was still being explored by private British interests, with the

[1] There is a broken record for Singapore before 1869, with some observations in the 1840s, and others from 1862-66.

Company's first resident being appointed at Sandakan in 1878. A third major source has been Singapore, where the *Straits Times* has been invaluable for the later dates[2]. The *Journal of the Straits Branch of the Royal Asiatic Society*, published from 1878, also contains accounts of meteorological conditions in Singapore.

One problem has been that Jakarta, Singapore, and Kuching were somewhat removed from the worst affected areas struck by drought or fire, and at times the knowledge base was simply lacking as colonial authorities simply had no presence in parts of their 'realm'[3]. Manila was even further from the forests of Mindanao or Palawan which might have burned in such an extreme drought[4]. For the 1877-8 event, Peninsular Malaysia is also lacking detail, as British administration on the peninsula was in its early stages: Hugh Low became Resident of Perak in 1877, but beyond remarking in his Journal on the prevalence of smallpox and cholera, together with failed rice crops and a lack of water for tin washing, there is little mention of a drought and none of fire (Low, 1954)[5]. Another problem is that even though urban dwellers might have been aware of unusual climatic conditions, their reports tended to focus on questions of water supply, crop failure and human disease, which affected them more directly than the state of the forests.

Equally useful source materials, and often more precise in locating the exact position of fires, have been accounts of traveller-scientists moving through forested districts either during or just after significant years of drought and fire. Whether a traveller actually discovers and describes a burned area, however, especially when some time has elapsed since the fire, must be regarded as a matter of chance, as travellers were not specifically looking for evidence of fires and were often surprised when they found it. Locations listed in this chapter should thus be seen as indicative only. This conclusion applies even more strongly to the cases in which a silviculturalist notes some unusual characteristics of a forest, which may lead to speculation about a burning event which occurred many years before. When we know that certain years experienced drought, it is tempting to assign any evidence concerning fire to those years, even though it may not really be possible to date the fire with such accuracy. Oral traditions may suffer from the same problem, this time that of precise recall. Such traditions have not yet been much tapped in respect of historical fires, but as Goldammer (1990) has shown, they can also yield useful information if approached with care. What all the fragmented observations add up to may only be a fraction of the areas actually affected, but they may be as good as we can get with current knowledge.

---

[2] Unfortunately I did not have access to the 1877-8 issues when compiling this chapter.

[3] For example, the Dutch scientific expedition to Central Sumatra 1877-9 was forbidden to enter the Sultanate of Jambi, although one of the express aims of the expedition was to explore the basin of the Batanghari, Jambi's most important river.

[4] It was noted by Goldammer (1988: 246) that there was a rain forest fire in Mindanao during 1983 which burned 20,000 ha, but it remained largely unreported, unlike the larger blazes in Kalimantan and Sabah. If the forest burned there in 1983, one might imagine it could also have burned in 1877.

[5] The fact that the Low Journal covers only the period from the end of April to mid-June may also be relevant, as according to Dale (1959), the dry season in central Perak is most marked from June to August.

From the beginning of the twentieth century there was a surge of scientific study of Indonesian rainfall and drought patterns, especially in the journal *Natuurkundig Tijdschrift voor Nederlandsch Indie*. The question of smoke and haze which moved beyond district or national boundaries was being raised by 1902, although in Singapore, which did experience such haze, there was considerable ignorance as to the likely source. This lack of easily accessible information was largely due to the language and cultural barriers between the colonies; Singapore in 1902, or at least the *Straits Times*, was more knowledgeable about drought conditions in Australia than in nearby Sumatra. By 1914, the second most severe burning event after 1877, ships' captains, rubber planters and scientists were furnishing more detailed reports and government administrators had access to rather better data. As institutional control over the dipterocarp forests was being established, professional foresters were turning their attention to questions of forest burning. Finally, the event of 1972 sounds very modern in its concerns, with haze occurring over Malaysia and Singapore, and Indonesia denying responsibility for its generation. Although the *Straits Times* once more provides useful data, for this event satellite photographs were for the first time being put to work to discover the precise locations of fires and to assign responsibility for them.

DROUGHT, FIRE AND HAZE IN INDONESIA AND EAST MALAYSIA: THE EVENT OF 1877-78

The following three paragraphs describing the situation in Indonesia are summarised in translation from a much fuller account in the *Javasche Courant*, covering the period from November 1876 to September 1878.

Like many other ENSOs, this particular climatic perturbation began with heavy rain and floods in several districts of Sumatra and South and East Borneo[6] in November/December 1876, extending eastwards to Sulawesi and Maluku by January. The rains were unreliable, with alternating flood and dry periods, difficult for crops to handle. By February unusually high temperatures began to appear, with the heat described as very great in West Borneo by May and insufferable in the capital, Pontianak. Hot days were often followed by cool nights, noted in Palembang and Banjarmasin.

Drought set in by June in parts of Sulawesi and Maluku[7]. In August there was a great forest fire in Ambon (Maluku), which also destroyed many fruit trees growing near the

---

[6] During Dutch times the Indonesian 2/3 of the island of Borneo was divided into two administrative districts, the 'Southern and Eastern divisions of Borneo' and the 'Western division of Borneo'. The first included the modern provinces of East, South and Central Kalimantan, administered from Banjarmasin, the second covered West Kalimantan with Pontianak as the main centre. Where possible, modern provincial and district names will be used in the ensuing discussion, unless the locations are too vague for such precision.

[7] The botanist Teijsmann, on a visit to Sulawesi between June and December 1877, also reported the dying and burning of coffee plantations and useful forest trees as a result of the

forest edge. By August, persistent drought was also being experienced in Sumatra, with accompanying health problems. Palembang residents had difficulty in collecting forest products because the rivers were so low. The pepper harvest failed in Lampung and the forests began burning, also houses and pepper vines. Palembang in September 1877 was often covered in the morning by what was described as "an extraordinary heavy mist, the result of forest fires". Frequent forest and village fires were recorded, consuming cultivation and houses alike. In South and East Borneo it was explained that the people needed to burn new upland fields (*ladangs*) as their harvests had failed; many forest fires were reported there in October. In Bengkulu, by November, all the animals were short of food, with wild tigers coming out to seek food near the houses. In South Sumatra much damage was done to gambier, durian and coffee plantations through bush and *alang-alang*[8] (grass) fires.

From the middle of November, rain was beginning to fall in parts of Sumatra. However, in districts further to the east the drought continued. In Ambon a fire destroyed sago, coconut and other trees. Fires also occurred in Timor, burning stocks of hay in the fields. During 1878, the rains in Sumatra were patchy and unreliable: much cultivation was attacked by pests and fires again broke out, for example burning young coffee plants in West Sumatra. In West Kalimantan, on the other hand, heavy rain and floods in August made it difficult for people to burn their ladangs, a task which they were able to begin by September.

The *Koloniaal Verslag* reported more briefly, for Palembang (South Sumatra) that bushfires had spread alarmingly in the lowlands. The capital was enveloped in smoke and great damage was done to cultivation and forests. Meanwhile meteorological measurements were being taken in the mountains of West Sumatra and in South Sumatra, as part of the research of the geographical expedition to Central Sumatra from 1877-79. The general dryness of the area around Padang (West Sumatra) and in the Padang uplands was noted from the second half of 1877 to the first months of 1878. Thick fog was also observed at a number of sites in the Padang uplands between May and November 1877 (Veth, 1882: 2, 53-4). Bushfires were reported to be widespread in the Palembang region when the expedition arrived there in February 1878 (Veth, 1882: 2, 57).

From these sources it seems clear that in the case of Sumatra, South Sumatra, especially the Palembang lowland, was most severely affected by fire as a result of the drought. The formation of haze is clearly described as resulting from such fires. When Forbes surveyed the lowland landscape from the summit of Mt. Dempo in 1880, he was amazed at the amount of land which was treeless and covered by *alang-alang* grass (*Imperata cylindrica*) (Forbes, 1885). It is likely that the fires of 1877-8 had encouraged the spread of grassland at the expense of forest. As most of the fires noted by the

drought and the strong east winds, which had destroyed everything (Teijsmann, 1879: 74, 93, 97-8).

[8] *Alang-alang* grass, or *Imperata cylindrica,* is highly combustible and spreads easily through burning. Its underground rhizomes are not destroyed by the passage of fire, and are in fact stimulated to send out new shoots.

Javasche Courant were described in conjunction with human cultivation, it seems reasonable to assume that many were indeed of human origin, perhaps ladang fires which had escaped, or attempts at clearing to extend areas under cash crops such as pepper or coffee. The failure of many crops in 1877 had made many farmers anxious to plant again as soon as possible, as was clear also for both divisions of Dutch Borneo (Kalimantan).

For more detailed information on Borneo, there are a number of scattered sources available for locations in both Kalimantan and Sabah. Because of its modern burning experience, one might expect East Kalimantan to be a prime contender for forest fires in 1877-8. While there is evidence of burning, it is indirect and only revealed through Tromp's discussion, published in 1890, and by Goldammer's analysis of local oral tradition, which suggested burning in the Mahakam lakes district near Muara Lawa (Goldammer, 1990: 19). Tromp argued that with good soil, the forest burned in Kutai in 1878 should already have recovered by 1890; the fact that it had not was to him a sure indicator of soil poverty. He also observed that the climate of the district was such that long droughts did occur, which had resulted in the forest fires of 1878 (Tromp, 1890: 732-733). The fact that the explorer Bock, who passed through the district in 1879, did not mention forest fires, even though large numbers of trees had been, as he put it, 'roasted' by the drought (Bock, 1988: 117), has led some later writers to conclude that there was no actual burning. While it was reported during the 1982-3 drought that some dipterocarp forests in Kutai National Park became deciduous under extreme dryness (Leighton, 1984), the dead trees observed by Bock along the Mahakam River could have simply been drought stressed, or could have burned the year before. Bock (1988: 117) described the scene as resembling an English landscape in early winter. One may compare the statement of Gerlach (1881) on the other side of the island. He wrote that the swamp forest around the Danau Sentarum lakes system in the Kapuas river (West Kalimantan), which had dried out in 1877, had here and there been burned by Dayaks, presenting a sad sight which made one think of the trees in a Netherlands winter flood (Gerlach, 1881: 293).

A further comparative description comes from Sabah, which was described by a rather derogatory article in the *Sarawak Gazette* (24 April 1878) as having suffered from drought for the previous six months:

> "During the period mentioned no rain has fallen; paddy farms, sago plantations and other cultivations are dying or dead, burnt up by the intense heat. Fires are bursting forth in all directions, and such a dense haze has been hanging over everything for months that traffic has been to a large extent put a stop to. One who saw it, compared it to a November English fog."

In the same month, April 1878, F.W. Burbidge sailed around the northern coast of Borneo, from Labuan to Sandakan. He noted that there had been an unprecedented five months of drought and mentioned that in several places he could see "jungle fires raging along the coast" (Burbidge, 1880: 183). Returning to Labuan three months later, the ship ran aground twice in thick fog.

From these passages, one is able to perceive some of the impacts of the 1877-8 ENSO in northern and eastern Borneo. In much of Sabah and East Kalimantan the drought appeared more severe in 1878 than in the previous year[9]. Kuching and the rest of Sarawak (and presumably also Brunei) experienced a more limited drought of about two months in August and September 1877, although the *Gazette* reported a more serious drought further north in the Bintulu region than in Kuching itself. The only burning noted was of new gambier plantations which had been made by Chinese immigrants in the forest near the capital. However, the most serious effect on Sarawak was a severe cholera epidemic brought about by inadequate water supplies, in which thousands died (*Sarawak Gazette*, 11 August 1877; 7 September 1877; 26 February 1878).

For southern Borneo so far only one description of burning has been found[10], an area in present Central Kalimantan between the Katingan and Cempaka rivers[11], which was traversed by Michielsen in 1880, three years after the great forest fire which had occurred there. The traveller remarked that it was difficult to see how the fire had spread, as he alternately passed dense forest, apparently unaffected, then completely burned patches, on which three years later only sparse low bush could be observed. On those burned patches, called *padang* by locals, burnt black or grey-black tree trunks, many uprooted, with their leafless branches, made a sorrowful display. Dead trees on the ground and dead trees in the air made a chaos of wood in which it was hard to find one's way (Michielsen, 1882: 83-4).

IMPACTS OF THE 1877-78 EVENT IN THE PHILIPPINES

For the Philippines the information is so far inadequate. While Manila observatory records show a four-month dry spell from January to April 1877 (Algue and Clos, 1900: 82-3), and observers from Sarawak were aware of a serious drought in Sulu, little useful data has so far been found. This may partly be due to the fact that the author does not have access to a comparable newspaper, equivalent to the *Javasche Courant* or *Sarawak Gazette*. However, the information is probably not available, for, as Foreman (1899) explained, very little of Mindanao was really under Spanish sovereignty, despite

---

[9] This effect was replicated just over a century later in 1982-3, when the most severe burning occurred in the early months of 1983 (see description in Brookfield *et al.*, 1995).

[10] Grabowsky (1890) travelling along the Batang Alai river in the Meratus Mountains in South Kalimantan noticed two treeless mountain peaks which he was told had been deforested by forest fire. However, he gives no indication as to when the event might have occurred. There is an earlier report by Dutch explorer von Gaffron of a very extensive fire in a swampy district of Kotawaringin, now Central Kalimantan, in 1846 (Pijnappel, 1860). That year was also listed as one with a strong ENSO by writers such as Quinn (1992), but no other information has been found about its impact on Indonesia.

[11] The area now lies on the main road between Central Kalimantan's capital Palangkaraya and Sampit. There was considerable burning again in this district in 1997, mainly a result of clearing for new plantation development.

occasional coastal bridgeheads and a little missionary activity. Although a treaty with Britain and Germany in 1877 recognised Spanish rights in the Sulu area, the Spanish presence there was very limited. Jesuit missionaries visiting Moro settlements in Mindanao in 1877 complained mainly of the extreme heat, especially when they climbed a mountain covered with *Imperata* grass (local name: *cogon*). The comment was that they "nearly suffocated under the scorching afternoon sun and the cogon grass, which seemed to be radiating flames of fire"(Schreuers, 1994: 1, 282).

THE DROUGHT OF 1902

Although there were a number of other dry events during the 1880s and 1890s, some of them accompanied by burning in particular districts[12], the drought of 1902 must be singled out, partly for its severe effect on food supplies in Indonesia, especially Java, but also for the haze which became widespread through the archipelago and was remarked upon in Singapore. It was described as a very abnormal year, with drought conditions lasting from April 1902 until February 1903, especially through July-October, where rainfall deficits were more than 50 per cent. In this exceptionally dry period, according to Braak (1915), a large number of reports were received of fog or haze, covering southeast Sumatra, southern Borneo, all of Sulawesi and extending across Maluku to the south coast of New Guinea. There were Dutch attempts to introduce regulations banning the burning of Imperata grassland in the Meratus mountains of South Kalimantan during this drought period (Wentholt, 1938), regulations described by the same writer as useless. So far no detailed accounts of exceptional fires burning at this time have come to light, except in the teak forests of Java (Boschwezen, 1903, 1904). Abandoned coffee gardens in Ulu Selangor on the Malay Peninsula were said to have been destroyed with the jungle in the dry weather (*Straits Times*, 3 September 1902).

While haze was experienced in Singapore, initially there was confusion as to its origin. On October 17 the *Straits Times* quoted the *Malay Mail* as saying: "there is a report current amongst the Natives near Seremban that a volcano in Sumatra in the Menang Kabu district is in a state of activity, and that the haze which has been prevalent lately is the smoke from that mountain.... Sumatra is very near at hand and the smoke could be very easily blown across the Straits". From that report it certainly sounded as

[12] The geologist Hooze, surveying the Meratus mountains of South Kalimantan in the late 1880s, like Forbes in Sumatra, described huge expanses of *alang-alang* grassland and commented: "This increasing deforestation is caused by the lay-out of dry rice fields (ladangs), for which over large areas the forest is felled and burnt and by the enormous field and bush fires, which in the dry season sometimes cover the whole divisions Martapura and Bandjermasin with smoke" (Hooze, 1893: 22). Endert (1925), exploring South and West Sumatran forests, quoted a report by Lever that in Sidjoengdjoeng, West Sumatra, a strong fire must have gone through in 1882, severely damaging the larger trees and altering the forest structure. This fire was near the Ombilin coalfield, and one suspects that underground coal seams may have caught alight, as has happened in recent times in East Kalimantan.

though there were fires in Sumatra, but whether or not they were of volcanic origin is presently unknown.

BURNING THE FORESTS OF THE FEDERATED MALAY STATES: THE TRANSITION TO RUBBER PLANTATIONS

One activity which has its counterpart in modern Indonesia, but which was unrelated to any drought event, was the calculated wide-scale burning in Peninsular Malaysia around 1909-10. This second rubber boom (the first was in 1905-6) saw the rapid transformation of large areas of rain forest into plantation monocultures. A graphic account in the *Straits Times* of August 24, 1910 describes the felling and clearing, usually carried out by a group of Chinese, who methodically stripped all the vegetation from a piece of forest then piled it into heaps for burning. After a few dry days and with the aid of kerosene, the fire "roars and blazes, covering the sky with a black pall of smoke, and turning to ashes all but the solid tree trunks". The subsequent description of the new plantations is easily recognisable, though not all would agree that their disciplined lines of trees could provide a potentially beautiful substitute for the forest:

> "Travelling through Malaya during the past year one came across scores of the black smoke columns which mark the chief stage of jungle clearing. But far more weirdly impressive is the aspect of the new plantations. Vast areas are seen on which there is no scrap of vegetation. Gaunt skeletons of the few trees which have been left to serve as landmarks stand out against the sky. The black earth is thickly strewn with the whitening trunks of felled trees, and they look like the rain-washed bones of great giants which have fallen in a dreadful battle. The old splendour of the jungle has passed away, as if a mighty cataclysm of nature had destroyed it, and the softer, more regular beauty of the plantation has still to come." (p. 12)

Such a description clearly demonstrated the attitude of the business interests of the time, and the belief that the dipterocarp forest had no value, so that even the largest trees were not worth salvaging.

DROUGHT AND FIRE IN THE PHILIPPINES: 1912 AND 1913

Two serious fires were reported in the forests of Negros in 1912, originating in areas cut over by two logging companies. Although incendiarism was suspected, it was also noted that there had been practically no rain at all in Northern Negros for ten months, which had made protection of the forests very difficult (Annual Reports, 1912: 31). In 1913 new rules were instituted to make timber companies responsible for preventing such fires over their concession areas: in particular they were required to remove all trash from the ground. The Director of Forestry observed: "It is really only within the logged over areas that difficulty will be encountered, as the virgin forests of the Philippine lowlands are immune to fire except at times of most severe drought, when their edges may be more or less damaged" (Annual Reports, 1913: 12).

DROUGHT, FIRES AND HAZE: INDONESIA AND SABAH IN 1914

The 1914 drought, judged to be only a moderate ENSO in other parts of the world, was nonetheless very severe in Indonesia and East Malaysia. Brookfield and Allen (1991) described it as the major regional dry event between 1877/8 and 1982/3. It was accompanied by much forest burning and well-documented examples of haze and mist.

While the Resident of South and East Borneo lamented the destruction of great forest complexes as a result of the 'long and fierce drought' (Rijckmans, 1916), forester Van der Laan (1925) specifically pointed to the burning of peat, which had caused the death of beautiful swamp forests in the same region. In East Kalimantan, Goldammer's Banjarese informant, whose grandmother had remembered the 1878 fires near Muara Lawa, had a story of burning in the same locality in 1914, and another informant described fires near the coast in Pasir (Goldammer, 1990: 19). Further evidence of burning was again observed some years after the event (Endert, 1927: 233), which had left patches of the vegetation thin-stemmed and poor in species. In West Kalimantan many tengkawang trees[13] were lost to bush fires (Boschwezen, 1914). Although the government forest service did not report other fires, they did institute prohibitions against burning of *Imperata* (Tapanuli, Sumatra); against the use of fire for hunting (Riau, North Sumatra); and against burning of ladang plots (Palembang, Bengkulu, Lampung) (Boschwezen, 1914). A dangerous fire in South Sumatra broke out at Muara Enim near the oil field installations of the Bataafsche Petroleum-Maatschappij, some of which were destroyed. The *Straits Times* (21 October 1914) reported: "The total damage is enormous...the conflagration still rages over a large area and great anxiety is felt for the petroleum fields in the neighbourhood."

One of the best fire descriptions came from Sabah, where Matthews, the Conservator of Forests, in 1917 came upon evidence of a large conflagration south of Mount Kinabalu. This fire, which had taken place two or three years earlier, had destroyed the forest over an estimated 100 square miles, including the area now known as the Sook Plain. Matthews was interested in the fact that this fire had been particularly widespread and destructive, and had "climbed the trunks of the trees to the crowns", which he considered very unusual in a tropical forest (Matthews, 1917). The Sook Plain forest has never grown back, as apparently the thin layer of peat covering an infertile sandy soil was destroyed. For many years the plain was clothed in *Imperata* grass (Cockburn, 1974) and more recently it has been included in government reforestation projects.

Dense haze and mist were observed during the 1914 drought in a number of locations in Sumatra. For two months Samosir Island could not be seen from the shores of Lake Toba and on the lowlands near Deli, "the sun appeared as a sharply delineated but dim

[13] *Tengkawang* or *Illipe* nut trees, which are a species of dipterocarp (*Shorea*) produce a nut rich in a kind of edible fat which may be used as a substitute for cocoa.

disk shining through the mist or smoke" (Braak, 1915: 131-2). A similar report came from the Kerinci highlands of Jambi. Both coasts of Sumatra were affected, with haze extending from the Malacca Straits to Borneo. The *Straits Times* of 31 October 1914 reported huge columns of smoke inconveniencing shipping in the Muntok and Karimata Straits from a large fire in West Central Borneo. Similarly, on the east coast near Samarinda and in Pasir, ship's captains were complaining of poor visibility caused by bush fires "which could be smelled and caused pain to one's eyes" while pieces of burning grass had landed on their decks (Braak, 1915: 133). Smoke originating from bush fires also increased the density of fog on the north coasts of Ceram (Maluku) and New Guinea.

Braak (1915) maintained that such fog and mist were not rare during the dry season over the archipelago, but that they became more pronounced with the greater intensity of the dry spells. They would always be dispersed with the arrival of the rains. Boerema (1917), in a general study of rainfall in the east monsoon (dry) months, set out to prove that 1914 was the driest year on record for Indonesia, using statistics from the network of 551 stations which then existed, some established after 1879 and others around 1912-13. Constructing a map of the rainless region, he showed that only the southernmost parts of Sumatra (South Sumatra and Lampung) would be included, plus the southern and eastern coasts of Borneo (penetrating further inland in East Kalimantan), together with most of Sulawesi and the remainder of Eastern Indonesia, including the western and southern coasts of New Guinea (see map 1). While such a map is reasonably accurate in plotting the locations of known forest fires, it omits the fires which certainly occurred in West Kalimantan, and in Sabah, although the latter admittedly has rather different seasons, experiencing its driest period early in the year. It is interesting also, given the distribution of haze over a large part of Sumatra, that much of that island was deemed to be excluded from participation in the most extreme drought conditions. From the descriptions in Braak, some burning did seem to be occurring further to the north. Peninsular Malaysia, West Kalimantan and Sarawak were equally excluded, though some of the Kalimantan forests were definitely burning.

Through an analysis of a large set of available rainfall data for Southeast Asia, Kripalani and Kulkarni (1997) have demonstrated that rainfall patterns in central and northern Sumatra and most of Peninsular Malaysia, together with Cambodia, southern Vietnam and the northern Philippines, are out of phase with the Indian Monsoon Rainfall (IMR). As drought periods in India tend to coincide with severe El Niño events, one might therefore expect those places to be less liable to such droughts. On the other hand, southern Sumatra, most of Borneo and Sulu, together with Eastern Indonesia, including the western coasts of Irian, are in phase with the IMR and are strongly influenced by

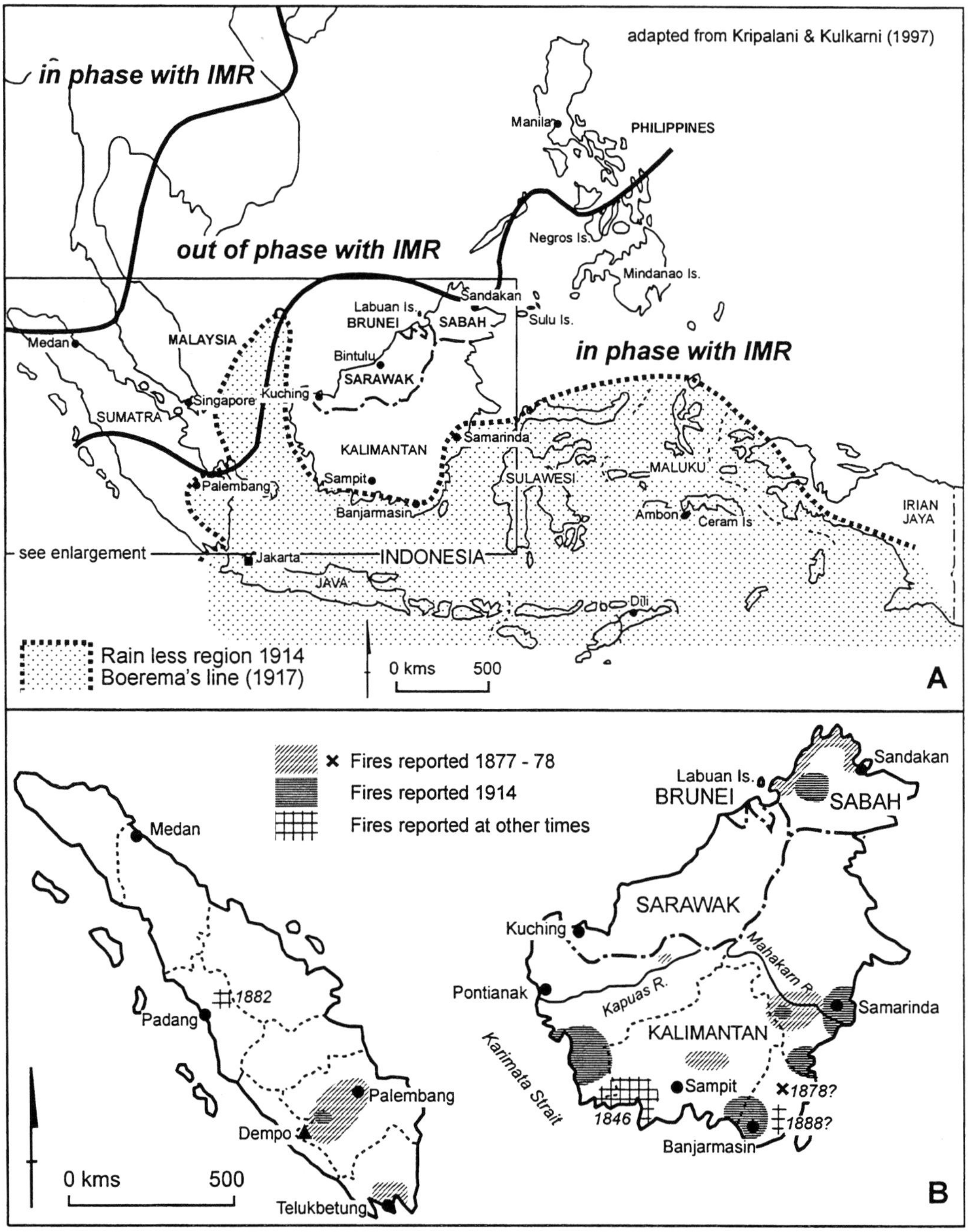

Map 1 A Regional climatic differentiation, **B** Indicative historical fire locations, Sumatra and Borneo

ENSO in a very similar way to India[14]. While Salavsky (1994) has noted the occurrenceof burning in West Kalimantan, it is also true that in most years the dry season is weaker there then further east, while King has observed the same for Sarawak, using rainfall for Bintulu (King, 1996). The difference may be clearly seen in the map compiled by Potter in Brookfield *et al.* (1995: 159). The greater extent of burning in 1914 may attest to the extreme severity of the drought in that year.

THE EVENT OF 1972

On 5 October 1972, Kuala Lumpur reported a thick haze, which lasted throughout the day. It was believed to consist of pollutants in the atmosphere caused either by a volcano or a forest fire. Singapore also reported haze, as did Medan airport, where, according to pilots, it extended up to 6000 feet. A week later the paper observed that the haze definitely did not have a source anywhere in Malaysia, and that satellite pictures had shown its origin to be Sumatra and southern Borneo (*Straits Times*, 13 October 1972). Jakarta denied that there had been any forest fire big enough to produce such a haze. The following day, as air traffic was disrupted in Palembang, the cause of the haze was being attributed to shifting cultivators clearing land for making ladangs. From various pilots' reports, it appeared that extensive fires were burning in South Kalimantan as well as South Sumatra. Crop failure and famine were at the same time reported from Papua New Guinea as a series of severe frosts hit highland agriculture[15].

With the now familiar pattern of fire in Indonesia during the dry season causing pollutants to be blown over its neighbours in the form of haze, one may inquire as to the trigger for these particular blazes. Blaming shifting cultivators is an easy option, but given that such people operate every year in the same manner and the haze does not occur every year, some other factor must have intervened. One suggestion is that such fires were connected with the young timber industry, just becoming established in parts of both Kalimantan and Sumatra. Clearing had to take place for the building of logging roads and camps; the use of fire for such clearing in what was obviously an ENSO drought year would be likely to lead to conflagration.

DISCUSSION

This paper has briefly examined a number of historical reports of forest fire and haze as these affected the 'maritime continent' of Malesia. There is strong evidence that severe

[14] Kripalani and Kulkarni (1997) produce a rather odd map (Fig 1, p. 1156) on which they indicate the location of the rainfall stations used. It is odd because they have called all of northern Borneo 'Brunei' and appear to believe that Malaysia means the Peninsula only. While this distorts some of their analysis, it does not affect their second map, already mentioned.

[15] Similar frosts had occurred in New Guinea during the droughts of 1902 and 1914 and also in 1941, another ENSO year which has not been discussed here.

drought conditions, usually accompanying an ENSO event, were very likely to produce fire in parts of the equatorial moist forest, especially if daily temperatures were higher than normal. While several of the fires reported were presumed to have taken place some years before they were noticed by an outside observer, the question of who was responsible for starting them, or indeed whether human action was involved at all, was generally not asked. In fact, it was only in 1972 that this question became really important. As the fire-haze situation was upgraded to fully international status, satellite imagery could pinpoint the location of a blaze and the Indonesian government was asked to explain the phenomenon. Since that time the identification of responsibility has become all-important, though even positive identification may not always lead to action.

In this chapter it has been shown that haze, as well as burning, was characteristic of the ENSOs of 1877-78 and 1914. The position in 1902, when drought and haze were well-established, but burning was either absent or less important, is so far anomalous, but might well repay more detailed research. Equally anomalous was the burning pattern in 1910 in the Federated Malay States, which demonstrated that the demands of the market, in that case the excitement of the rubber boom, could quickly produce a transformation in the vegetation which was far greater than that attributable to natural causes. Such a transformation has its modern counterparts, which are not limited to ENSO years. If one compares the present maps of fire occurrence in 1997-8 with the drought map of Boerema (1917), one can still see the impact of lower rainfall and ENSO-type factors, but the areas actually burned have gone far beyond Boerema's limits. Deliberate burning behaviour has obviously become more widespread and more market-driven than was the situation in 1914, although the Malaysian rubber boom produced extensive forest clearing at that time also.

The meaning of the drought-fire rich historical record, in terms of its impact on the forests, especially its relation to diversity and speciation, is still open to speculation. Goldammer, examining the patterns of recovery in the East Kalimantan forests, has made some interesting observations about the dominance of ironwood (*Eusideroxylon zwageri*), one of the most drought and fire resistant of the species involved, yet one of the fastest to recover. He wonders whether the dominance of ironwood in East Kalimantan might signal a forest prone to the stresses of drought and fire (Goldammer, 1990: 27). Ironwood (*ulin*) is similarly widespread in the Meratus mountains of South Kalimantan (though much reduced from its former abundance by clearing), and was once equally important in the South Sumatra lowlands and on the islands of Bangka and Billiton. Its range extended to Sulu in the extreme south of the Philippines, and to Sabah. These again are areas included in the map of districts most subject to drought. However, *belian* the ironwood tree is also widespread in West Kalimantan and Sarawak, which are not especially drought-prone, and there are pure ironwood forests in Jambi (Soerianegara and Lemmens, 1994: 214). If indeed ironwood is some kind of indicator of drought or fire, then more studies are needed to establish the circumstances under which this would occur. It is true that many of the dipterocarp species have also shown themselves well-adapted to cyclic droughts, even in apparently aseasonal Sarawak, where localised water stress may appear, especially on sandy soils (Brunig, 1969). The dipterocarps appear to need this stimulus for mast fruiting, a phenomenon of equal importance for their associated animal populations. The increased areas of *Imperata* grassland towards the end of the nineteenth century might also be seen as indicative of regions experiencing cyclic fire, as such grasslands are likely to expand rapidly after major bushfires.

The historical data supplied in this chapter provide a context for the modern studies of drought, fire and deforestation. Their intent is to make the reader aware of the differential history and the rapidly changing characteristics of much of the forest estate, an estate which is disappearing far too quickly throughout all of Malesia. While the record is incomplete and there is more to be learned about historical burning patterns, the links with the present are clear to see.

REFERENCES

1. Algue, J. and J. Clos, 1900. *El Archipelago Philippino*, Vol II, Washington DC, Government Printing Office.
2. Allen, B. 1989. Frost and drought through time and space, part 1: the climatological record. *Mountain Research and Development* **9** (3): 252-278.
3. Allen, B., H.C. Brookfield, and Y. Byron, 1989. Frost and drought through time and space, part II: the written, oral and proxy records and their meaning. *Mountain Research and Development* **9** (3): 279-305.
4. Annual Reports of the Director of Forestry of the Philippine Islands, 1912, 1913.
5. Bergsma, P.A. 1879. Regenwaarnemingen in Nederlandsch-Indie (Rainfall observations in the Netherlands Indies). *Natuurkundig Zeitschrift voor Nederlandsch Indie* **38**: 232-237
6. Bock, C. 1881. *The headhunters of Borneo: a narrative of travel up the Mahakam and down the Barito, also journeying to Sumatra.* Sampsom Low, Marston, Searle and Rivington. London. Reprinted, Singapore, Brash, 1988.
7. Boerema, J. 1917. Regenval in de oostmoessonmaanden en uitbreiding van het regenlooze gebied gedurende droge oostmoessons in Nederlandsch Indie. (Rainfall in the east monsoon months and the extent of the rainless region during dry east monsoons in the Netherlands Indies). *Natuurkundig Tijdschrift voor Nederlandsch Indie* **76**: 46-63.
8. Boschwezen, 1903, 1904, 1914. *Verslag van den Dienst van het Boschwezen in Nederlandsch Indie over het Jaar 1903, 1904, 1914* (Annual Report of the Forest Service of the Netherlands East Indies 1903, 1904, 1914), Landsdrukkerij, Weltevreden.
9. Braak, C. 1915. De sterke nevel in de oostmoesson van 1914 (The dense fog in the east monsoon of 1914). *Natuurkundig Tijdschrift voor Nederlandsch Indie* **74**: 131-142
10. Brookfield, H.C., L.M. Potter, and Y. Byron, 1995. *In Place of the Forest: Environmental and Socio-economic Transformation in Borneo and the eastern Malay Peninsula.* United Nations University Press, Tokyo.

11. Brunig, E.F. 1969. On the seasonality of droughts in the lowlands of Sarawak (Borneo). *Erdkunde* **23** (2): 127-133.
12. Burbidge, F.W. 1880. *Gardens of the Sun: or A Naturalist's Journal on the Mountains and in the Forests and Swamps of Borneo and the Sulu Archipelago.* John Murray, London.
13. Cockburn, P. 1974. The origin of Sook Plain, Sabah. *Malaysian Forester* **37**: 61-63.
14. Crutzen, P. J. and J.G. Goldammer, 1993. *Fire in the Environment: the Ecologic, Atmospheric, and Climatic Importance of Vegetation Fires.* Wiley, London.
15. Dale, W. 1959. The rainfall of Malaya. *Journal of Tropical Geography* **13**: 23-37.
16. Endert, F., 1925. Boschbouwkundige Aanteekeningen over een Reis in Atjeh, Sumatra's Oostkust, Tapanoeli en Sumatra's Westkust. (Silvicultural notes from a trip to Atjeh, Sumatra's East Coast, Tapanoeli and Sumatra's West Coast). *Tectona* **XVIII** (1): 1-116.
17. Endert, F., 1927. Floristisch Verslag (Report on Flora). *Midden-Oost Borneo Expeditie*, 1925 (Expedition to Central-East Borneo, 1925) D. Buijs, H. Witcamp, F. Endert and T. Siebers, Weltevreden,G. Kolff and Co., pp. 200-283.
18. Forbes, H.O. 1885. *A Naturalist's Wanderings in the Eastern Archipelago: a Narrative of Travel and Exploration from 1878 to 1883.* Harper & Bros., New York.
19. Foreman, J. 1899. *The Philippine Islands: a Political, Geographical, Ethnographical, Social and Commercial History of the Philippine Archipelago and its Political Dependencies, Embracing the whole Period of Spanish Rule*, 2nd. edition. Scribner's Sons, New York.
20. Gerlach, L.W.C. 1881. Reis naar het Meergebied van den Kapoeas in Borneo's Westerafdeeling (Journey to the Lake District of the Kapoeas in Borneo's Western Division). *Bijdragen tot de Taal- Land- en Volkenkunde van Nederlandsch Indie* **29**: 285-322.
21. Goldammer, J.G. 1988. Rural land-use and wildland fires in the tropics. *Agroforestry Systems* **6** (3): 235-252.
22. Goldammer, J.G. 1991. Tropical wildland fires and global changes: Prehistoric evidence, present fire regimes and future trends. In: *Global Biomass Burning: Atmospheric, Climatic, and Biospheric Implications*, ed. J. Levine, MIT Press, Cambridge, MA, pp. 83-91.
23. Goldammer, J. and B. Seibert, 1989. Natural rain forest fires in Eastern Borneo during the Pleistocene and Holocene. *Natuurwissenschaften* **76** : 518-520.
24. Goldammer, J.G. and B. Seibert, 1990. The Impact of Droughts and Forest Fires on Tropical Lowland Rain Forest of East Kalimantan. In: *Fire in the Tropical Biota: Ecosystem Processes and Global Challenges*, ed. J.G. Goldammer, Springer-Verlag, Berlin. pp.11-31.
25. Grabowsky, F. 1890. Streifzuerge durch die malayischen Distrikte Suedost-Borneos II. *Globus* **57**: 219-221.
26. Harger, J.R.E. 1995a. Air Temperature Variations and ENSO Effects in Indonesia, the Philippines and El Salvador. ENSO Patterns and Changes from 1866 to 1993. *Atmospheric Environment* **29** (16): 1919-1942.

27. Harger, J.R.E. 1995b. ENSO Variations and Drought Occurrence in Indonesia and the Philippines. *Atmospheric Environment* **29** (16): 1943-1955.
28. Hooze, J. A. 1893. Topografische, geologische, mineralogische en mijnbouwkundige beschrijving van een gedeelte der Afdeeling Martapoera in de residentie Zuider- en Oosterafdeeling van Borneo (Topographic, geological, mineralogical and mining description of a section of the district of Martapura in the residency of South and East Borneo). *Jaarboek van het Mijnwezen* **22**, pt. 2: 22.
29. *Javasche Courant* (Daily newspaper, Batavia), 1877-78.
30. *Journal of the Straits Branch of the Royal Asiatic Society*, 1878-1923.
31. Kiladis, G.N. and H.F. Diaz, 1986. An Analysis of the 1977-78 ENSO Episode and Comparison with 1982-3. *Monthly Weather Review* 114: 1035-1047.
32. Kiladis, G. N. and H. F. Diaz, 1989. Global climatic anomalies associated with extremes in the Southern Oscillation. *Journal of Climate* **2**: 1069-1087.
33. King, V.T. 1996. Environmental change in Malaysian Borneo: fire, drought and rain, In: *Environmental Change in Southeast Asia*. Eds. M. Parnwell and R. Bryant, Routledge, London. pp. 165-189.
34. *Koloniaal Verslag* (Colonial Report), 1878, 1879, 1903, 1915.
35. Kripalani, R.H. and A. Kulkarni, 1997. Rainfall variability over South-East Asia – connections with Indian monsoon and ENSO extremes: new perspectives. *International Journal of Climatology* **17**: 1155-1168.
36. Leighton, M. 1984. The El Niño-Southern Oscillation event in Southeast Asia: effects of drought and fire in tropical forest in eastern Borneo. WWF-US No 293 project report (typescript).
37. Low, H. 1955. *The Journal of Sir Hugh Low, Perak 1877*. Ed. E. Sadka, Government Printing Office, Singapore.
38. Matthews, D.M. 1917. Conservator of Forests, British North Borneo to Chairman North Borneo Company, 20 April 1917. CO 874/714. *British North Borneo Company Papers*, Singapore University Library.
39. Michielson, W.J.M.. 1882. Verslag eener reis door de bovendistricten der Sampit en Katingan rivieren in Maart en April 1880. (Report of a journey through the upper districts of the Sampit and Katingan rivers in March and April 1880). Tijdschrift voor Indische Taal, *Land-en Volkenkunde* **28**: 1-87.
40. Nicholls, N. 1992. Historical El Niño/Southern Oscillation variability in the Australasian region. In: *El Niño: Historical and Paleoclimatic Aspects of the Southern Oscillation,* Eds. H.F. Diaz and V. Markgraf, Cambridge University Press, Cambridge. pp. 151-172.
41. Nicholls, N. 1993 ENSO, drought and flooding rain in Southeast Asia. In: *South-East Asia's Environmental Future: the Search for Sustainability*. Eds. H.C. Brookfield and Y. Byron, Oxford University Press and United Nations University Press, Singapore and Tokyo, pp. 154-175.
42. Pijnappel, J. 1860. Beschrijving van het Westelijke Gedeelte van de Zuider en Ooster-Afdeeling van Borneo (De Afdeeling Sampit en de Zuidkust). *Bijdragen tot de Taal-Land-en Volkenkunde* **3**: 243-346.

43. Potter, L. M. 1999. Forest and grassland, drought and fire: past and present environmental transformations in Borneo. In: *Environment and Conservation in Borneo*. Ed. P. Eaton, Borneo Research Council. pp. 9-35.
44. Quinn, W. H., D. O. Zopf, K.S. Short, and T.W. Kuo Yang, 1978. Historical Trends and Statistics of the Southern Oscillation, El Niño, and Indonesian Droughts. *Fishery Bulletin* **76** (3): 663-678.
45. Quinn, W.H. 1992. A study of Southern Oscillation-related climatic activity for A.D. 622-1900 incorporating Nile River flood data. In: *El Niño: Historical and Paleoclimatic Aspects of the Southern Oscillation*. Eds. H.F. Diaz and V. Markgraf, Cambridge University Press, Cambridge. pp.119-148.
46. Ropelewski, C.F. and M.S. Halpert, 1987. Global and regional scale precipitation patterns associated with the El Niño/Southern Oscillation. *Monthly Weather Review* **115**: 1606-1626.
47. Rijckmans, L.F.J. 1916. *Memorie van overgave van de Zuider-en-Ooster Afdeeling van Borneo*, (Handing-over memorial of the Southern and Eastern Division of Borneo) No 271, Algemeen Rijksarchief (ARA), The Hague, (unpublished).
48. Salavsky, N. 1994. Drought in the rain forest: effects of the 1991 El Niño-Southern Oscillation event on a rural economy in West Kalimantan, Indonesia. *Climatic Change* **27**: 373-396.
49. *Sarawak Gazette*, daily newspaper, Kuching, Sarawak, 1877-8, 1902-3.
50. Schreurs, P. 1994. *Mission to Mindanao 1859-1900*. From the Spanish of S.J. Pablo Pastells, Vol 1, Cebu City, University of San Carlos Publications.
51. Soerianegara, I. and R.H.M.J. Lemmens (Eds.), 1994. Plant Resources of South-East Asia, No. 5 (1) Timber Trees: Major Commercial Timbers. PROSEA, Bogor.
52. *Straits Times*, daily newspaper, Singapore, 1902-3, 1910, 1914-15, 1972.
53. Tromp, S.W. 1890. Mededeelingen uit Borneo (Information from Borneo). *Tijdschrift voor de Aardrijkskundig Genootschap van Nederlandsch Indie* **7**: 728-763.
54. Van der Laan, E. 1925. De Bosschen van de Zuider- en Ooster Afdeeling van Borneo, (The forests of the South-Eastern Residency of Borneo) *Tectona* XVIII: 925-952.
55. Veth, P.J. 1882. Midden-Sumatra. Reizen en Onderzoekingen der Sumatra-Expeditie, Uitgerust door het Aardrijskundig Genootschap, 1877-1879. Tweede Deel Aardrijkskundige Beschrijving (Middle Sumatra. Journeys and Investigations of the Sumatra Expedition, facilitated by the Geographical Society, 1877-79. Part 2, Geographical Description). Brill, Leiden.
56. Wentholt, J. 1938. *Memorie van overgave, Afdeeling Martapoera, Zuider-en Ooster Afdeeling van Borneo* (Handing-over memorial, division of Martapura, Southern and Eastern Divisions of Borneo). Algemeen Rijksarchief, The Hague.
57. Whetton, P., D. Adamson, and M. Williams 1990. Rainfall and river flow variability in Africa, Australia and East Asia linked to El Niño-Southern Oscillation events.*Geological Society of Australia Symposium Proceedings* **1**: 71-82.

Chapter 4

ANTHROPOGENIC FIRES IN INDONESIA: A VIEW FROM SUMATRA

M. Roderick Bowen[1], Jean Marie Bompard[2], Ivan P. Anderson[1], Philippe Guizol[3] and Anne Gouyon[4*]

[1]Natural Resources Institute, University of Greenwich Central Avenue, Chatham Martime, Kent ME4 4TB, UK

[2]c/o CIRAD-Forêt, Campus International de Baillarguet, B.P. 5035, 34032 Montpellier cedex 1, France

[3]CIRAD-Forêt attached to CIFOR, P.O. Box 5696 JKPWB, Jakarta 10065, Indonesia

[4]Idé-Force, 8 rue de l'Isly, 75008 Paris, France

INTRODUCTION

It is safe to regard all vegetation fires that occur on the main forested islands of Indonesia – Kalimantan, Sumatra and Irian Jaya – as anthropogenic, i.e. they are started by man and it is man, not fire, that destroys the forests. Only vested interests in the logging and plantation industries suggest that lightning strikes are responsible, and the small number of vegetation fires in Sumatra and Kalimantan that are ignited by exposed burning coal seams is insignificant within the overall picture.

We need to define our terms with care. The words 'forest fire' are emotive and may conjure up a picture of a tropical primary rain forest in flames. This was not the general case during the 1997-98 fires in Indonesia (at least in Sumatra) where it is rare for fires to pass beyond the margin of virgin forest even in drought years except in certain forest

* At the time the work was carried out the authors were attached to the European Union-Forest Fire Prevention and Control Project, Palembang, South Sumatra in conjunction with the Indonesian Ministry of Forestry and Estate Crops. The opinions expressed here are those of the authors.

types (e.g. exposed up-hill sites and ridge tops with shallow soils, heath forests – *kerangas* – and dried swamps). It should be noted also that relatively undisturbed primary forest is rare in areas most affected by fires. However, the term is popular with both conservationists and donors as it attracts attention and with it, money.

A second reason why the term is widely used is that areas that were governed under the Basic Forestry Law of 1967, and since September 1999 under the New Forestry Law, are known as Forest Land. Any fires that occur on them are thus classified by government as forest fires. (Fires that occur outside the forest estate are often referred to as land fires). As over 140 Mha, or nearly 75 percent of the entire land area of Indonesia, is under the jurisdiction of the Ministry of Forestry and Estate Crops, it is easy to see why the appellation forest fire has stuck.

The main, and to us over-riding, objection to the use of the phrase, is that the great majority of fires burn grassland, scrub and stretches of logged-out forest that have been, or are about to be, felled, often for conversion to agriculture. In general, they do not burn forest in the commonly understood sense of, 'large tracts of land covered with closed-canopy trees'. Thus while we recognise that heavily logged forest, disturbed forest in nature reserves and national parks, and locally relatively untouched peat swamp forest and heath forest do suffer considerable fire damage during periods of drought, we prefer to apply the single term 'vegetation fire' irrespective of the main fuel or the location of the fire.

The expression 'wildfire' is used only in the sense of: 'any fire that is not meeting its management objectives and thus requires suppression' (Nicolas and Beebe, 1999). In non-drought years we believe that there are few wildfires of any consequence in Indonesia. In 1999 in Sumatra only one was identified with certainty and one with high probability; both are strongly believed to have arisen from commercial land clearance. A similar situation is reported in Kalimantan with no evidence of wildfires during the year (Nicolas, pers. com.). In El Niño years the picture is dramatically different and thousands of fires run way beyond their intended limits.

Nobody knows with certainty how much forest is left in Indonesia. Whitmore (1990) noted that more than 50 percent of the primary rainforest in Sumatra that existed in 1980 had been cleared using fire by 1985, and that by the end of the decade the dry-land forests of the lowlands remained only as relics. The National Forest Inventory published in 1996, based on 1986 – 1991 satellite data and generally held to be reliable, suggests that 121 Mha were 'forest covered' at the time of the survey. Of these, 22 Mha were montane forest (on land over 1000 m), and 28 Mha were classed as conservation and protection forest, none of which is supposed to be commercially exploited. The remaining 71 Mha is open to selective logging.

Of the 71 Mha of forest set aside for logging, 22.5 Mha is officially "Conversion Forest" and is liable to be 'developed'. In the Indonesian context, development can be often equated to opening access to large tracts of sparsely populated Forest Land for other uses. This is particularly the case in the outer islands of Sumatra and Kalimantan where huge areas have already been unlocked by large-scale logging, by clear felling for agro-industrial crops, by oil and gas exploration, and in pursuit of the transmigration programme.

Fraser (1999) suggests that conversion to agriculture could continue for 25 years at around 900 000 ha a year, and of this 600 000 ha a year is likely to be for oil palm estates (Wakker, 1999). Most will be cleared through the systematic burning of logged forest. There is unlikely to be any primary lowland forest left in Indonesia within ten years.

The examples we have chosen to illustrate the use of fire by man and the pattern of vegetation fires that their use produces come from Sumatra (Map 1). Causes and effects, however, are similar throughout the country. References to field-based studies similar to our own in East Kalimantan include, amongst others, Schindler (1998) and Hoffmann *et al.* (1999). No similar assessments are known from Irian Jaya, although according to Ministry of Forestry and Estate Crops statistics, its Forest Land totals 41 Mha (Sumatra 31 Mha, Kalimantan 45 Mha) and its logging and estate industries are increasingly active. The gap in coverage needs filling.

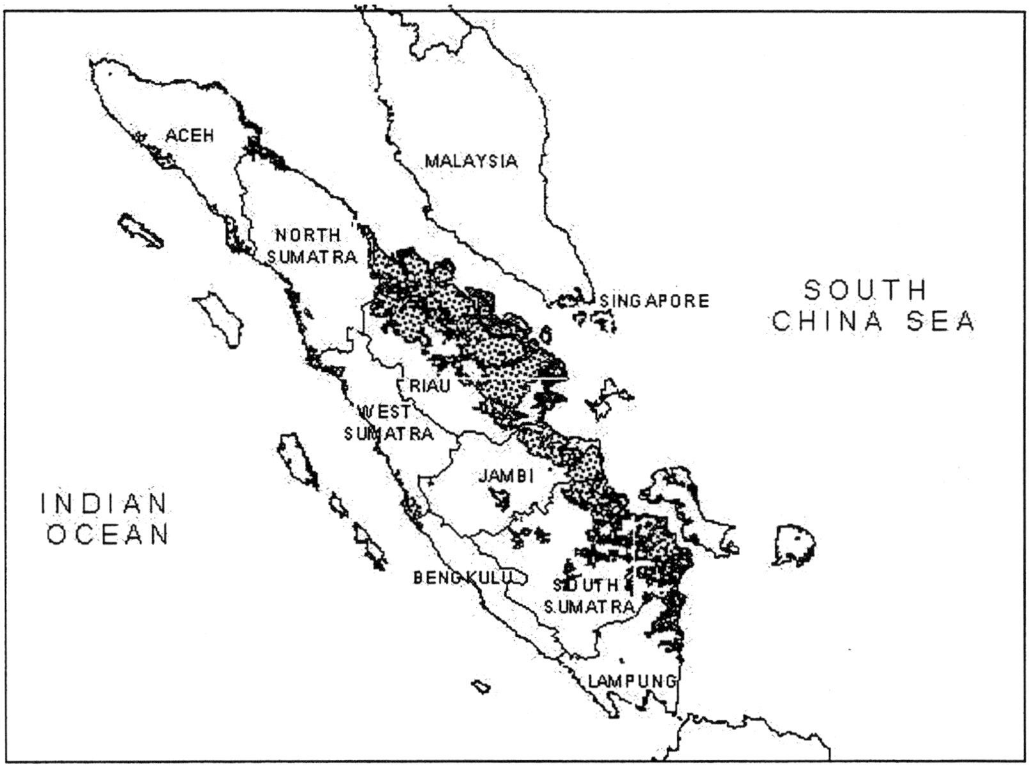

Map 1 Sumatra showing the provincial boundaries: areas with peat swamp soils are shaded grey.

The Use of Fire in Land Management

The agro-industrial development of Sumatra began at the turn of the twentieth century when the Dutch established plantations – mainly rubber – on the fertile, easily accessible coastal plains of what is now North Sumatra province (Fisher, 1964). Known as the *culuurgebied*, this was one of the first areas outside Java to be permanently settled by the Dutch and the use of fire to clear land for rubber estates around Medan in the twenties is vividly described by Laszlo Szekely (1936) in his semi-autobiographical novel *Tropic Fever*.

Forest exploitation, followed by conversion to agriculture, next started in Lampung province in the far south of the island, a development promoted by its proximity to land-hungry Java. From there, timber harvesting, followed by settlement and land conversion, spread northwards through the provinces of South Sumatra in the eighties and early nineties, Jambi in the early and mid-nineties and finally reached Riau within the last four years to meet up with the original development in North Sumatra. The transformation of much of Sumatra took just twenty years and has been accompanied by fire throughout.

Three provinces have so far escaped relatively lightly from the development process, and thus from fires. Bengkulu has a small west-coast plain but is mostly within the mountain spine of the Bukit Barisan which deters logging. West Sumatra province similarly lies partly within the mountains, although some logging and conversion to agriculture and estate crops has taken place on its coastal plain, most recently in wetlands. In Aceh, in the far north of the island, the loss of primary forest has been somewhat less than in other regions ("Only 30 percent between 1980 and 1985", Whitmore, 1990), deterred by the more difficult terrain and perhaps also by political unrest. Aceh has also seen considerable local resistance to transmigration, particularly during the 1980s.

We have mentioned that development is to all intents, synonymous with opening up of land. The process leads to a rash of large fires until development is complete and a new 'fire equilibrium' is reached. The parts played by logging, forest conversion, the petroleum industry, small-holders, transmigration, the invasion of the wetlands, and arson provide convenient headings under which fire-use and fire-risk can be examined, although in reality all are closely inter-linked and the artificial separation can mislead.

Commercial Logging

Fire has followed the chain saw. Logging *per se* starts no fires but the indirect effects have been, and continue to be, devastating. As pointed out by Nicolas and Beebe (1999), "Where [logging] roads penetrate primary forest, fire danger is considerably increased. Humidity is lower, wind speeds are greater and there is always a ready supply of drier fuel available: both from the dumping of debris during road construction and from the grass, creeping vines and small shrubs that invade along the road line". Wildfires started by carelessness or by neighbouring land clearance are frequent and spread readily over long distances adjacent to the road.

The degree of danger to the forest after logging basically depends on the felling regime and the time that has passed since the coup. A heavily logged forest – the great majority in Indonesia – has some 50 percent of the ground flattened by bulldozer extraction and contains an ample load of light fuels from crown off-cuts, broken small trees and invading shrubs and herbs that dry easily under the opened canopy. In the few areas where low impact logging is practiced, the immediate danger is considerably reduced but the fire risk remains well above that in undisturbed forest for three to five years. Regardless of the logging regime, a second fire, if sufficiently intense and following the first within a few years, will permanently destroy the forest and the area will revert to scrub and grassland.

Forest Conversion and New Plantations

The agro-industry arrives and takes over the area once the forests have been logged. It is this clearing of the remaining logged forest for oil palm estates, and at present to a much lesser extent for rubber and wood-crop plantations, that is unquestionably the prime cause of the great majority of the large vegetation fires that occur with regularity each year. The total area burnt is huge and smoke haze pollution can be severe when clearing wetland sites. In a typical conversion operation, all valuable timber is cut and extracted before the remainder of the trees is felled. Bulldozers are then used to pile (windrow) the debris that is left to dry before being set alight. Properly managed, although it rarely is, the burning is controllable.

Ideally the new estate crop is established at the start of the next rainy season to allow it to gain a rapid control of the site and suppress weed growth. But often far greater areas are cleared at one time than are planted. This is partly poor planning and partly a desire by the company to demonstrate possession of the land. In either case the open ground encourages the rapid colonization by herbs particularly the grass *Imperata cylindrica (alang-alang)*. The area then becomes highly vulnerable to repeat fires in the next dry period.

Even where new crops are established promptly, they are often inadequately weeded and thus at high risk of fire for three to six years until canopy closure, which is itself delayed by weed competition. After canopy closure oil palm and rubber plantations become low fire risks. However some pulp and medium density fiberboard species may experience considerable leaf loss during long dry spells: leaf-litter accumulates in the somewhat open plantations, particularly in periods of drought, and invasive wildfires readily become destructive crown fires.

There are 261 registered oil palm estates in Riau province alone, with concessions estimated to cover between one and two million hectares out of the total of five million within the province officially listed as Conversion Forest. A sizeable percentage of the one to two million hectares already has been, and eventually all will be, cleared by fire.

Elsewhere in Sumatra, controlled burning on this scale is, fortunately, rare, although blocks of fire linked to commercial land clearance – mostly for oil palm - are occasionally seen in all Sumatra provinces except in the more settled agricultural regions in the north and the south of the island. The smoke haze which periodically affects Singapore and peninsular Malaysia (most recently in April and July 1999) derives from peat set on fire while clearing for new estates in Riau, and it is likely to recur for some time yet.

The Palm Oil Industry

Indonesia is the second largest supplier of palm oil in the world – Malaysia is the largest - and in 1999 is expected to produce 5.4 Mt (World Bank, 1999). Current prices stand at \$550 t^{-1} and the Bank forecast suggests that they will only decline slightly over the next ten years. A middle-of-the-road estate yields 4.8 t ha^{-1} y^{-1} of oil, giving an income well above the break-even production cost of \$150 t^{-1} of a company with average capitalisation costs (Vincent, pers. com.). It therefore appears that in simple economic terms, conversion to oil palm plantations will remain attractive in the foreseeable future. That a similar conclusion has been reached by government and the plantation industry is clear from the large tracts earmarked for conversion.

The Petroleum Industry

Commercial logging is the main but not the only industry that opens up the forest and makes it directly and indirectly vulnerable to fire. The oil and gas industry in both Kalimantan and Sumatra has developed an extensive network of all-weather roads that lead to numerous wells, storage areas and pumping stations and run alongside the interconnecting pipework. Riau and South Sumatra are major gas and petroleum exploration provinces and the industry has had a profound effect on the landscape. In Riau these roads have greatly facilitated the influx of both illegal loggers and migrants into formerly remote areas. A similar instant response to new roads is also reported by estate crop companies who claim that illegal loggers frequently fell timber in anticipation of the road's arrival, tipped off by junior employees or government officials (Khoo, pers. com.).

Smallholders

Indigenous hunter-gatherers who live within the forest now form a tiny percentage of Indonesia's population of over 200 million. Ethnically diverse, they are increasingly found only in the remotest regions of Sumatra and the less accessible parts of Kalimantan and Irian Jaya. Their use of fire to clear land is negligible - a practice new to their traditional hunter-gatherer activities, brought about by land shortage, economic pressure and coercion to participate in resettlement programmes. There is a tendency by the ignorant, however, to equate such peoples with slash-and-burn farmers, a broad grouping often painted in the past – less so now - by government and international agencies as the villains.

To many people the phrase 'slash-and-burn farmer' means one who wantonly fells an area of virgin forest, burns it, plants subsistence crops, harvests, and moves on to repeat the process, and, in doing so, destroys large areas of forest. It is now recognised that in Indonesia this is not the reality. In South Sumatra province, farmers live in settled rural communities. Individually they burn only limited areas within the known permanent boundaries of their own farms to clear land on a cyclic basis to plant the food and the income-earning crops that they need to support their families. We thus prefer to use the less prejudicial and far more accurate term 'smallholders'. While smallholders light the great majority of fires that occur in South Sumatra, Nicolas (1998) notes that "Farmers have burnt land to clear it for centuries and have developed the skills to control the fires. They have an excellent knowledge of fire behaviour in their environment. Burning is usually carried out by small groups to control the fire within limited areas of 2 to [a rare maximum of] 25 ha, and the fires are set [in any one village] over only a few days."

Transmigration Schemes

Indonesia's Transmigration Program is one of the largest resettlement schemes in the world and has been much criticized. Supporters point to the orderly movement as alleviating pressure on Java, Bali and Madura (the inner islands) and contributing significantly to the development of the outer islands; in this instance referring largely to Kalimantan and Sumatra. According to the World Bank (1994), more than 3.6 million people were resettled between 1903 and 1990, the great majority between 1968 and 1990. Government figures put the number at approximately 8.5 million between 1969 and 1990 (GoI, 1999). Detractors point to considerable resources wasted – World Bank support totalled $ 560 million up to 1994 – in settling people who have been unable to progress beyond subsistence levels, and also to the concomitant extensive damage to the environment, as well as to the great detriment of local people living in the transmigration areas.

Sections of a World Bank report on the programme (World Bank, 1994) certainly provide ammunition for the detractors. "Probably no single transmigration site, treecrop area, logging concession, or industrial timber plantation has caused a loss of biodiversity that is significant at the national level [a contentious statement]. But when they are viewed in combination their serious impact can be appreciated." The same report also observes that land clearing failed to comply with guidelines, and on a number of occasions commercial timber was left partly burned in the fields.

There have been many widely varying 'guestimates' but no agreement, on how much land has been directly cleared for transmigrants. If we assume as Dick (1991), that 2 ha per family were cleared for each family-farm plus an additional hectare for infrastructure, and also that a family is five people, then based on the World Bank figure of transmigrant numbers, this equates to something over 2 million hectares. All was cleared by bulldozer and fire at an average rate of around 90 000 ha y^{-1}.

To add to the problem, many of the areas initially settled were totally unsuitable for farming and have by now been largely abandoned. They have subsequently been regularly reburnt by wildfires and will remain a highly fire-susceptible reservoir for years ahead. Dense smoke haze can also be expected to continue as many of the schemes were in peat swamp areas.

The Invasion of the Wetlands

A substantial literature exists on the peatlands and swamps of Indonesia, and while their role in carbon sequestration remains under debate, there is unanimous agreement on their great ecological importance and on their fragility. Extensive coastal and inland wetlands are found in all three main forestry islands. Rieley and Page (1997) and Rieley *et al.* (1997) between them provide a comprehensive, nationwide overview of the geography, vegetation, forestry and land-use of these zones.

The wetlands of Sumatra lie largely along the eastern coastal regions in the provinces of Riau, Jambi and South Sumatra (Map 1). There are also limited coastal areas in Lampung, and some inland peats in other provinces. Peat and swamps together cover some 8.3 Mha (RePPRrot, 1990) of which 4.3 Mha of peat soils up to 10 m deep lie in Riau and account for 27 percent of Indonesia's total (PEMDA, 1999).

These areas have never been 'fire-proof' and there is a limited history of fire occurrence before the eighties. Since then frequency, numbers and severity have risen steadily (Meijaard and Dennis, 1997), exacerbated by unsustainable land-use and exploitative forest management practices that began in the late sixties (Danielsen and Verheugt, 1990; Brinn, 1992). Legal and illegal logging, the transmigration programme and conversion to estate and pulpwood plantation crops have all fragmented the forest and the present vegetation is a mosaic of grassland, shrubland, open woodland and rare relicts of the natural closed forest cover. All are vulnerable to fire.

South Sumatra province has received large influxes of transmigrants since 1969 and, around 30 000 ha of pristine coastal peat swamp forest were 'reclaimed' by chainsaw, drainage and fire to accommodate many of the new arrivals (Dis, 1986). At least 28 000 people were settled on tidal swamp schemes to the north and east of Palembang (Danielsen and Verheugt, 1990). Intended for tidal rice cultivation, the schemes were conspicuously unsuccessful and in many cases failed completely. The drainage has caused a marked reduction in the dry season water table over large areas, and this, together with the grassland and shrub vegetation that replaced the forest, has substantially increased the fire risk that is low in the undisturbed ecosystem.

Extensive legal and illegal logging have devastated widespread areas and increased fire risk in other parts of the swamps. Some of the most heavily exploited areas are now being planted with pulp and timber species. The use of fire to clear the land is standard, and in 1997 one logging company that had applied to government for authorization to convert to a pulpwood estate, systematically burnt 40 000 ha of peatland in anticipation of approval (Nicolas and Bowen, 1999). Areas that have previously burnt are particularly fire-vulnerable even in years of average rainfall: thus in 1999 a wildfire that lasted from June to September ranged over some 15 000 ha of the 40 000 ha first torched in 1997.

Coastal peats in Sumatra and inland peats in Kalimantan are also increasingly being used to establish oil palm plantations as more suitable land becomes scarce. Maintenance costs are higher – regular applications of boron, zinc and copper are needed with heavy liming – but the land is cheap and with careful water management, profits remain high. Having cleared the land by fire, it is paradoxically necessary to maintain a high watertable for the palm during periods of drought (Khoo, pers. com.) and this may serve to protect the peat from further fires, and its resultant dense smoke haze.

The growing of *sonor* rice is a tradition of people living along rivers in the wetlands of South Sumatra and has now been taken-up by transmigrants. It is practiced only at the end of at least a three months drought when long-straw local rice varieties are direct-sown into a rising water table. The crop is harvested by boat. The danger comes from the fires set to clear the land. The farmers 'prospect' to find the most fertile soils and clear up to five times more land than is sown; wildfires are an inevitable consequence. Ramon (pers. com.) suggests that a minimum of 75 000 ha were cleared for *sonor* in South Sumatra in 1997. The additional area burnt by the wildfires that resulted is impossible to quantify but is likely to be far higher. (Nicolas and Bowen (1999) estimated 500 000 ha of coastal wetland was damaged by fires from all sources). *Sonor* cannot, and should not, be eliminated by authority but steps can be taken to limit the fire risks. Bompard and Guizol (1999) give a full description.

When access to the fire location is possible, surface fires in peat areas are extinguishable with hand tools and back-pack pumps although the fires are liable to spread quickly and overwhelm resources. Sub-surface fires are far more difficult to control and need to be isolated within trenches dug to the full depth of the peat that commonly exceeds two metres, an impossible task when such fires are numerous in drought years. Prevention is far preferable to failed suppression.

Accidental Fires

Vayda (1999) cautions against the tendency to be dismissive of "so-called accidental causes" of vegetation fires and singles out Byron and Shepherd (1998) for their statement that fires from carelessly discarded cigarettes are rare. However, while correctly pointing out the lack of evidence one way or the other, Vayda then goes on to suggest, with no evidence presented, that probably much more common as a source of wildfires are small fires lit by hunters and collectors within the forest and by drivers repairing vehicles alongside roads! As we mentioned earlier, serious wildfires, in this sense synonymous with accidental fires, in non-drought years are rare. In El Niño years there is indeed no evidence for or against accidents but it is undoubted the great majority have been deliberately lit and then escaped their intended limits.

Arson

Arson is often mentioned, with no supporting evidence, as a source of fire in commercial estates; official sympathy seems to lie with the companies rather than with the communities that feel they have lost their land rights. Evidence is indeed, and understandably, hard to come by and the nearest that villagers come to admitting to the use of fire as a weapon is to talk of *tangan liar* (wild hands) (Bompard and Guizol, 1999). Companies are equally reluctant to draw attention to any question of arson as to do so also calls attention to their dubious claims to some lands. Even less frequently mentioned is that village communities in turn, strongly suspect the companies of arson and fireing smallholder plots as a way to reduce the compensation they need to pay when they eventually acquire the land.

Reliable figures are impossible to obtain but Gouyon (1999) records that in 1998 in South Sumatra, 104 communities were in legal dispute with 40 oil palm estates who had reported 3 500 ha burnt – out of 79 000 ha they controlled – in the 1997 drought. The fires may have been arson or accident but the number of disputes gives an indication of the magnitude of the potential problem in one province.

In Riau, relationships between local residents and Javanese transmigrants and Batak spontaneous migrants have been embittered by the manipulation of funds and facilities designated for transmigration projects (Hoshour, 1995). And conflicts over land rights, to the point of five recent deaths (Suwandi, pers. com.), have further deteriorated with the arrival of the oil palm estates. The frequency of fires along the province border with North Sumatra may, in part, be a symptom of this tension. The stage of land allocation, however, varies greatly from province to province and thus the relative importance of arson within the overall 'fire-scene' will also be markedly different. But, even where tensions are high, arson is towards the bottom of the list of causes of vegetation fires.

The Record of Vegetation Fires in Indonesia

An Overview

The forests of the Far East can be broadly classified into three formations – dryland rain forest, wetland rain forest and monsoon forest. The key differentiating factor is the availability of water, irrespective of whether this is controlled by climate or soil-water relations (Whitmore, 1990). All three forest types have been degraded by man over millennia through exploitation for forest products, cultivation, grazing, and by burning, the last of which has had the most profound effects.

The history of vegetation fires in Indonesia is as old as the history of mankind. There is evidence from soil charcoal of major fires within the rain forest over the past several thousand years (Goldammer, 1993), although these were rare events that coincided with known climatic dry phases. Written records are, of course, few but it is said that one Seventeenth century European explorer seeking Borneo was guided to the island by sailing up-wind into smoke smelt miles out to sea.

Monsoon forests, with a seasonally dry climate, are the most prone to fire, as eloquently put by Steenis and Schippers-Lammertse (1965) when referring to Java that has long since lost all its primary forest under pressure from a rapidly increasing population, "Fire for hunting, for pleasure, for pestering neighbours or neighbouring villages, by carelessness, for clearing land, for making land passable, for converting forest into pasture land, in short for innumerable purposes, has played havoc within the monsoon forest" [Quoted by Whitmore, 1975].

The Major Fire Years Since 1980

In Sumatra, Dutch colonists recorded a number of wildfires during prolonged droughts from the turn of the twentieth century until the nineteen forties, and it was also during this time that the final conversion of Java into a settled agricultural economy was completed by the use of fire. But since the early eighties more frequent and more severe droughts, coupled with the irresponsible use of land and natural resources, have greatly increased both the numbers and the severity of the fires throughout Indonesia.

There have been five major fire events since 1980; in 1982-83, 1987, 1991, 1994 and 1997-98. These have coincided with El Niño periods, and had their greatest influence in Kalimantan and the south-eastern regions of Sumatra, northwards as far as Jambi (Map 1). The double-year events, e.g. 1982-83, occur as the droughts, and thus the fires, began and ended in the one calendar year in Sumatra but reached Kalimantan later in that year and extended into the first months of the following year. All the fires are well recorded by time and island, although much of the data quoted on areas burnt is at best of dubious quality. Data originating from government agencies were particularly weak and in some cases seem to be deliberately designed to shield commercial companies from accusations of negligent fire prevention and suppression. The history of the fires and the figures for the areas burnt given in the next paragraphs are taken from a number of sources including ECE/FAO (1997), Dennis (1998), EUFREG (1998), Makarim *et al.* (1998), Chandrasekharan (1998) and Schweithelm (1999), who have, in turn, compiled data from many further secondary sources. As the authors themselves often acknowledge, the figures must therefore be read with circumspection.

Kalimantan, with Brunei Darussalam and the Malaysian states of Sabah and Sarawak that together make up the island of Borneo, bore the brunt of the 1982-83 fires. The more recently logged areas were particularly hard hit, as were secondary forests not yet fully recovered after earlier timber extraction. The fires also entered the inland and coastal peat swamps - areas as the name implies, previously felt to be far too wet to burn - and caused considerable damage. In total, some 3 Mha of forest land were affected.

The 1987 fires were considerably less damaging in terms of area burnt - officially only 66 000 ha but probably ten times more - but this time stretched from Sumatra in the west, through Kalimantan to Timor in the east. Moderately and heavily disturbed forests are noted in a number of reports as the most seriously affected by fire. Widespread fires again broke out in 1991 on the same islands as in 1987. Official estimates showed 500 000 ha burned and there were reports of dense but fairly localized smoke haze.

Indonesia was once more gripped by prolonged drought in 1994 with Sumatra and Kalimantan swept by fires. The resultant smoke haze – as before mainly from peat swamp areas - drifted over East and Peninsular Malaysia and reduced visibility to 500 m in Singapore by the end of September. It was this extensive transboundary pollution that finally alerted the international donor community and led to the setting up of three, province-based, forest fire prevention and management projects. A startling 31 percent of the fire-affected area was said to have been burnt by slash-and-burn practices and a further 58 percent was noted in government reports, quoted by ADB (1999), as 'farmland'. This graphically highlights the difficulty brought about by the term 'forest fire'. On the one hand official figures state that 4.9 Mha of forest burned, while the same data suggest that agriculture, transmigration schemes, plantations and timber estates together accounted for 99 percent of the burnt area.

The 1997-98 drought was the severest in the last twenty years and fire numbers in all 27 provinces from Aceh in the north west of Sumatra, to Merauke in the south east of Irian Jaya, suffered numerous fires. While many – perhaps the majority – of these were wildfires beyond the control of local resources, many others were set by commercial companies and smallholder farmers who took advantage of the exceptionally dry periods to clear more extensive estate land and farming plots than normal. This allowed them to plant larger areas than they had previously planned in the 1997 rainy season, and also to 'bank' pre-cleared land for planting in the 1998 rains.

Estimates of areas burnt in 1997-98 by province and vegetation type vary sharply between institutions (see amongst others EUFREG (1998), Stolle (1998), Schindler (1998a)). All authors acknowledge that they were greatly hampered by poor land use maps and by conflicting figures from different government agencies when preparing their estimates. ADB (1999) combined data from a number of secondary sources and has suggested that 9 745 000 ha burnt between May 1997 and March 1998 (Table 1). Sub-divided by island, this equates to around 12 percent of the land in Kalimantan, 4 percent in Sumatra and 2.5 percent in Irian Jaya: figures which are probably a fair reflection of the scale of forestry and agricultural expansion in each of the islands.

Table 1 Estimated areas burnt in 1997-98 (x 1000 ha) (Data from ADB, 1999)

Island	Montane Forest	Lowland Forest	Peat and Swamp Forest	Scrub and Grass	Tree Plantation	Agriculture	Estate Crops	Total
Kalimantan		2375	750	375	116	2830	55	6501
Sumatra		380	300	260	70	670	60	1740
Java		25		25		50		100
Sulawesi		200				200	1	401
Irian Jaya	100	300	400	100		100	3	1003
Total	100	3280	1450	760	186	3850	119	9745

Prolonged transboundary smoke haze, which again drew much international scrutiny, originated from the relatively small but damaging proportion of long-duration fires that developed in degraded peat swamps and peat soil zones in east coastal Sumatra and southern Kalimantan. Serious peat fires also occurred in Irian Jaya but were little noticed as the smoke stayed within Indonesia and the island is outside the range of the main satellite stations used to monitor fires (Legg and Laumonier, 1999).

The Non-Drought Years in Sumatra

The five 'fire years' since 1980 have distracted attention from the equally serious numbers of vegetation fires that now form a routine part of land management throughout Indonesia. This is well illustrated by the fire history of Sumatra in 1996 and 1998 (Anderson *et al.*, 1999) and 1999, years of average annual rainfall with a 'standard' distribution pattern. Over 13 000 hot spots (Table 2a) were detected in each of these years in the provinces monitored using NOAA satellite imagery (see box).

Table 2a Total number of hot-spots detected using NOAA satellite data in the provinces of Sumatra from 1996 to 1999 (No data are available for Aceh, North Sumatra and West Sumatra from 1996 to 1998).

Province	1996	1997	1998	1999
Aceh	NA	NA	NA	276
North Sumatra	NA	NA	NA	1791
Riau	6067[a]	160883	13064	8251
Jambi	2946	12392	961	1840
South Sumatra	3371	60763	1043	7250[b]
West Sumatra	NA	NA	NA	1143
Bengkulu	532	3299	145	189
Lampung	351	21069[c]	119	1351
Total	**13277**	**113606**	**15 332**	**15551**

This overall trend in fire occurrence in Sumatra, as throughout Indonesia, is controlled – as must be expected - by the cycle of seasonal rainfall, though this is overlaid, and heavily influenced, by the markedly different types of land-use in each of the eight provinces. In spite of the fact that there are considerable regional differences in mean annual rainfall in Sumatra, the climate is essentially humid and 85 percent of the island has a dry season (average monthly rainfall less than 100 mm) of less than two months. Based on meteorological data from the provincial capitals, the month of least rainfall in the eastern plains of the northern provinces of Aceh and North Sumatra occurs in February. The plains of the central province of Riau are driest in June. The month of minimal rainfall in the southern provinces of Jambi, South Sumatra and Lampung is later and is more variable occurring in the period July to September. As elsewhere in the world rainfall generally increases with altitude and the western mountain spine, the Bukit Barisan, rarely experiences drought. The narrow coastal plain to the west of the mountains has a wetter climate than its eastern counterpart (Holmes, 1998). And, as rainfall is the primary inhibitor and suppressant of vegetation fires, the mountains and western plains seldom experience prolonged serious fires.

Table 2b Number of hot-spots detected per 100 km^2 in the same provinces and years

Province	Area (ha)	1996	1997	1998	1999
Aceh	5539 200	NA	NA	NA	0.5
North Sumatra	7078700	NA	NA	NA	2.5
Riau	9514300	6[a]	17	14	9
Jambi	4 901 300	6	25	2	4
South Sumatra	10 226400	3	59	1	7[b]
West Sumatra	4 977800	NA	NA	NA	2
Bengkulu	2116800	2.5	16	0.7	1
Lampung	335200	1	63[c]	0.4	4

[a]Detection of fires in Riau province did not start until May 1996.
[b]25 percent accounted for by a single large wildfire in the Pampangan wetlands.
[c]Anomalously high owing to the inclusion of false-fires caused by high soil temperatures during the drought.

Satellite Fire Detection

Before the mid-eighties the reporting of wildfires relied on the local population and on the use of watch-towers where commercial or state resources were at risk. Since then techniques have been developed that allow fires with a flaming front of 30 m or more to be detected through the acquisition and processing of data from the NOAA series of meteorological satellites.

The method has the considerable advantages that the data are free, transmitted continuously, and are available to any user with a receiving system. Three, sometimes four, images of Sumatra are obtained each day from NOAA 12 and 14. The data have a coarse spatial resolution (1.1 km) but are well suited to locate and monitor the larger fires that persist for longer than a few hours and which are thus potentially dangerous.

The NOAA mid-infrared sensor integrates all radiation emitted from each 1.1x1.1km sampled. This may come from either one or from several fires and thus gives an underestimate of fire numbers. Numbers are also underestimated when thin cloud (and smoke haze) which is common over much of Indonesia, absorbs and scatters the radiant energy from the fires. The weakened signal is rejected as a hot-spot and thus as a fire during processing. (Thick cloud and dense smoke haze are opaque and no hot-spots are seen). Sub-surface peat fires, although common, are also unlikely to be detected because of the absence of flames.

However, a small hot fire regularly produces more than one hot-spot in a contiguous array (possibly owing to heated gases and suspended particles from the fire emitting additional radiation). This effect gives an overestimate of fire numbers. Thus one hot-spot location does not equal one fire, although in everyday circumstances it is a fair approximation. A full discussion of capture, data processing and interpretation of NOAA data and sources of error are given by Anderson *et al.* (1999a , 1999b).

High-resolution satellite images from LANDSAT and SPOT provide more detailed and accurate data but are expensive and their repeat cycles of more than two weeks preclude their use for regular fire monitoring.

In each of the four years 1996 to 1999, including the drought year of 1997, a wave of land clearance fires moved from the north to the south of the island (Figure 1). Fire numbers peaked in Aceh, North Sumatra and Riau provinces in the months before and including July, in Jambi from July to September, in South Sumatra in September, and in Lampung province from August to October. These fire peaks broadly correlate with months of lowest rainfall in each of the provinces.

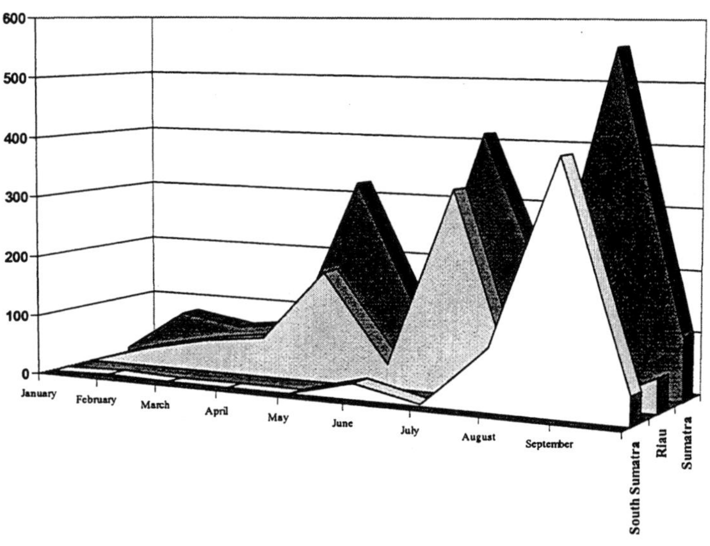

Figure 1 The numbers of fires detected on one day during each month from January to October 1999 in the central province of Riau and the southern province of South Sumatra compared to the total fire numbers on the selected days in the island of Sumatra as a whole. The data used are from an afternoon NOAA 14 pass on days when Sumatra was generally free of cloud.

The strong influence of land-use, and thus of man and man's policies, on the pattern of fire distribution in two Sumatran provinces, Riau and South Sumatra, is shown in Map 2. NOAA satellite data from June 1998 (Map 2a) show the typical pattern of fires associated with large-scale land clearance. Four large clusters of hot-spots approximately define the blocks of land that will be planted in the next rainy season. (The accuracy of the NOAA data was confirmed by reference to the archive of SPOT high resolution satellite images held by the Centre for Remote Imaging, Sensing and Processing at the National University of Singapore (CRISP, 1999) and by field visits).

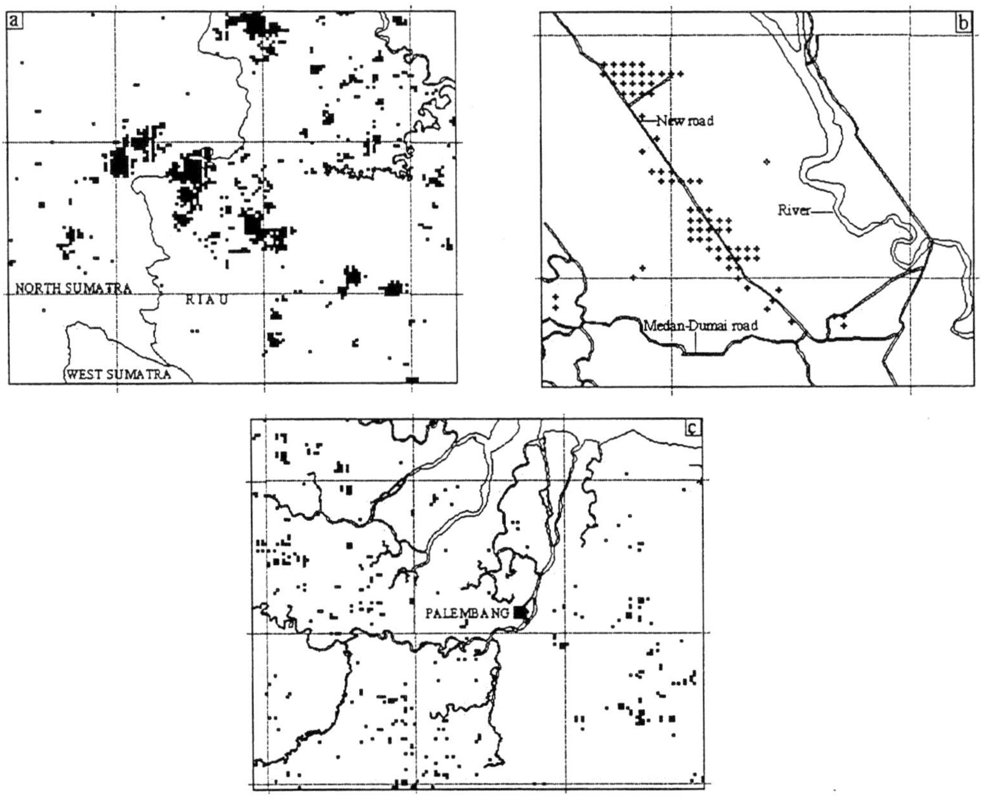

Map 2 The effects of different land-use on the pattern of vegetation fires as detected by NOAA satellite imagery. (a) fire clusters associated with commercial-scale land clearance in Riau province in June 1998, (b) linear fires that often mark the course of new roads – Riau, June 1999, and (c) widely dispersed fires typical of plot burning by smallholders in South Sumatra province in the first two weeks of September 1999. Approximate scales (a) and (c) 1:2.8 million, (b) 1:900 000.

The effects of different land-use on the pattern of vegetation fires as detected by NOAA satellite imagery. (a) fire clusters associated with commercial-scale land clearance in Riau province in June 1998, (b) linear fires that often mark the course of new roads – Riau, June 1999, and (c) widely dispersed fires typical of plot burning by smallholders in South Sumatra province in the first two weeks of September 1999. Approximate scales (a) and (c) 1:2.8 million, (b) 1:900 000.

Block clearance fires on this scale often persist for up to ten successive days and have been detected along the Riau – North Sumatra border periodically during dry spells from February to July and again in November each year since 1996. They are caused by often illegal, company-driven land clearing for oil palm. From the NOAA and SPOT satellite data it is estimated that in June 1998 alone 20 000 ha were cleared by fire for oil palm estates in the Riau - North Sumatra border area, much of it on peat soil areas.

A narrow linear pattern of fires (Map 2b) seen on NOAA imagery of Riau province in June 1999 prompted a field visit. The forest alongside a new road and pipeline had been intermittently cleared since early 1998 for distances of between 100 m and 500 m on each side over a total length of 30 km. The area was in a recently logged swamp forest but the clear-felling and burning had been carried out by new-comers (spontaneous migrants) from the neighbouring province of North Sumatra who said that they rented the land from local individuals. Pepper, beans and maize were already growing for the early use of the migrants and the main crop of oil palm seedlings, bought by the plot owners, was being planted.

In contrast to the large blocks of fire seen in Riau, the typical pattern of fires detected in the southern provinces of South Sumatra and Lampung (Map 2c), is random and ephemeral. The fires are lit mainly by smallholders clearing their farm plots in anticipation of the start of the rains. Fire numbers peak in August or September but field observations show that they are carefully controlled within their intended limit that seldom exceeds two hectares at any one time. Burning usually starts in the early afternoon and the fires last until dusk: satellite data show that the same fire is not often detected on two consecutive images captured two to three hours apart. This is by far the most frequent type of fire throughout Sumatra, and accounts for some 90 percent of all fires detected in a year of average rainfall.

The pattern of fires now seen in Jambi with fairly frequent block fires combined with numerous transient small fires, reflects the transitional status of the province situated between the more or less settled agricultural economy in the south, with few large fires, and the rapid exploitation of large tracts of forest in Riau. It is interesting that far fewer large-scale fires connected with commercial land clearing, were detected in 1998 and 1999 than in 1996. This suggests that conversion of Jambi's forests into estate crops is now largely complete.

THE ECONOMIC IMPLICATIONS

There have been a number of studies on the direct and indirect costs of the 1997-98 fires. Some, such as Dieterle and Heil (1998) and Levine (1998), assessed the impact of the transboundary smoke haze pollution on the health of the region. Others attempted a much wider cost evaluation and to put figures on damage to agriculture and forestry and to assign monetary values to the loss of biodiversity, erosion and siltation, firefighting, etc., as well as on health.

Of the latter studies, ADB (1999) placed a total value of $9.3 billion on the cost to Indonesia of the 1997-98 fires, while EEPSEA (1998) gives a figure of $4 billion for the 1997 fires alone. As the continuation of the fires into 1998 was severe in Kalimantan, these figures may not be so far apart as they first appear. Direct comparison between the two studies is difficult as they use different cost headings. There are however, some notable variations. For example the EEPSEA study that covers the shorter period, puts health damage at $924 million while the ADB figure, which may be expected to be higher, places loss of health at 'only' $145 million. Conversely losses to agriculture are estimated by ADB to be six-times higher than those by EEPSEA. Such comparisons, while they point out the difficulties of costing fires, should not be used to downplay their damaging effects. There appear to be no comparable studies for 'routine' fire years such as 1996 and 1998.

Land Management Institutions and Laws

We have so far looked at the history and pattern of vegetation fires in Indonesia in general and Sumatra in particular and have attempted to identify the major sources of fires, their direct causes and their relative importance within the overall picture. It is, however, the national and provincial land management policies and laws, and the way that these are implemented at field level, that shape the attitudes and the behaviour of individuals, communities, companies and government agencies in how they regard, and hence manage, their national resources. And it is these policies that have done much to indirectly encourage the use of fire as a tool in the management of land throughout the country, while the decrees and regulations that exist to control the use of fire have been ignored with impunity.

The Institutions

Three institutions play a major, and three a minor part in national land management. Their overlapping functions, together with the diffusion of responsibility, contribute greatly to the fire problem. Over the last thirty years planning in all sectors has been strongly 'Jakarta centred' and directly and tightly controlled by the President. Despite much discussion since the movement for *reformasi* (comprehensive change) began in 1998, it remains to be seen if greater powers will be given in fact, not simply in theory, to the provinces.

The Ministry of State for National Development Planning is responsible to the president, and formulates policy and coordinates the planning activities of all government institutions, in particular those of the National Development Planning Agency (BAPPENAS) and its regional wing, BAPPEDA. The President is further assisted in his control of land by the National Land Agency (BPN) which is in a charge of land registration, land measurement, land-mapping, land certification and the granting of land: a powerful remit.

The Ministry of Forestry and Estate Crops (MoFEC), controls the great majority of the country's land and thus has considerable potential to influence the use of fire. The Ministry has two distinct roles in managing its huge territory. As the Forest Authority it is responsible for licensing forest exploitation and for policing the activities of the licencees. In this MoFEC has to enforce the Forestry Law (see below) and be relatively inflexible, operating straightforward procedures and enforcing regulations without favour or discrimination. But as the overseer of six State Owned Forest Enterprises (Inhutani I-VI), MoFEC finds itself responsible for commercial companies that carry out logging, reforestation, agroforestry and work with transmigrants. All these duties require flexible, efficient and accountable management. The two roles are in many ways incompatible as MoFEC is called upon to control and police itself. That it has not always done so, has had a strong negative impact in the quality of land management in Indonesia.

Of the three less influential institutions involved in land management, the regional offices of the State Ministry for the Environmental (BAPEDAL) are in charge of environmental rehabilitation schemes and vegetation fire prevention. The Ministry is also given control when a national disaster is declared as it was during the 1997 fires; a fire management role for which it has no regular training.

The main task of the Ministry of Agriculture is to develop production and processing methods for agriculture, horticulture, fisheries and livestock. It has limited land development powers but retains an interest in fire-use and fire prevention. The once powerful Ministry of Transmigration and Forest Squatters Resettlement (Ministry of Transmigration and Population since October 1999) continues to manage existing schemes, and thus sizeable areas of fire-prone land. However new programmes are considerably reduced after the - often predictable - failure of most projects and the consequent reassessment of policy and reduced support from the World Bank.

Cooperation between the various ministries and government agencies to find a joint approach to reduce the number of vegetation fires, and to suppress those that do occur, is at best weak and is often non-existent. The lack of clear responsibilities is well illustrated by the 'conference round': from mid-1998 to September 1999, fifteen national and international seminars, workshops and conferences took place in Indonesia and several more in Singapore, Malaysia and Brunei Darussalam, all aimed at solving Indonesia's (and thus its ASEAN neighbours') fire and smoke haze problems. Some twenty donor organisations were involved, with support being offered in fairly equal measure through MoFEC, BAPPENAS and BAPEDAL but also through the Ministry of Transmigration, through the State Ministry of Human Settlement and through the government Agency for the Assessment and Application of Technology.

The Land Management Laws

Indonesian laws are based on the 1945 constitution that specifies that all branches of production which are important to the state and which affect the lives of the majority are controlled by the state. Accordingly, the Basic Forestry Law of 1967 stipulated that, "The government determines the use of forests according to the interest and benefit of the state and population, the government determines the legal relationship between people and legal entities with forests". However all the procedures specified under the law were 'Jakarta centred'. And the administrative approach allowed little involvement of the local people and communities in whose "interest and benefit" the forest was exploited. The people's disinterest in fire prevention and suppression that followed is unsurprising.

The lack of local involvement in forest management was partially recognised when drafting the New Basic Forest Law passed in September 1999, although the statute that enforces total state control over Forest Land is unmodified and the condescending paradigm towards local communities is maintained. For example, "Forestry extension shall be intended to increase the knowledge and skill and to change the mental attitude of the communities so they shall be capable to support the forestry development and be aware of the importance of forest resources to human life."

The new law does, however, allow that State Forest can be in the form of traditional (*adat*) forest, that long-standing community rights will prevail, and that management plans shall take into account the aspirations of the local communities. Nevertheless opportunities for genuine local management are restricted and likely to be tightly controlled by MoFEC.

Nineteen presidential, ministerial and gubernatorial decrees that deal directly with fire on forest land supplement the Basic Law. Most were passed during, or in the wake of, the 1994 and 1997-98 fires and many lay down strict regulations requiring forest logging and forest plantation concession-holders to set-up fire prevention and control teams. Others deal with the establishment of Local Fire Control Centres and Land and Forest Fire Control Task Force teams to be housed in provincial and district forestry offices.

A number of forest plantation companies in Sumatra and Kalimantan have taken their responsibilities seriously and are well equipped to protect against fires and to control those which do occur. Many others, and most estate crop companies, are not so well prepared to protect their investments. Few of the official government local fire control centres exist in anything but name, and those that do are generally ill-funded, ill-equipped and ill-trained.

It is ironic that so much effort has gone into laying down control mechanisms for the relatively few wildfires while, as a matter of policy, the routine use of fire to clear millions of hectares of land proceeds unchecked with almost no enforcement of burning permit requirements and even less enthusiasm to investigate, let alone prosecute, companies suspected of breaking fire regulations.

THE FUTURE

Shortcomings and outright failures in land resource policies over the last twenty years have contributed to the number and severity of the vegetation fires throughout

Indonesia. If there is to be any improvement, then there must first be a genuine will to reform, followed by rapid and decisive action to integrate fire prevention and fire protection in to all land management decisions and activities (Bompard and Guizol, 1999).

Support must be given to the provincial, district and sub-district offices that are most closely involved in land management, rather than to the central, Jakarta-based institution. There is a need to strengthen local capacity to plan wisely and to foresee the likely effects of decisions both on the direct use of fire and on how these decisions may also increase or decrease the indirect fire-risks. BAPPEDA, working with BPN and MoFEC, need to upgrade staff skills within the provinces and districts. They also, as suggested by Scott and Lusli (1998), need to establish local land management data centres to identify and map areas known to be at high risk of fires.

Mature and open cooperation between village communities and government agencies, private companies and parastatal enterprises is essential if fire numbers and severity are to be reduced. Village communities must be fully recognised as the dominant partner when land management decisions are made. This is the only level at which there can be workable control and accountability for the use of natural resources. And sub-district and district are the only levels where fire prevention can be effective and fire suppression is physically possible.

The rash of land disputes between government and companies on the one side, and communities on the other, lead to the use of fire as a weapon. Mediation mechanisms need to be developed to help prevent new, and solve existing, conflicts. Only land in which local people have a close political and a financial interest will be protected from fire. As suggested by, amongst others, Potter and Lee (1998) and Gouyon (1999), NGOs may be best placed to act as intermediaries between the warring parties. Co-management of land by communities with companies and government would be a major step forward, although it will entail a detailed, village by village revision of the boundaries of state forest land, and consensus boundary setting with MoFEC and BAPPEDA.

Priority must clearly be given to fire prevention in wetland areas. These areas are in imminent danger of suffering irreversible damage: logging and conversion to estate crops should cease, not only to preserve these ecologically valuable areas but also to greatly reduce the noxious smoke haze pollution that results when peat burns.

The mechanical clearance of land is widely promoted by many who see it as a preferable alternative to clearance by fire. Such a course is not without its own risks as contractors are paid by the hectare and pay scant heed to preserving topsoil, avoiding soil compaction and to not filling gullies with debris. And, whatever proponents of mechanical clearance would like to suggest, direct costs are far higher. Only if a subsidy, perhaps in the form of direct cash inputs from the reforestation fund or as tax incentive schemes, is put in place 'to repay' the indirect environmental costs of fire, will commercial companies cease to burn: and only then if stiff fire penalties are enforced.

To expect smallholders not to use fire is both impracticable and unnecessary: they have neither the financial resources nor the manpower available to clean their plots by hand and their fires are usually well controlled and relatively non-polluting.

The solution to Indonesia's fire problem lies largely in much improved local level land-use planning and strengthened local management that together fosters local initiative and allow local autonomy. National control from the centre must, concomitantly, become weaker. Any continuation of a top-down, bureaucratic approach to fire management that focuses on fire suppression will fail in the field during the next El Niño drought as it did in 1997. As things stand now, fire-prone areas continue to expand and the costs of fire damage and fire management will continue to increase over the years.

REFERENCES

1. ADB (1999). "Planning for fire prevention and drought management project" Report, Asian Development Bank, Manila.
2. Anderson, I.P., Bowen, M.R., Imanda, I.D. and Muhnandar (1999). *Vegetation fires in Indonesia: the fire history of the Sumatra provinces 1996 – 1998 as a predictor of future areas at risk.* EU-Forest Fire Prevention and Control Project with Kanwil Kehutanan dan Perkebunan Palembang. European Union and Ministry of Forestry and Estate Crops, Jakarta. 22 pp plus 16 maps and tables.
3. Anderson, I.P., Imanda, I.D. and Muhnandar (1999a). *Vegetation fires in Indonesia: the interpretation of NOAA derived hot spot data.* EU-Forest Fire Prevention and Control Project with Kanwil Kehutanan dan Perkebunan Palembang. European Union and Ministry of Forestry and Estate Crops, Jakarta. 19 pp plus 3 maps.
4. Anderson, I.P., Imanda I.D. and Muhnandar (1999b). *Vegetation fires in Sumatra, Indonesia: the presentation and distribution of NOAA-derived data.* EU-Forest Fire Prevention and Control Project with Kanwil Kehutanan dan Perkebunan Palembang. European Union and Ministry of Forestry and Estate Crops, Jakarta. 8 pp plus 6 maps.
5. Bompard, J.M. and Guizol, P. (1999). *Land Management in South Sumatra Province, Indonesia Fanning the flames: the institutional cause of vegetation fires.* EU-Forest Fire Prevention and Control Project with Kanwil Kehutanan dan Perkebunan Palembang. European Union and Ministry of Forestry and Estate Crops, Jakarta. 40 pp plus annexes.
6. Brinn, P.J. (1992). *Recommended Development Area 99, South Sumatra. Rapid appraisal report.* Regional Physical Planning, Map Improvement and Training Project. Land Resource Planning Component. Ministry of Transmigration, P.T. Nasuma Putra Consulting Engineers and Natural Resources Institute, UK. 43 pp.
7. Byron, N. and Shepherd, G. (1998). *Indonesia and the 1997-98 El Niño: fire problems and long-term solutions.* Natural Resources Perspectives Number 28, Overseas Development Institute, London. 1-4.
8. Chandrasekharan, C. (1998). The mission on forest fire prevention and management to Indonesia and Malaysia (Sarawak). Paper, 'International Cross Sectoral Forum on Forest Fire Management in South East Asia'. Indonesian National Planning

Agency, Japan International Cooperation Agency and International Tropical Timber Organization. 7-8 December 1998. Jakarta.

9. CRISP (1999). Centre for Remote Imaging, Sensing and Processing, National University of Singapore. (http://www.crisp.nus.edu.sg/crisp.html;http://www.crisp.nus.edu.sg/crisp.html)
10. Danielsen, F. and Verheugt, W.J.M. (1990). *Integrating conservation and land-use planning in the coastal region of South Sumatra, Indonesia*. The Directorate General of Forest Protection and Nature Conservation, Ministry of Forestry and Estate Crops and Asian Wetland Bureau. Jakarta.
11. Dennis, R. (1998). 'A review of fire projects in Indonesia'. Prepared for International Center for Forestry Research, International Centre for Research in Agroforestry, UNESCO and EU. Bogor/Jakarta.
12. Dick, J. (1991). 'Forest land use, forest use zonation, and deforestation in Indonesia: a summary and interpretation of existing information.' Background paper to United Nations Conference on Environment and Development for the Indonesian State Ministry for Population and Environment and the Environmental Impact Management Agency.
13. Dieterle, G. and Heil, A. (1998). *Impacts of large scale forest and land fires in Indonesia in 1997 on regional air pollution*. Paper, 'International Cross Sectoral Forum on Forest Fire Management in South East Asia'. Indonesian National Planning Agency, Japan International Cooperation Agency and International Tropical Timber Organization. 7-8 December 1998. Jakarta.
14. Dis, van M.M.U. (1986). 'The art of lowland development. Theme 1. Land and water management.' In *Symposium on Lowland Development in Indonesia*. Jakarta, Indonesia, 24-31 July 1986. International Institute for Land Reclamation and Improvement, Wageningen, The Netherlands. 34-53.
15. ECE/FAO (1997). *Forest fire statistics 1994-1996*. Timber Bulletin Number 4. ECE/TIM/BULL/50/4.United Nations, New York, Geneva. 19 pp.
16. EEPSEA (1998). *The Indonesian fires and haze of 1997: the economic toll*. Economy and Environment Programme for South East Asia and World Wide Fund for Nature. (http://www.wwf.or.id/EEPSEARE.html; http://www.wwf.or.id/EEPSEARE.html) 6pp.
17. EUFREG (1998). *Assessment of the 1997 fires in Indonesia*. European Union Fire Response Group with he Ministry of Forestry and Estate Crops, Jakarta, Indonesia. 33 pp.
18. Fraser, A.I. (1999). 'Skenario-skenario untuk masa depan sektor kehutanan dan kebun di Indonesia. [Possible scenarios for the future of forestry and estates in Indonesia.]. Seminar, Sustainable Forest and Estate Crops Management for the Maximum Prosperity of People. Ministry of Forestry and Estate Crops, Department for International Development (UK) and Consultative Group for Indonesia Forestry. 8 September 1999, Jakarta, Indonesia. 6 pp.
19. GoI (1999). *National population program profile*. Badan Koordinasi Keluarga Berencana Nasional. 17 pp. http://www.bkkn.go.id/english/b43.htm

20. Goldammer, J.G. (1993). Historical biogeography of fire: tropical and subtropical. In, *Fire in the environment: the ecological, atmospheric, and climatical importance of vegetation fires*. Editors, P.J. Crutzen and J.G. Goldammer. John Wiley and Sons, Chichester, UK. 297-314.
21. Gouyon, A. (1999). *The sustainable development of tree crops and the prevention of vegetation fires in South Sumatra, Indonesia: jungle rubber*. EU-Forest Fire Prevention and Control Unit with Kanwil Kehutan dan Perkebunan Palembang. European Unit and Ministry of of Forestry and Estate Crops, Jakarta, 55 pp. plus annexes.
22. Hoffmann, A,A., Siegert, F. and Hinrichs,A. (1999) 'Fire damage in East Kalimantan in 1997/98 related to land use and vegetation classes based on satellite analysis with ERS SAR-2 and
NOAA-AVHRR. GTZ Integrated Forest Fire Management Project, Samarinda, East Kalimantan. (In preparation).
23. Holmes, D.A. (1998), Rainfall and droughts in Indonesia. A Study for the World Bank. Volumes 3A; Sumatra. World Bank, Jakarta. 77pp.
24. Hoshour, C. (1995). *Resettlement and Politicization of ethnicity in Indonesia.* Conference; Riau in Transition', International Institute for Asian Studies, 4-6 October 1995, Leiden, The Netherlands.
25. Khoo, L.L. (pers.com). Oil Palm plantations consultant.
26. Legg.C and Laumonier Y. (1999). '1997 Fires in Indonesia: a remote sensing perspective'. *Ambio*, **28**, 479-485.
27. Levine, J. (1998). 'Gaseous and particulate emissions released to the atmosphere during forest fires: a case study of Kalimantan and Sumatra, Indonesia in 1997'. Conference; 'Health guidelines for forest episodic events'. World Health Organization, 6-9 October 1998, Lima, Peru.
28. Makarim, N., Arba 'I, Y.A., Deddy, A. and Brady, M. (1998). 'Assessment of 1997 land and forest fires in Indonesia: national coordination.' Environmental Impact Management Agency, Jakarta, 7 pp.
29. Meijaard, E. and Dennis, R. (1977). *Forest fires in Indonesia: bibliography and background information*. World Wide Fund for Nature, Amsterdam.
30. Nicolas, M.V.J. (1998). 'Fighting the forest fires: the South Sumatra experience.' EU- Forest Fire Prevention and Control Project, Palembang, Indonesia. Internal report. 22pp.
31. Nicolas, M.V.J. (pers.com). Forest Fire Management Expert. GTZ Integrated Forest Fire Management Project, Samarinda, East Kalimantan, Indonesia.
32. Nicolas, M.V.J. and Beebe, G.S. (1999). *Fire Management in the logging concessions and plantation forests of Indonesia*. Forest Fire Prevention and Control Project (South Sumatra) and Integrated Forest Fire Management Project (East Kalimantan). European Union and GTZ with Kanwil Departemen Kehutanan dan Perkebunan, Propinsi Sumatra Selatan. European Union and Ministry of Forestry and Estate Crops, Jakarta, Indonesia. 38 pp.
33. Nicolas, M.V.J. and Bowen, M.R. (1999). *A field level approach to coastal peat and coal-seam fires inSouth Sumatra, Indonesia*. Forest Fire Prevention and

Control Project (South Sumatra). European Union with Kanwil Departemen Kehutanan dan Perkebunan, Propinsi Sumatera Selatan. European Union and Ministry of forestry and Estate Crops, Jakarta. 39pp.

34. PEMDA (1999). *Pemerintah Daerah Riau.* [The provincial government of Riau]. (*http://www.riau.go.id/pemda/indez2.html*)
35. Potter, L. and Lee,J. (1998) *Oil Palm in Indonesia: its role in forest conversion and the fires of 1997/98*Report for World Wide Fund for Nature-Indonesia programme. University of Adelaide, Adelaide, Australia.
36. Ramon, J. (pers.com). Social Forestry Expert, EU-Forest Fire Prevention and Control Project, Palembang, 1995-1998.
37. RePPRrot (1990). A national overview from the regional Physical Planning Programme for Transmigration. UK Overseas Development Administration and Directorate BINA Programme, Ministry of Transmigration and Forest Squatters Resettlement, Jakarta, 32pp.
38. Reieley, J.O.and Page, S.E. (editors) (1997), *Biodiversity and Sustainability of Tropical Peatlands.* Proceedings, 'International Symposium on Biodiversity, Environmental Importance and Sustainability of Tropical Peatlands', Palangkaraya, Central Kalimantan , 4-8 September 1995.Samara Publishing Limited, Tresaith, Cardigan, UK. 369pp.
39. Rieley, J.O., Page, S.E., Limm, S.H. and Winarti, S. (1997). 'The peatland resources of Indonesia and the Kalimantan Peat Swamp Forest Research Project.' In, Rieley, J.O. and Page, S.E. (editors). *Biodiversity and Sustainability of Tropical Peatlands.* Samara Publishing Limited, Tresaith, Cardigan, UK. 37-45.
40. Schindler, L. (1998). 'Fire Management in Indonesia – quo vadis?' Paper, International Cross Sectoral Forum on Forest Fire Management in South East Asia. Indonesian National Planning Agency, Japan International Cooperation Agency and International Tropical Timber Organization. 7-8 December 1998. Jakarta.
41. Schindler, L. (1998a). 'The Indonesian fires and SE Asian haze 1997/98: review, damages, causes and necessary steps.' Paper presented at the, Asia-Pacific Regional Workshop on Transboundary Atmospheric Pollution, 27-28 May 1998, Singapore.
42. Schweithelm, J. (1999). The fire this time: an overview of Indonesia's forest fire in 1997/1998. (*http://www.wwf.or.id*) 44pp.
43. Scott, I. And Lusli, S. (1998). 'Land use planning and the IFRIS database.' Forest Inventory and Monitoring Project, European Union and Ministry of Forestry and Estate Crops, Jakarta, (Draft,December 1998). 42 pp plus annex.
44. Steenis, van C.G.G.J and Schippers- Lammertse, A.F. (1965*).* 'Concise plant geography of Java.' In *Flora of Java* (editors C.A. Backer and R.V. van der Brink Bakhuizen), Volume 2. Leiden.
45. Stolle, F. (1998). 'The 1997 Sumatra fires.' Paper presented at the Workshop on Fire Hazards, Transboundary Haze and Sustainable Forestry in East Asia and the Pacific. Environment and Natural Resources Division - World Bank, Surabaya, Indonesia. 9-12 December 1998.
46. Suwandi,A. *(pers.com)* . Head of Natural Resouces Conversation, Provincial Forestry and Estate Crops Office, Pekanbaru, Riau province.

47. Szekely, L. (1936). *Tropic fever: the adventures of a planter in Sumatra*. Hamish Hamilton, London. 352pp.
48. Vayda, A.P. (1999). Finding causes of the 1997-98 Indonesian forest fires: problems and possibilities. WWF Indonesia. (*http://www.wwf.or.id*) 51pp
49. Vincent, A. (pers.com.). Oil palm plantations manager.
50. Wakker, E. (1999). *Forest fires and the expansion of Indonesia's oil-palm industry*. WWF Indonesia Forest Fire Project. World Wide Fund for Nature Indonesia. Jakarta. 19 pp.
51. Whitmore, T.C. (1975). *Tropical rain forest of the far east*. Clarendon Press, Oxford. 282 pp.
52. Whitmore, T.C. (1990). *An introduction to tropical rain forests*. ELBS with Oxford University Press. 226 pp.
53. World Bank (1994). *Transmigration in Indonesia*. OED Precis Number 72, September 1994.8pp. (htpp://www.worldbank.org/html/oed/pro/2/htm)
54. World Bank (1999). *Global commodity markets*. World Bank Commodities Team, March 1999. 32pp.

Chapter 5

REMOTE SENSING OF FIRE AND HAZE

Soo Chin Liew, Leong Keong Kwoh, Oo Kaw Lim, and Hock Lim
Centre for Remote Imaging, Sensing and Processing, Faculty of Science, National University of Singapore, Lower Kent Ridge Road, 119260 Singapore

INTRODUCTION

Remote sensing generally refers to the activities of recording/observing/perceiving (i.e. sensing) objects or events at far away (i.e. remote) locations. In contrast to *in situ sensing*, the sensors used in remote sensing are not in direct contact with the objects or events being observed. In this chapter, we will restrict our discussions to the applications of satellite remote sensing in observation of fire and haze. Satellite remote sensing refers to the technology of acquiring information about the earth's surface and atmosphere using sensors mounted onboard satellites orbiting the earth. In comparison to the conventional methods of information gathering, satellite remote sensing provides the following advantages: large area coverage, frequent and repetitive coverage of an area of interest, quantitative measurement of ground features using radiometrically calibrated sensors, synoptic views of events in relation to the environment, semiautomated computerised processing and analysis, and relatively low cost per unit area of coverage.

Satellite remote sensing is able to provide key information useful in fire monitoring and management as well as damage assessment. The acquisition of this information will be examined in relation to the types of signals observable from space before, during and after a fire event.

When solar radiation hits the earth surface, it may be transmitted, absorbed or reflected. Different land cover types reflect and absorb differently at different wavelengths. This property serves as a unique signature for identification of the land cover types in an optical remote sensing image. The land cover maps derived from remote sensing images are useful in ascertaining the types of land affected by fires during a fire event for effective fire fighting and fire management, and for damage assessment.

During a fire event, the heat emitted by active fires can be detected by infrared sensors onboard a satellite. This procedure is commonly known as 'hot spot detection'. Fire hot spots can be detected and monitored during a fire event to provide information on the general locations, spatial distributions and temporal evolution of fire. Light emitted by fires at night may be detected using sensitive optical sensors onboard a satellite. Besides heat and light emission, smoke plumes can be observed emanating from the sites of active fires. The locations of the active fire sites can be determined accurately in high resolution remote sensing images, by measuring the locations of these smoke plumes. Large smoke plumes can be observed from relatively low-resolution sensors such as the ones onboard a meteorological satellite. Smoke plumes are aggregations of fine aerosol particles which absorb and scatter light. Algorithms have been developed to derive the optical thickness (which relates to the aerosol concentration) and an indication of the particle size from the smoke plumes observed in remote sensing images. The distribution of these smoke plumes, together with the prevailing wind conditions, provide early warnings for the possible spreading of smoke haze to the neighbouring regions. Fires alter the reflectance characteristics of the affected land. The burned areas are void of live vegetation with deposition of black char. The fire affected areas can be discriminated in a remote sensing image due to the strong contrast with the unburned areas. Burned scar maps derived from remote sensing images provide information on the spatial extent and distribution of the fire affected areas, and the total area burned. The burned area, together with land cover information, provide the basis for estimating emissions of carbon dioxide and other gases implicated for global climate change, and for assessment of fire damage to the economy and environment.

AVAILABLE SATELLITE-SENSOR PLATFORMS

Several remote sensing satellites are currently available, providing imagery suitable for forest fire research and fire monitoring operations. Each of these satellite-sensor platforms is characterised by the wavelength bands employed in image acquisition, spatial resolution of the sensor, the coverage area and how frequently a given location on the earth surface can be imaged by the imaging system. The satellite-sensor systems commonly used in fire detection and monitoring are listed in Chapter 9 together with their characteristics (see Table 6 in Chapter 9).

Spatial Resolution and Coverage

Spatial resolution refers to the size of the smallest object that can be resolved on the ground by an imaging system. The intrinsic resolution of an imaging system is determined primarily by the instantaneous field of view (IFOV) of the sensor, which is a measure of the ground area viewed by a single detector element in a given instant in time. However this intrinsic resolution can often be degraded by other factors which introduce blurring of the image, such as improper focusing, atmospheric scattering and target motion. A "High Resolution" image refers to one with a small resolution size. Fine details can be seen in a high resolution image. On the other hand, a "Low Resolution" image is one with a large resolution size, i.e. only coarse features can be observed in the image. Due to the constraint in downlinking the remote sensing data from the satellites back to earth, a high resolution sensor usually has a smaller coverage area, and the converse is true for the low resolution sensors. The SPOT-HRV (20 m resolution in multispectral mode, 10 m in panchromatic mode) and the Landsat-TM (30 m resolution) are examples of high resolution satellite-sensor systems. The sensors on meteorological satellites are usually low resolution. For example, the resolution of NOAA-AVHRR is about 1.1 km. but the AVHRR sensor has a ground swath of about 2400 km. In comparison, a SPOT scene only covers an area of about 60 km x 60 km on the ground.

Both low-resolution and high-resolution satellite imaging systems have their respective roles to play. They complement each other in a fire monitoring operation. The low-resolution wide-coverage system (such as the NOAA-AVHRR) provides a synoptic view of the fire situation (e.g. the general locations and distributions of active fires), while the high resolution system is used to pinpoint the exact locations of fires, to determine the types of land cover affected, and to assess the damage due to fires.

Temporal Coverage

The ability to acquire images of any given location on the earth surface at regular intervals with high revisit frequency is essential in a fire monitoring operation. This requirement can be met in several ways. If a satellite is launched in such a way that it follows an orbit parallel to the equator and travels in the same direction as the earth's rotation with the same period of 24 hours, the satellite will appear stationary with respect to the earth surface. This orbit is known as a geostationary orbit. A satellite following a geostationary orbit always views the same area on the earth. Thus, it is possible to monitor continuously any location within the field of view of its sensor. A large area of the earth can also be covered by the satellite. However, satellites in the geostationary orbits are located at a high altitude of 36,000 km from the earth's surface. At this distance, only low resolution images are acquired. The geostationary orbits are commonly used by meteorological satellites. Currently, the available geostationary satellites are the Japanese GMS (Geostationary Meteorological Satellites) over the Asia-Pacific region (140°E), the US GOES (Geostationary Operational Environmental Satellites) over the American continents (75°W and 135°W) , and the European METEOSAT over Europe and Africa (0°E). These satellites are primarily meant for observing meteorological conditions, but their use in fire detection and monitoring has also been reported (Prins and Menzel, 1992; Prins *et al.*, 1998; Weaver *et al.*, 1995).

High-resolution remote sensing satellites are usually launched into a low-earth orbit (LEO), about 700 to 800 km above the earth surface. In order to achieve global coverage, a near-polar orbit is usually employed. In this orbit, a satellite travels in a near circular orbit, with the orbital plane inclined at a small angle to the earth's axis of rotation. The satellite takes about 100 minutes to revolve once around the earth, along a general north-south direction. The satellite traces out a path on the earth surface, called its ground track, as it moves across the sky. As the earth below is rotating, the satellite traces out a different path on the ground in each subsequent cycle. If the satellite's altitude and the orbital inclination angle meet certain conditions, the satellite repeats its path after a fixed time interval. This time interval is called the repeat cycle of the satellite. For example, the Landsat satellite has a repeat cycle of 16 days. Thus, Landsat is not suitable for routine monitoring of active fires. The Landsat-TM sensors can nevertheless be used for detecting, mapping and assessing the conditions of the burnt areas. The SPOT satellites have a longer repeat cycle of 26 days. Unlike Landsat-TM, SPOT's sensors can be programmed to view obliquely. With this oblique viewing capacity, high resolution images of a location on the earth's surface can be acquired more frequently. The frequency varies with latitude. At the equator, a given area can be imaged 7 times during a 26-day orbital cycle, i.e. once in 3 to 4 days. A higher frequency of observation can be achieved using more than one satellite of the same family. For example, there are currently three operational SPOT satellites (SPOT 1, 2 and 4) orbiting the earth. With these three satellites, any location can be imaged by at least one SPOT satellite per day. The NOAA series of polar orbiting environment satellites are also launched into near-polar low earth orbits. With the wide swath of the NOAA-AVHRR sensors, any location can be imaged at least twice a day by a particular NOAA satellite. The imaging frequency increases further with three NOAA satellites (NOAA 12, 14 and 15) currently in operation.

Wavelength Bands Used in Remote Sensing

The layer of atmosphere covering the Earth's surface limits the wavelength bands useable in satellite remote sensing. Specifically, the electromagnetic radiation used by the sensor needs to be able to penetrate the atmosphere. The atmosphere consists of molecules and particulate matters which absorb and scatter radiation. Each type of molecules has its own set of absorption bands in various parts of the electromagnetic spectrum. As a result, only the wavelength regions outside the main absorption bands of the atmospheric gases can be used for remote sensing. These regions are known as the atmospheric transmission windows. The wavelength bands used in remote sensing systems are usually designed to fall within these windows to minimise the atmospheric absorption effects. These windows are found in the visible, near-infrared, certain bands in thermal infrared and the microwave regions (see Figure1).

Most of the solar energy reaching the earth's surface is contained in the wavelength region from about 400 nm to about 3 μm. Optical remote sensing makes use of visible, near infrared (NIR) and short-wave infrared (SWIR) sensors to form images of the earth's surface by detecting the solar radiation reflected from the earth's surface. For example, the NOAA-AVHRR sensors have one band in the visible region and one in the NIR region. The first three bands of Landsat-TM cover the blue, green and red regions of the visible spectrum. The fourth band is in the NIR region and the fifth band detects SWIR.

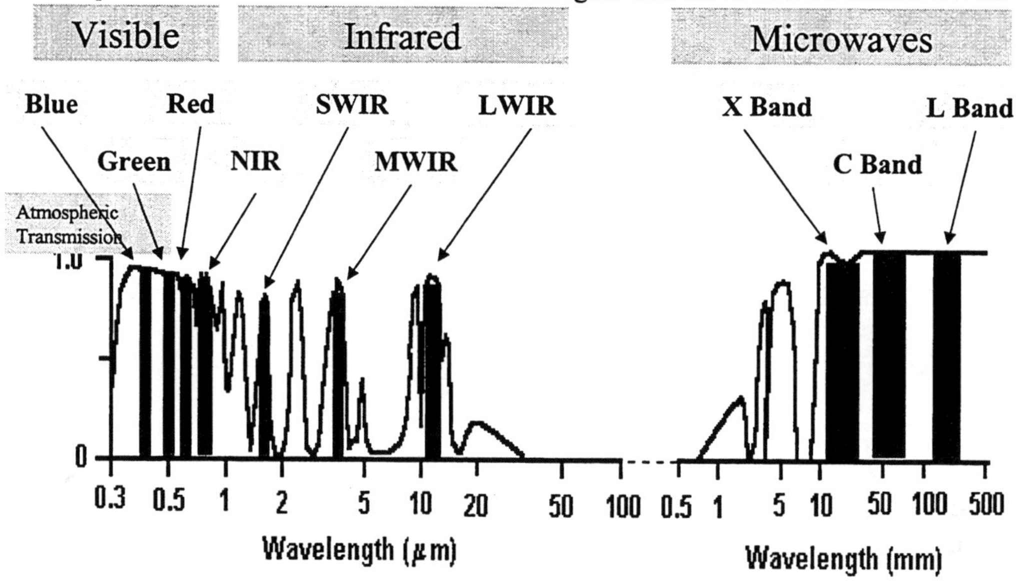

Figure 1 Wavelength bands used in remote sensing.

The SPOT HRV sensors on-board the SPOT 1 and SPOT 3 satellites have three bands covering the green, red and NIR regions. The sensors on-board the recent SPOT 4 satellite has a fourth SWIR band. Different land cover types can be distinguished by measuring the solar radiation reflected from the earth surface in these spectral bands because different materials reflect and absorb differently at different wavelengths.

Green plants have unique reflectance signatures which distinguish them from non-vegetation. Typical vegetation has low reflectance in the red and blue bands, due to absorption of light by chlorophyll. There is a small peak in the vegetation reflectance spectrum at the green band which gives rise to the green colour of vegetation. Vegetation reflects the NIR radiation strongly (much stronger than the reflectance at the visible bands), due to the cellular structure of leaves. Many vegetation indices have been composed by using combinations of different wavelength bands for detection of vegetation in optical remote sensing images, and for deriving biophysical and physiological parameters characterising the conditions of the vegetation (Huete *et al.*, 1997). One such combination is the ratio of the near-infrared band to the red band. This ratio is known as the Ratio Vegetation Index (RVI). Since vegetation has a high NIR reflectance but low red reflectance, vegetated areas have higher RVI values compared to non-vegetated areas. Another commonly used vegetation index is the Normalised Difference Vegetation Index (NDVI) computed by NDVI = (NIR - Red)/(NIR + Red). The NDVI is related to the RVI by the relation NDVI = (RVI - 1)/(RVI + 1).

The reflectance of vegetation in the SWIR region (e.g. band 5 of Landsat TM and band 4 of SPOT 4 sensors) is more varied, depending on the types of plants and the plant's water content. Water has strong absorption bands around 1.45, 1.95 and 2.50 μm.

Outside these absorption bands in the SWIR region, reflectance of leaves generally increases when leaf liquid water content decreases. This property can be used for identifying tree types and plant conditions from remote sensing images. The SWIR band 5 on Landsat TM has been used in detecting plant drought stress and delineating burnt areas and fire-affected vegetation. This SWIR band is also sensitive to the thermal radiation emitted by intense fires, and hence can be used to detect active fires, especially during night-time when the background interference from SWIR in reflected sunlight is absent .

The middle-wave infrared (MWIR) and long-wave infrared (LWIR) are within the thermal infrared region. These radiations are emitted from warm objects such as the Earth's surface. They are used in satellite remote sensing for measurements of the earth's surface temperature. The MWIR and LWIR bands (e.g. Band 3 centred at 3.7 μm and Band 4 at 10.6 μm of NOAA-AVHRR) have been extensively used in active fire detection, by detecting the thermal radiation emitted from the fires.

One major limitation of optical/infrared remote sensing lies in the inability of optical/infrared radiation in penetrating clouds, especially in the tropical regions which are frequently under cloud covers throughout the year. Cloud-penetrating microwave radiation can be used to generate high resolution images of the earth's surface. One common microwave imaging system in satellite remote sensing is the synthetic aperture radar (SAR). In SAR imaging, microwave pulses are transmitted by an antenna towards the Earth's surface. The microwave energy scattered back to the spacecraft is measured. The SAR makes use of the radar principle to form an image by utilising the time delay of the backscattered signals. Due to the cloud penetrating property of microwaves, SAR is able to acquire "cloud-free" images in all weather. However, interpreting SAR images is not a straightforward task. It very often requires some familiarity with the ground conditions of the areas imaged and knowledge of the interaction mechanisms of microwaves with the land cover types of interest (Henderson and Lewis, 1998). The backscattered signals depend not only on the land cover types, but also on the wavelength and polarisation of the microwave used. SAR images cannot be used for detection of active fires as microwave is not sensitive to the heat and light emitted by active fires. Another direct signature of fires, i.e. smoke, is not visible in SAR images. SAR images are mainly used in detection of land cover change which affect the radar backscattering properties of the earth's surface. Some of the microwave remote sensing satellites are ERS-1 and 2 (C band, VV polarisation), RADARSAT (C band, HH polarisation) and the now defunct JERS (L band, HH polarisation). The orbits of the ERS 1 and ERS 2 satellites are identical except that ERS-2 lags ERS-1 by one day. This configuration of the two satellite orbits provides a unique opportunity for performing interferometric SAR in the tandem mode; i.e. a location on earth can be imaged by the two satellites with identical geometry, at one-day apart. The pairs of images acquired in tandem can be used to generate interferometric coherence maps which are useful for tropical forest applications (Wegmuller and Werner 1995; Stussi *et al.*, 1997; Liew *et al.* 1999).

Choice of Satellite-Sensors for Remote Sensing of Fire

Given the multitude of the satellite-sensor platforms available for remote sensing, it is natural to ask which one is the most suitable for remote sensing of fires. Each of these sensors has its advantages and limitations. The choice of sensors depends on the intended applications. At present, there is no one single satellite-sensor platform that can serve all the requirements for fire monitoring and mapping. Perhaps the best strategy is to engage a multisensor approach; i.e. data from all available satellite-sensor platforms are used (Eva and Lambin, 1998).

Several options are available when choosing satellite-sensors for remote sensing of fire. First, there is the selection of the attributes of fires to monitor: active fires, smoke haze or burnt areas. Next, the requirements of spatial resolution, coverage area, revisit frequency and timing of image acquisition (e.g. daytime or night-time, morning, noon or evening) have to be considered.

REMOTE SENSING OF ACTIVE FIRES

Three manifestations of active fires can be detected by satellite-sensors: heat, light and smoke plumes.

Heat Emission

The heat emitted by active fires is the signal most commonly used for detection of active fires (Robinson, 1991). The AVHRR sensors on board the NOAA Polar Orbiting Environment Satellites are commonly used to detect heat emission from fires for the construction of fire hotspot distribution maps (Matson *et al.*, 1987). The physical principles involved and the limitations of fire detection using AVHRR will be described here.

The amount of thermal radiation emitted at a particular wavelength from a warm object depends on its temperature. If the earth's surface is regarded as a blackbody emitter, its apparent temperature (known as the brightness temperature) and the spectral radiance are related by the Planck's blackbody formula:

$$L_\lambda = \frac{c_1 / \lambda^5}{\exp(c_2 / \lambda T) - 1}$$

where λ is the wavelength, T is the brightness temperature and L_λ is the spectral radiance (power emitted per unit surface area per solid angle per wavelength interval) at

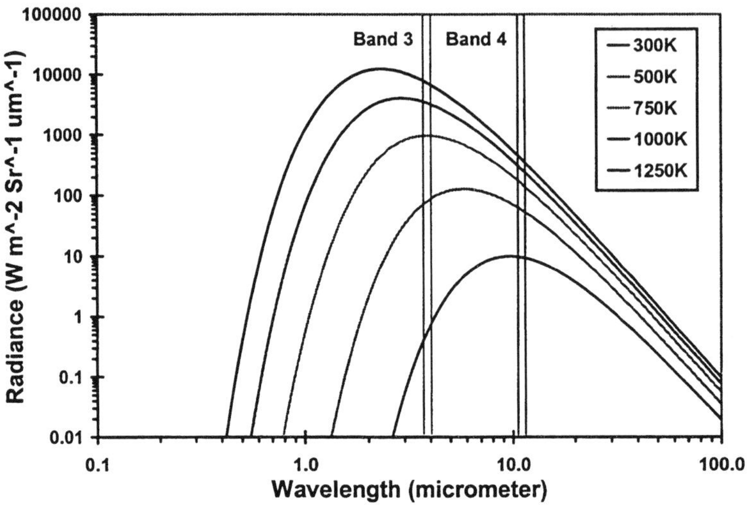

Figure 2 Blackbody Radiance at different brightness temperatures

For typical fire temperatures from about 500 K (smouldering fire) to over 1000 K (flaming fire), the radiance versus wavelength curves peak at around 3.8 μm, which coincides with Band 3 of AVHRR. For this reason, Band 3 of AVHRR is often used for detection of 'fire hotspots'.

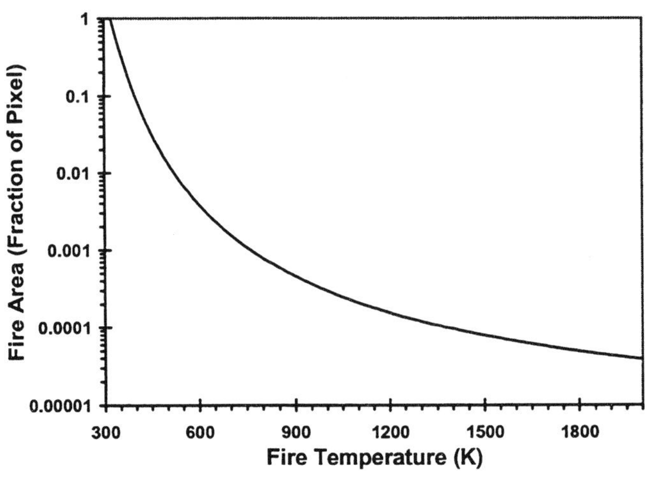

Figure 3 The fraction of a pixel required to be covered by fire at a given temperature in order to saturate Band 3 of the AVHRR sensor. The background land surface temperature is assumed to be 300 K.

wavelength λ. This relation is plotted in Figure 2. For a surface at a brightness temperature around 300 K, the spectral radiance peaks at a wavelength around 10 μm. The peak wavelength decreases as the brightness temperature increases.

Band 3 and other thermal bands of AVHRR are actually designed for detecting night-time clouds and for measurement of sea surface temperature, and thus are not optimised for detection of active fires. The dynamic range of AVHRR Band 3 has been designed such that the sensor saturates at a radiance corresponding to a brightness temperature of about 320 K (47 °C), well below the typical fire temperature. There are several consequences, in relation to fire detection, due to this low saturation temperature. While terrestrial non-fire targets rarely exceed this temperature, any fire of a respectable intensity will easily saturate the sensor. Thus, it is not possible to determine the fire temperature and hence AVHRR is not able to discriminate between smouldering fire and flaming fire. It is also not necessary for a fire to cover the whole pixel area (about 1 km^2 for AVHRR) before it can be detected as a fire hotspot. In Figure 3, the fraction of a pixel required to be covered by fire in order to saturate Band 3 of AVHRR is plotted against the fire temperature. For example, a smouldering fire at 500 K needs to cover more than a hundredth of a pixel area (i.e. about 100 m x 100 m) before it saturates Band 3. On the other hand, a flaming fire at 1000 K will saturates Band 3 even if it covers an area as small as 20 m x 20 m. Hence, a saturated pixel may be due to a single intense fire covering a small area anywhere within the 1 km^2 pixel area, a large fire or several fires affecting a much larger area within the pixel. Consequently, the low saturation temperature prohibits the discrimination between large and small fires. Due to the coarse resolution of the AVHRR sensors, the exact locations of fires cannot be determined and measurement from hot spot images generally overestimates the area burnt but underestimates the total fire count.

Numerous algorithms exist for detection of fire hotspots from AVHRR images (Robinson, 1991; Kaufman and Tucker, 1990; Setzer and Malingreau 1996). Most of the algorithms are based on thresholding the Band 3 and Band 4 pixel values. The differences between these algorithms lie mainly in how the threshold values are determined. Since most active fires would saturate Band 3, a straightforward way of detecting hotspots is to impose a threshold on the Band 3 data. Any pixel with equivalent Band 3 temperature exceeding this threshold would be labelled as a hotspot pixel. However, this method would introduce many false alarms, for Band 3 would also be saturated by sun glints from clouds and water, as well as by thermal emission from warm land surfaces. Sun glints from open water can be removed by using a water mask of sea and inland waters to mask out the water areas before further analysis. The other two types of false alarms can be eliminated with the help of Band 4 (11 μm). Solar radiation is practically nil in the 11 μm band and since clouds are significantly cooler than land surfaces, clouds can be easily detected using Band 4. Removal of false alarms due to warm land surfaces is more tricky. Band 4 is less sensitive to an alleviation of temperature than Band 3. A fire hotspot usually registers a lower apparent temperature in Band 4 than in Band 3, whereas non-fire land surfaces appear to have similar temperatures in Band 3 and Band 4. Hence, a potential hotspot pixel (i.e. one with Band 3 temperature exceeding the threshold) will not be classified as a hotspot if the difference between its Band 3 and Band 4 temperatures is less than a certain threshold value (usually around 10K).

The AVHRR sensor is carried on NOAA's Polar Orbiting Environmental Satellites (POES), beginning with TIROS-N in 1978. Currently, there are three operational NOAA satellites: NOAA-12 with morning and evening passes, NOAA-14 with afternoon and night-time passes and NOAA-15 with morning and evening passes. Eventually, NOAA-15 will replace NOAA-12. Together, the NOAA satellites provide at least four image acquisitions per day for any location on the earth. AVHRR data are acquired in three formats: High Resolution Picture Transmission (HRPT), Local Area Coverage (LAC), and Global Area Coverage (GAC). HRPT data are full resolution image data transmitted to a ground station as they are collected. LAC are also full resolution data, but recorded with an on-board tape recorder for subsequent transmission during a station overpass. GAC data provide daily subsampled global coverage recorded on the tape recorders and then transmitted to a ground station. Many weather stations around the world operate ground stations that routinely receive real-time HRPT data. LAC data are available from NOAA's Satellite Active Archive (SAA) through the world-wide web (http://www.saa.noaa.gov). Since on-board tape facility is limited, only a limited number of scenes are archived. GAC are subsampled on-board to about 4-km pixel separation at nadir for global fire monitoring. GAC data are also available from NOAA SAA.

Figure 4 shows an example of a cumulative hotspot distribution map for Borneo derived from daily NOAA-14 day time AVHRR images during the fire period from May to October, 1997. This cumulative hotspot map shows the general areas affected by fires during this period.

Besides the NOAA-AVHRR, other satellite sensors also have thermal bands equivalent to the Band 3 (around 4 μm) and Band 4 (around 11 μm) of AVHRR that can be used for detecting fire hotspots. The family of the Along Track Scanning Radiometer (ATSR) instruments on board the ERS-1 and ERS-2 satellite has 4 thermal bands (1.6, 3.7, 10.8 and 12.0 μm), 3 of which are equivalent to Bands 3, 4 and 5 of AVHRR. The other 1.6 μm is also present in the newer generation of AVHRR on the recent NOAA-15. The ATSR also has a pixel resolution of 1 km, but with a better pixel location accuracy than AVHRR. However, its swath width is about 1000 km, in comparison to the larger swath of over 2000 km for AVHRR. Furthermore, it has a less frequent temporal coverage than the AVHRR.

The NASA's Terra (formerly known as EOS AM-1) polar orbiting satellite launched on 18 December 1999 carries the MODIS instrument which has 36 spectral bands for observation of land, ocean and atmosphere (Running *et al.*, 1994). Among the 36 bands, Band 20 (3.66 - 3.84 μm) and Band 31 (10.78 - 11.28 μm) are equivalent to Bands 4 and 5 of AVHRR useful for fire monitoring (Kaufman *et al.*, 1998). The MODIS thermal bands have similar spatial resolution (i.e. 1 km) as the AVHRR. However, the MODIS bands are specially designed for fire monitoring. They saturate at higher temperatures (450 K for the 3.7 μm band and 400 K for the 11 μm band) and the data are digitised at 12 bits, instead of 10 bits for AVHRR. Except for very widespread fires that almost cover

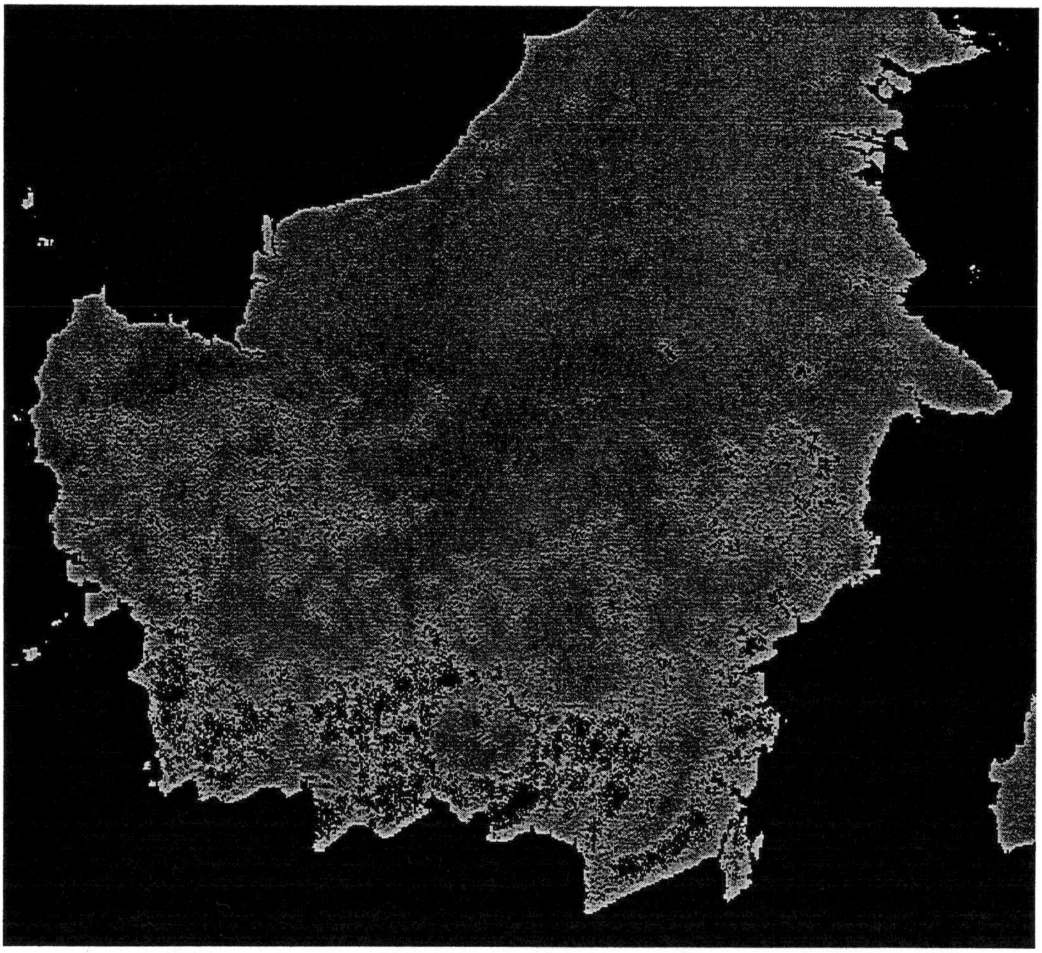

Figure 4 Cumulative hotspot image of Borneo derived from day time NOAA-14 AVHRR imagery during the period May to October, 1997. The areas coloured red have two or more hot spots detected during this period while the areas with one hot spot detected are coloured dark green. The AVHRR data were provided by the Meteorological Service Singapore (MSS) and processed at the Centre for Remote Imaging, Sensing and Processing (CRISP).

the area of a whole pixel, the MODIS bands are not expected to saturate. Hence, it is possible to derive the fire temperature and fire area using an algorithm similar to the one developed by Dozier (1981) for AVHRR. The Terra-MODIS sensor provides two observations per day during the morning and evening passes. Another similar satellite carrying the MODIS instrument (Aqua, formerly EOS PM-1) is being planned to launch in the near future. This second satellite will provide observations in the afternoon and night-time passes. These satellites are expected to enhance the fire monitoring activities around the world when their data are available.

The higher resolution satellite-sensors such as the Landsat-TM and SPOT-HRV do not have the 4 μm thermal band useful for fire detection. The short wave infrared (1.6 μm) band present in the Landsat-TM and the new HVIR sensor on board the SPOT 4 satellite has the capability to detect hot surfaces due to active fires burning during the time of imaging. An example of such application is shown in Figure 5. In the false colour image (left panel), the vegetated areas are shown in green while the reddish areas are exposed soils. In this high resolution image, a smoke plume due to an active fire can be seen emanating from the edge of a linear grid-like structure, indicating that the fire occurred in a plantation or agricultural area. The plot of land adjacent to the smoke plume had just been cleared by the fire. It was so hot that the radiation emitted from the hot land surface saturated the 1.6 μm of the SPOT4-HVIR sensor, as shown in the right panel of Figure 5.

Light Emission

The visible light emitted from active fires is another form of signal which permits the detection of fires by satellite-borne sensors. The visible and near infrared bands (bands 1 and 2 respectively) of AVHRR are not sufficiently sensitive to detect light emission from fires. The US Air Force (USAF) Defence Meteorological Satellite Program (DMSP) operates a series of satellites which carry very sensitive light sensors known as the Operational Linescan System (OLS) that can detect light emission from the earth surface at night (Elvidge *et al.*, 1997). The digital data from DMSP-OLS were archived at NOAA's National Geophysical Data Center (NGDC) since 1992. The USAF initially imposed a 72 hours embargo on the DMSP-OLS before they could be released by NOAA. Since 21 Dec 1999, the 72 hours hold has been relaxed to 3 hours.

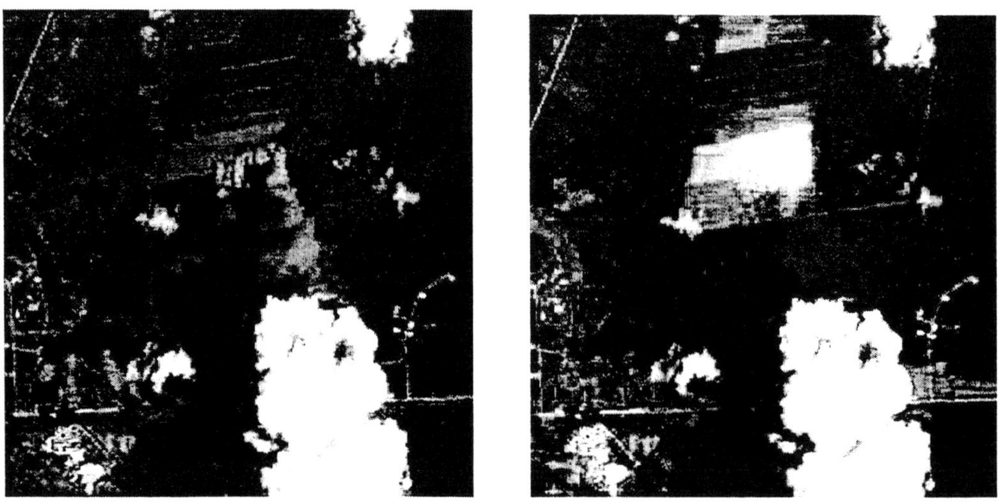

Figure 5 A false-colour SPOT4 HVIR image (Red: band 4, Green: band 3, Blue: band 2) (left panel) and the corresponding band 4 image (right panel) of a plantation fire near the town of Pekanbaru, Sumatra, Indonesia on 3 February, 2000. The area shown is about 6 km x 6 km. © CNES 2000, acquired and processed by CRISP.

The DMSP-OLS has a ground swath of about 3000 km. It has two broad spectral bands, one covering the visible-near infrared region (0.5 - 0.9 μm) and the other is in the thermal infrared region around 10 μm. The OLS data are acquired in two spatial resolution modes: 'fine' and 'smoothed'. The full resolution fine data have a nominal spatial resolution of 0.56 km. On board averaging of five by five blocks of fine data produces "smoothed" data with a nominal spatial resolution of 2.7 km. Most of the data received by NOAA-NGDC is in the smoothed spatial resolution mode. The DMSP-OLS data are also available from NOAA SAA. The DMSP-OLS has a unique capability to observe faint sources of visible- near infrared emissions present at night on the earth's surface, including cities, towns, villages, gas flares, heavily lit fishing boats and fires (Croft, 1978; Elvidge *et al.*, 1997). By analysing a time series of DMSP-OLS images, it is possible to define a reference set of "stable" lights, which are present in the same location on a consistent basis. Fires are identified as lights detected on the land surface outside the reference set of stable lights (Elvidge *et al.*, 1996). The identified fire pixels are also commonly referred to as the 'fire hotspots'.

Due to the coarse resolution of the smoothed OLS data, they have similar limitations as the AVHRR data for fire detection. For example, it is not possible to determine the number of fires, locations of individual fires and fire size within an identified fire hotspot. Furthermore, smouldering fire or fire burning below the canopy will not be detected by OLS.

Figure 6 shows a cumulative fire hotspot distribution map derived from DMSP-OLS images for parts of Indonesia during the period July to December 1997. Comparison with the AVHRR cumulative hotspots distribution map for Kalimantan (Figure 4) shows that the two maps generally agree regarding the areas affected by the 1997 fires.

Smoke Plumes

Smoke plumes in optical remote sensing images are direct telltale signs of active fires. They can be visually identified in an image as some fuzzy bright strips, usually conical in shape converging to a point on the ground. In high resolution images such as those acquired by the SPOT satellites, the precise locations of the active fires can be determined from the points of origin of the individual smoke plumes.

An example of a high resolution image of a smoke plume from a plantation fire is illustrated in Figure 5. This image was acquired by the SPOT 4 HVIR sensor. The smoke plume is seen as a bluish conical stripe. It spreads out to a width of about 3 km after travelling 4 km towards the south. The adjacent grid-like structures reveal that the fire is burning in a plantation. Another example of smoke plumes detected by a SPOT satellite is shown in Figure 7. In this case, the smoke plumes are numerous and intense that they can even be seen in the quicklook image. Quicklook images are routinely processed and archived at satellite receiving ground stations for quality control and to aid users in selecting suitable scenes. These quicklook images are subsampled from the full resolution scenes at 100-m pixel interval while preserving the radiometric characteristics of the

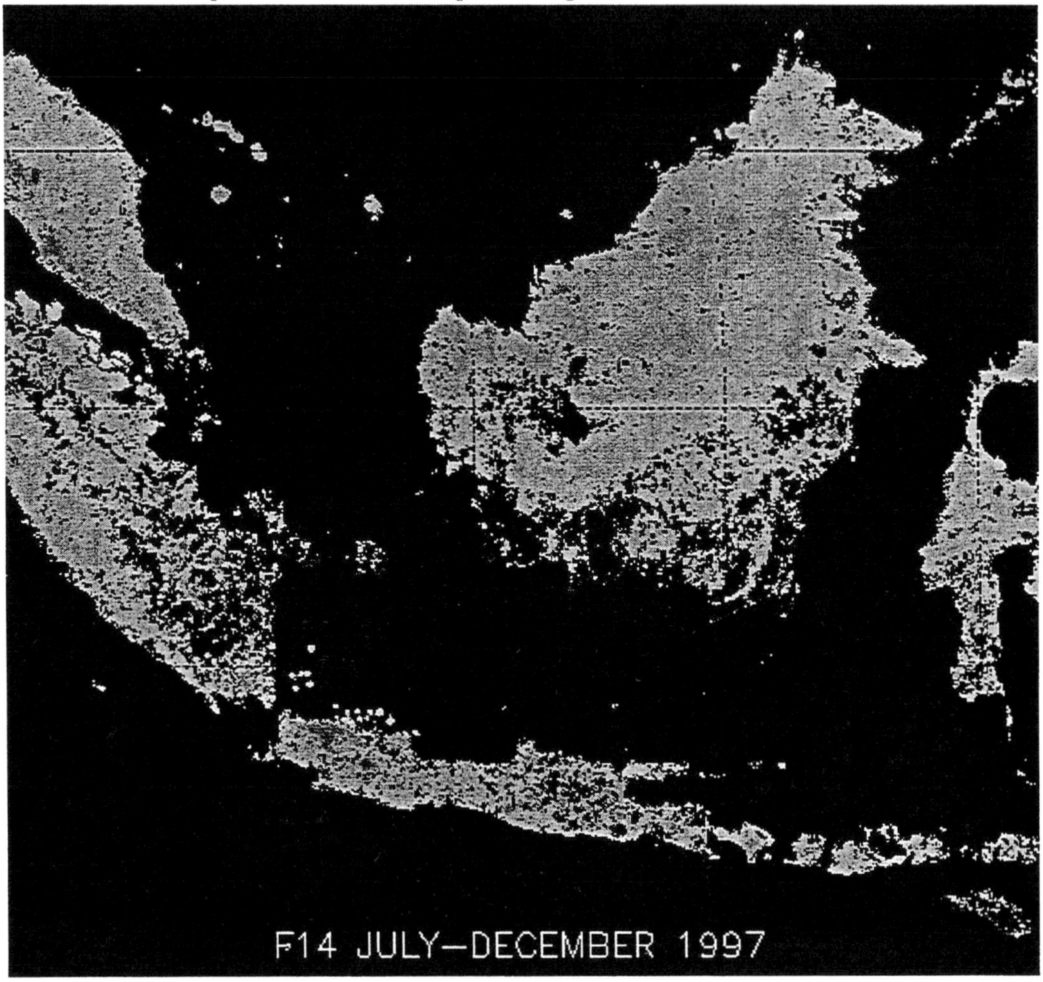

Figure 6 Cumulative fire hotspots derived from night-time DMSP-OLS images from July to December 1997 (Cyan: stable light, Red: fire hotspots). Source: NOAA National Geophysical Data Center (NGDC).

original pixel values. The area shown in Figure 7 is part of the 'Mega Rice Project' area of Central Kalimantan where about 1 Mha of peat swamp was converted for rice cultivation This area was one of the regions affected by fires during 1997. In this image, long streaks of smoke plumes can be seen, some extending over 30 km away from their sources.

Smoke plumes can also show up in the NOAA AVHRR images if the smoke plumes are intense and spread out far away from the site of active fires.

The combustion products during a forest fire exist in the forms of gaseous and particulate emissions. Near to the vicinity of active fires, the particulate emissions are noticeable as smoke plumes in satellite images. If these emissions spread and persist in the atmosphere for a considerable length of time, they create the phenomenon of smoke haze, which reduces visibility. Depending on the height reached by the combustion products and the strength and direction of wind, these emissions can be carried by wind and spread to other neighbouring regions.

Remote sensing of smoke haze usually depends on the light scattering properties of the fine particulate matters (typically less than 1 μm in size) suspended in the atmosphere (known as aerosols). When light from the sun traverses through an aerosol, some of the light is absorbed and some is scattered by the particulates, while the remaining light is transmitted through the aerosol. In a very thick haze, light undergoes multiple scattering and spreads almost uniformly in all directions. In this case, the direct transmitted light is negligible compared to the scattered light. An aerosol is characterised by several optical parameters.

An important parameter is the aerosol optical thickness (τ), which depends on the concentration of aerosol particles and the effective thickness of the aerosol layer in the atmosphere. The optical thickness is defined as:

$$\tau = -\log\left(\frac{I}{I_0}\right)$$

where I_0 is the intensity of light at the top of atmosphere incident along the vertical direction, and I is the intensity of the directly transmitted light (excluding scattered light) at the bottom of the atmosphere. The optical thickness is wavelength dependent and depends on the types of particles present in the atmosphere. Typically, for a given type of particles, the optical thickness is larger for shorter wavelength and smaller for longer wavelength. The wavelength dependence can be empirically fitted to a power law relation:

$$\tau(\lambda) = \tau(\lambda_0)\left(\frac{\lambda_0}{\lambda}\right)^{\alpha}$$

where $\tau(\lambda)$ is the optical thickness measured at wavelength λ, λ_0 is a reference wavelength, and α is a parameter known as the Angstrom exponent. The Angstrom exponent is related to the effective size of the particles. It ranges from 0 to 4 and increases as the particle size decreases. For very small particles (much smaller than the wavelength of light) such as gaseous molecules, the Angstrom exponent is equal to 4. For submicron particles, α typically has a value equal to 1.5 or larger. The optical thickness gives an indication of the total extinction (i.e. the sum of absorption and scattering) undergone by an incident light beam.

Figure 7 A SPOT quicklook image of part of the 'Mega Rice Project' area in Central Kalimantan, acquired on 8 September 1997, showing burnt areas and smoke plumes arising from sites of active fires. The image covers an area of 60 km x 60 km. © CNES 1997, acquired and processed by CRISP.

The ratio of scattering to extinction is known as the single scattering albedo (ω_0). The value of the single scattering albedo ranges from 0 for total absorption to 1 for total scattering. For smoke haze arising from biomass burning, the single scattering albedo may give hints to the type of burning. The single scattering albedo depends on the amount of graphitic carbon in the aerosol, which is a strong absorber of light in the visible spectrum. High intensity flaming combustion results in release of higher concentration of graphitic carbon but only moderate amount of trace gases. On the other hand, smouldering conditions result in inefficient burning and release of large amount of trace gases and particles which contain a higher percent composition of organic matter but a much smaller amount of graphitic carbon (Ward and Hardy, 1984; Ward, 1986). Therefore, ω_0 can be as high as 0.97 for smouldering and as low as 0.66 for flaming conditions in laboratory measurements (Patterson and MacMahon, 1984). Gras *et al.* (1999) studied the optical properties of smoke aerosol in Indonesia and tropical Australia during the 1997 dry season fires. Their data indicated relatively small absorption in the Indonesian smoke, suggesting that the Indonesia fires are predominantly due to smouldering fires. Nakajima *et al.* (1999) studied the optical properties of biomass burning aerosols in the 1997 Indonesian forest fires by ground based sky radiometry and satellite remote sensing with AVHRR and TOMS data. The single scattering albedo was found to be approximately 0.9 in the September-October 1997 period. Their data also suggested the predominance of small particles with a peak radius of 0.25 μm in the smoke haze.

Operational monitoring of smoke haze by satellite remote sensing is often hindered by cloud covers, especially in the tropical regions. Interference from ground reflection also complicates the retrieval of aerosol properties over land. Very often, visual qualitative analysis of remote sensing images is performed to delineate the region covered by smoke haze, and to forecast the transport of haze to other regions given the prevailing wind conditions. The matter is a bit simpler over oceans, since the ocean surface typically has low reflectance in the visible and near infrared wavelength bands. An operational algorithm to retrieve the aerosol optical thickness over ocean using the AVHRR band 1 and band 2 data has been developed at NOAA (Rao *et al.*, 1989; Stowe *et al.*, 1997; Husar *et al.*, 1997). In this algorithm, the algorithm has been implemented to produce operational global aerosol thickness products over ocean.

Aerosols arising from biomass burning are usually accompanied by an alleviation of ozone. This property has been exploited to indirectly detect and quantify the amount of aerosol in the atmosphere using the Total Ozone Mapping Spectrometer (TOMS) (Herman *et al.*, 1997; Hsu *et al.*, 1996). TOMS is an instrument built and operated by the National Aeronautics and Space Administration (NASA). The instrument uses backscattered ultraviolet radiance to infer total column ozone measurements. The data consists of daily gridded averages of total ozone covering the entire globe. The original Nimbus-7 TOMS operated from November 1978 until May 1993. Meteor-3 TOMS was launched in August 1991 and operated until December 1994. After a gap of one and a half years, two new TOMS instruments began operation in 1996; Earth-Probe TOMS was launched on 2nd July 1996 and started to produce data on 25th July. ADEOS TOMS was launched on 17th August 1996 and started producing data on 11th September. The satellites were originally placed in different orbits, giving complete global coverage with the ADEOS data, while Earth-Probe had complete coverage at the poles with an increased ability to measure UV-absorbing aerosols in the troposphere. ADEOS failed in June 1997 and Earth-Probe was subsequently placed in a higher orbit to give global coverage. Earth-Probe TOMS continues to operate successfully. Global data sets of TOMS aerosol products are available. Since the observation dated back to the seventies, historical changes in biomass burning activities can be monitored.

REMOTE SENSING OF FIRE AFFECTED AREAS

Remote sensing images are useful for assessing the extent and impact of fires after a fire event. Fire burn scars can be detected quite easily in the high resolution SPOT-HRV or Landsat-TM images. Recent fire scars appear dark due to the deposition of black char which suppresses the reflectance in the visible and near infrared wavelength bands. This phenomenon is clearly illustrated in Figure 7, a SPOT scene showing burnt areas and smoke plumes. The 1.6 μm short wave infrared band (as in Landsat TM and SPOT 4) is also useful for the detection of burn scars. Burnt areas generally show a higher reflectance in the 1.6 μm band compared with the surrounding unaffected forest. Maps of fire affected areas can be generated by delineating burn scars observed in remote sensing images (Liew *et al.*, 1998). These maps provide information on the spatial extent and distribution of the fire affected areas, and the total area burned. The burn scar maps (Liew *et. al.*, 1998) produced for the 1997 forest fire in Southeast Asia have been used for assessing the economic damage due to the fires (EEPSEA/WWF, 1999), estimated to be more than 4 billion US dollars. These burn scar maps together with the figure for the total burn areas were also used in estimating the gaseous and particulate emissions resulting from the 1997 fires (Levine, 1999).

SATELLITE MONITORING OF FOREST FIRES IN SOUTHEAST ASIA

Biomass burning has been the traditional method of clearing land for shifting cultivation in Southeast Asia. However, the small scale clearing of land is increasingly being replaced by modern large-scale conversion of forests into plantations/agricultural land, usually by fires (Goldammer, 1997). In periods of extreme drought, especially those associated with the El Niño Southern Oscillation (ENSO) phenomenon, the fires get out of control resulting in severe transboundary air pollution in the form of smoke haze. The 1982-83 forest fire event in Borneo has been reported to affect 5 million hectares of primary and secondary rain forests (Goldammer and Seibert, 1990). Despite daily satellite observations by meteorological satellites over the area during the fire event, the extent and severity of the fires were little noticed. The history of the burning was reconstructed using the 4-km GAC data as none of the local stations archived the HRPT data (Malingreau *et al.*, 1985).

Forest fire events occurred during the dry seasons in 1987, 1991, 1994 and 1997. The recent one in 1997 was especially severe, causing many countries in Southeast Asia to be affected by thick smoke haze. Prompted by the recurrent haze events, the Association of South East Asian Nations (ASEAN) Environment Ministers had in June 1995 agreed on an ASEAN Co-operation Plan on Transboundary Pollution. The Co-operation Plan contains broad policies and strategies to deal with transboundary pollution. In response to the latest haze episode in 1997, the ASEAN Environment Ministers agreed on a Regional Haze Action Plan, which sets out co-operative measures needed amongst ASEAN member countries to address the problem of smoke haze in the region arising from land and forest fires. Under this plan, Singapore has taken up the responsibility of coordinating the regional land and forest fire monitoring mechanisms.

The Centre for Remote Imaging, Sensing and Processing (CRISP) and Meteorological Service Singapore (MSS) jointly assist the Ministry of the Environment of Singapore to monitor the regional fires with remote sensing images. MSS acquires daily NOAA AVHRR and GMS images, and processes "hotspot" maps from the AVHRR data. The images and hotspot maps are posted on an intranet for access by authorised users. CRISP uses SPOT images to zoom in onto the hotspot areas to accurately locate the individual fires and to determine where possible the nature of the fires. Daily, passes of SPOT images are inspected and reports of fires with annotated images are sent to the Ministry of the Environment. A few hardware and software developments have enabled CRISP to efficiently perform this fire monitoring operation in near real-time (Lim *et al.*, 1998). Fire detection is carried out by visual inspection of the high resolution (20 m) image strips, looking out for tell tale signs of fires (smoke plumes and burn scars). The fire report lists for each detected fire: the accurate location (longitude and latitude), smoke plume condition (length, width and thickness), wind direction, the type of land cover on fire (forest, plantation, peat swamp, etc.) and the extent of burnt scar associated with the fire. The fire reports and annotated images are ready for delivery within five to six hours after the initial SPOT data reception.

The ground station at CRISP has acquired numerous SPOT images of the Southeast Asian region during this 1997 forest fire episode with almost complete coverage of the Sumatra and Borneo islands. SPOT quicklook images in CRISP's archive covering Sumatra and Kalimantan in Indonesia from January to December 1997 were used to generate two composites for each of the two regions, one before the fire episode from January to June and the other during the fire episode from July to December 1997 using a total of 766 SPOT scenes. The burnt areas in 1997 were delineated on the mosaics during the fire period (July to December, 1997) by visual inspection. Manual interpretation and classification was performed to classify the burnt areas into three classes: plantations/agricultural lands, peat swamp forests and other vegetation types which were mainly secondary forests and bushes. The plantations/agricultural lands could be distinguished by the presence of linear grid-like features which were probably canals or road networks. Peat swamp forests were characterised by their deep red appearance (due to lower near infrared reflectance than other forest types but with overall high vegetation index) in the before fire mosaics. The burnt areas were concentrated in the peat swamp areas along the northern coast of the Province of South Sumatra, near the town of Palembang. Much of the previous peat swamp forests had been converted to plantations/agricultural lands. The burnt scars stretched from Palembang along the coast to the Province of Jambi. Isolated fire scars were also found on the lowland areas parallel to the northern slope of the Barisan Mountains. In Kalimantan, the fire burnt scars were concentrated in two main areas: the peat swamp area in southern part of the central and the southwest coastal area. Some burn scars were also found in East Kalimantan. The burnt areas in central Kalimantan were in the 'Mega Rice Project' area. In both Sumatra and Kalimantan, the burnt areas were mostly located in low land areas close to roads or river basins, indicating correlation of the fires with human activities. The highland rain forests seemed to be untouched by the fires.

The total area of burn scars was initially estimated to be 1.50 Mha in Sumatra and 3.06 Mha in Kalimantan (Liew *et al.*, 1998). These figures were later refined after eliminating swamps that were misclassfied as burnt scars, and including burnt scars along the southern coast of Kalimantan that were missed in the initial attempt, as well as areas in the Riau province, Sumatra, that have been burnt in early 1997. The revised estimates for the burnt areas during the 1997 fires are 2.08 Mha in Sumatra and 3.58 Mha in Kalimantan. In Sumatra, about 45% of the burnt area was associated with plantations/agricultural land, 30% was in coastal peat swamp areas. In Kalimantan, 20% of the burnt area was associated with land clearings in peat swamp areas.

The 1997 fires subsided in November, with the coming of the rainy season which did not last long. Fires raged again from February to April 1998, affecting mainly the East Kalimantan province of Indonesia, and the northern part of the Borneo island, i.e. Brunei and Sabah and Sarawak states of Malaysia. The burnt areas estimated from SPOT images are 2.51 Mha in East Kalimantan, 530 kha in Sabah and 90 kha in Sarawak and Brunei.

In 1999, fires were detected in March, mainly concentrated in the Riau Province in Sumatra, as the weather became drier. From the archived SPOT images, the statistics on the number of active fires detected at different land cover types were compiled (Table 1). The land cover types were classified into five classes: big plantations (>100 ha), small agricultural fields and plantations, timber logging areas, forests and others. The statistics show that more than 40% of the total detected active fires occurred in big plantations and 82% of the active fires are associated with human activities.

Table 1 Statistics of active fires detected in 1999 SPOT images of Sumatra and Kalimantan

	Sumatra	Kalimantan	Total	Percentage of all active fires
Big Plantations	116	16	132	41.3
Small agricultural fields	89	15	104	32.5
Logging areas	25	2	27	8.4
Forests	35	7	42	13.1
Others	11	4	15	4.7

REFERENCES

1. Croft, T.A. 1978. Nighttime images of the earth from space. *Scientific American* **239**: 68-79.

2. Dozier, J. 1981. A method for satellite identification of surface temperature fields of subpixel resolution. *Remote Sensing of Environment* **11**: 221-229.
3. Elvidge, C.D., H.W. Kroehl, E.A. Khin, K.E. Baugh, E.R. Davis and W.M. Hao, 1996. Algorithm for the retrieval of fire pixels from DMSP Operational Linescan System. In: *Global Biomass Burning*. Ed. Levine, J.S., MIT Press, Cambridge, Massachusetts, pp. 73-85.
4. Elvidge, C.D., K.E. Baugh, E.A. Kihn, H.W. Kroehl and E.R. Davis, 1997. Mapping city lights with nighttime data from the DMSP operational linescan system. *Photogrammetric Engineering and Remote Sensing* **63**(6): 727-734.
5. Eva, H. and F. Lambin, 1998. Remote sensing of biomass burning in tropical regions: sampling issues and multisensor approach, *Remote Sensing of Environment* **64**: 292-315.
6. Goldammer, J.G. 1997. The role of fire on greenhouse gas and aerosol emissions and land use and cover change in Southeast Asia: Ecological background and research & development needs. Paper presented at the *International Conference on Science and Technology for the Assessment of Global Environmental Change and its Impacts on the Indonesian Maritime Continent*, Jakarta, Indonesia, 10-12 November 1997.
7. Goldammer, J.G. and B. Seibert, 1990. The impact of droughts and forest fires on tropical lowland rain forest of East Kalimantan. In *Fires in the Tropical Biota :Ecosystem Procecesses and Global Challenges*. Ed. Goldammer, J.G., Springer-Verlag, Berlin. pp. 11-31.
8. Gras, J.L., J.B. Jensen, K. Okada, M. Ikegami, Y. Zaizen and Y. Makino, 1999. Some optical properties of smoke aerosol in Indonesia and tropical Australia. *Geophysical Research Letters* **26** (10):1393-1396.
9. Henderson, F. and A. Lewis (Eds.), 1998. *Principles and Applications of Imaging Radar, Manual of Remote Sensing*, Vol. 2, J. Wiley,New York.
10. Herman, J.R., P.K. Bhartia, O. Torres, C. Hsu, C. Seftor and E. Celarier, 1997. Global distribution of UV-absorbing aerosols from Nimbus 7/TOMS data. *Journal of Geophysical Research* **102**(D14): 16911-16922.
11. Hsu, N.C., J.R. Herman, P.K. Bhartia, C.J. Seftor, O. Torres, A.M. Thompson, J.F. Gleason, T.F. Eck and B.N. Holben, 1996. Detection of biomass burning smoke from TOMS measurements. *Geophysical Research Letters* **23**(7): 745-748.
12. Huete, A.R., H.Q. Liu, K. Batchily and W. van Leeuwen, 1997. A comparison of vegetation indices over a global set of TM images for EOS-MODIS. *Remote Sensing of Environment* **59**: 440-451.
13. Husar, R.B., J.M. Prospero and L.L. Stowe, 1997. Characterization of tropospheric aerosols over the ocean with the NOAA advanced very high resolution radiometer optical thickness operational product. *Journal of Geophysical Research* **102**(D14): 16889-16909.
14. Kaufman, Y.J. and C.J. Tucker, 1990. Remote sensing of biomass burning in the tropics. *Journal of Geophysical Research* **95**(D7): 9927-9939.

15. Kaufman, Y.J., C. Justice, L. Flynn, J. Kendall, E. Prins, D.E .Ward, P. Menzeland A. Setzer, 1998. Potential global fire monitoring from EOS-MODIS. *Journal of Geophysical Research* **103**: 32215-32238.
16. Liew, S.C., O.K. Lim, L.K. Kwoh and H. Lim, 1998. Study of the 1997 forest fires in South East Asia using SPOT quicklook mosaics. *Proceedings of the 1998 International Geoscience and Remote Sensing Symposium*, Vol. 2, pp. 879-881.
17. Liew, S.C., L.K. Kwoh, K. Padmanabhan, O.K. Lim and H. Lim, 1999. Delineating land/forest fire burnt scars with ERS interferometric synthetic aperture radar. *Geophysical Research Letters* **26**(16): 2409-2412.
18. Lim, O.K., S.C. Liew, L.K. Kwoh and H. Lim, 1999. Forest fires under close surveillance. *SPOT Magazine* No. 30, April 1999, pp. 22 - 24.
19. Malingreau, J.P., G. Stephens and L. Fellows (1985), Remote sensing of forest fires: Kalimantan and North Borneo in 1982-3. *Ambio* **14**(6): 314-315.
20. Matson, M., G. Stephens and J. Robinson, 1987. Fire detection using data from the NOAA-N satellites. *International Journal of Remote Sensing* **8**(7): 961-970.
21. Nakajima, T., A. Higurashi, N. Takeuchi and J.R. Herman, 1999. Satellite and ground-based study of optical properties of 1997 Indonesian forest fire aerosols. *Geophysical Research Letters* **26**(16): 2421-2424.
22. Patterson, E.M. and C.K. McMahon, 1984. Absorption characteristics of forest fire particulate matter. *Atmospheric Environment* **18**: 2541-2551.
23. Prins, E.M. and W.P. Menzel, 1992. Geostationary satellite detection of biomass burning in South America. *International Journal of Remote Sensing* **13**(15): 2783-2799.
24. Prins, E.M., J.M. Feltz, W.P. Menzel and D.E. Ward, 1998. An overview of GOES-8 diurnal fire and smoke results for SCAR-B and 1995 fire season in South America. *Journal of Geophysical Research* **103**(D24): 31821 - 31835.
25. Rao, N., L. Stowe and P. McClain, 1989. Remote sensing of aerosols over ocean using AVHRR data: Theory, practice and applications. *International Journal of Remote Sensing* **10**: 743-749.
26. Robinson, J.M. 1991. Fire from space: global fire evaluation using infrared remote sensing. *International Journal of Remote Sensing* **12**(1): 3-24.
27. Running, S.W., C.O. Justice, V Salomonson, D. Hall, J. Barker, Y.J. Kaufmann, A.H. Strahler, A.R. Huete, J.P. Muller, V. Vanderbilt, Z.M. Wan, P. Teillet and D. Carneggie, 1994. Terrestrial remote sensing science and algorithms planned for EOS MODIS. *International Journal of Remote Sensing* **15**(17): 3587-3620.
28. Setzer, A.W. and J.P. Malingreau, 1996. AVHRR monitoring of vegetation fires in the tropics: Toward the development of a global product. In: *Biomass Burning and Global Change*. Ed. Levine, J.S., The MIT Press, Cambridge, Massachusetts, pp. 25-39.
29. Stowe, L.L., A.M. Ignatov and R.R. Singh, 1997. Development, validation, and potential enhancements to the second generation operational aerosol product at the National Environmental Satellite, Data, and Information Service of the National

Oceanic and Atmospheric Administration. *Journal of Geophysical Research* **102**(D14): 16923-16934.

30. Stussi, N., S.C. Liew, L.K. Kwoh, and H. Lim, 1997. Landcover classification using ERS-SAR/INSAR data over tropical areas. *Proceedings of the 1997 International Geoscience and Remote Sensing Symposium*, pp. 813-815.
31. Ward, D.E. and C.C. Hardy, 1984. Advances in the characterization and control of emissions from prescribed fires. Paper presented at the *77th Annual Meeting of the Air Pollution Control Association*, San Francisco, California.
32. Ward, D.E. 1986. Field scale measurements of emissions from open fires, technical paper presented at the *Defense Nuclear Agency Global Effects Review*, Def. Nucl. Agency, Washington DC.
33. Weaver, J.F., J.F.W. Purdom and T.L. Schneider, 1995. Observing forest fires with the GOES-8, 3.9-μm imaging channel. *Weather and Forecasting* **10**(4): 803-808.
34. Wegmuller, U., and C.L. Werner, 1995. SAR interferometric signatures of forest. *IEEE Transactions on Geoscience and Remote Sensing* **33**: 1153-1161.

Chapter 6

AIR POLLUTION MEASUREMENTS AND MODELING IN SOUTHEAST ASIA DURING EL NIÑO OF 1997

Mahesh J. Phadnis[1], Gregory R. Carmichael[2] and Sze F. Lim[3]

[1]Woodrow Wilson School of Public and International Affairs, Princeton University, Princeton, NJ 08544. U.S.A

[2]Center for Global and Regional Environmental Research, The University of Iowa, Iowa City, IA 52242. U.S.A

[3]Malaysian Meteorological Service, Jalan Sultan, 46667 Petaling Jaya, Selangor D.E., Malaysia.

INTRODUCTION

Biomass burning, mostly initiated by human beings, occurs regularly during the dry season in the tropics. Most important sources of biomass burning are the savanna fires in tropical Africa, forest fires and savanna fires in tropical America, and forest fires in tropical Asia (Hao and Liu, 1994). Although studies and field measurements of atmospheric pollutants have been carried out elsewhere, Southeast Asia remains relatively unexplored. This situation was magnified during the forest fires in Indonesia during the fall of 1997 that created a severe regional scale air pollution episode.

The severe drought over Indonesia as a result of El Niño was responsible for fires that burned out of control over the islands of Sumatra and Kalimantan. An estimated 300,000 hectares of forest was destroyed by fires. The total carbon released in 1997 from forest burning in Indonesia was estimated to be as much as that from fossil fuel emissions in Europe (Lewan, 1997). In September 1997, Air Pollution Indices (APIs) (see Chapter 10 for definition of API) reached hazardous levels for many days in Malaysia and Singapore while the visibility remained very low. In Southeast Asia, an estimated 20 million people were treated for illnesses such as asthma, bronchitis, emphysema, and eye, skin and cardiovascular diseases; a passenger plane crashed over Sumatra and ships collided in the Strait of Malacca killing many (Simons, 1998). Socioeconomic losses resulting from the forest fires and exacerbated air pollution episodes were staggering.

Scattered information is available on observations of the ambient levels of atmospheric pollutants from the forest fires. Sampling of atmospheric aerosols and ozone carried out during the fall of 1997 focussed mostly on the most severe month of September. At Kuching in Sarawak, Malaysia, biomass smoke pollution resulted in elevated levels of particulate matter. Measurements showed the 24-hour PM10 (Particulate Matter < 10 μm) levels reached as high as 930 μg/m3, more than 15 times higher than normal levels (Brauer and Hisham-Hashim, 1998) (Figure 1). For high concentration episodes observed in the afternoon due to downward flux from aloft, ratio of PM10 to TSP (Total Suspended Particulates) approached 90% (Geernaert and Waahlin, 1998). Aircraft sampling for atmospheric trace gases and aerosols indicated the dominance of particles with smaller diameters (0.03-0.1 μm) over northern Australia, whereas concentrations of larger particles were high over Kalimantan (Zaizen *et al.*, 1998). Aerosol and NOx levels were found to decrease from surface to upper elevations, but ozone did not. While ozone was high in the boundary layer over Australia, over Kalimantan it showed maximum concentration (~80 ppb) in the middle layer of the smoke haze and low concentrations below (Tsutsumi *et al.*, 1999). Vertical profiles of ozone over Petaling Jaya, Malaysia, corroborate with the above observations (Lim *et al.*, 1998).

In this chapter, we investigate the impacts of the gaseous and particulate emissions from the forest fires in Indonesia on the regional air quality of Southeast Asia during the fall of 1997. A three-dimensional numerical transport model is used to study a month-long episode of air pollution in October 1997. A brief analysis of observed ozone and particulate matter data over Southeast Asia will be presented in the next section. Emissions of trace pollutants from the forest fires and results from numerical modeling of the regional air pollution will be presented thereafter.

Data Analysis for Tropospheric Ozone and Particulate Matter over Southeast Asia

Satellite Observations

Long-term measurements of the total atmospheric column ozone are available from the Nimbus 7 (November 1978 to May 1993) and Earth Probe (July 1996 to present) total ozone mapping spectrometer (TOMS). Tropospheric column ozone can be derived from these satellite measurement as shown by Fishman *et al.* (1990), Hudson and Thompson (1998) and Ziemke *et al.* (1998). In addition to ozone, Herman *et al.* (1997) have shown that large-scale distribution of optically absorbing and non-absorbing aerosols can also be retrieved from the satellite data.

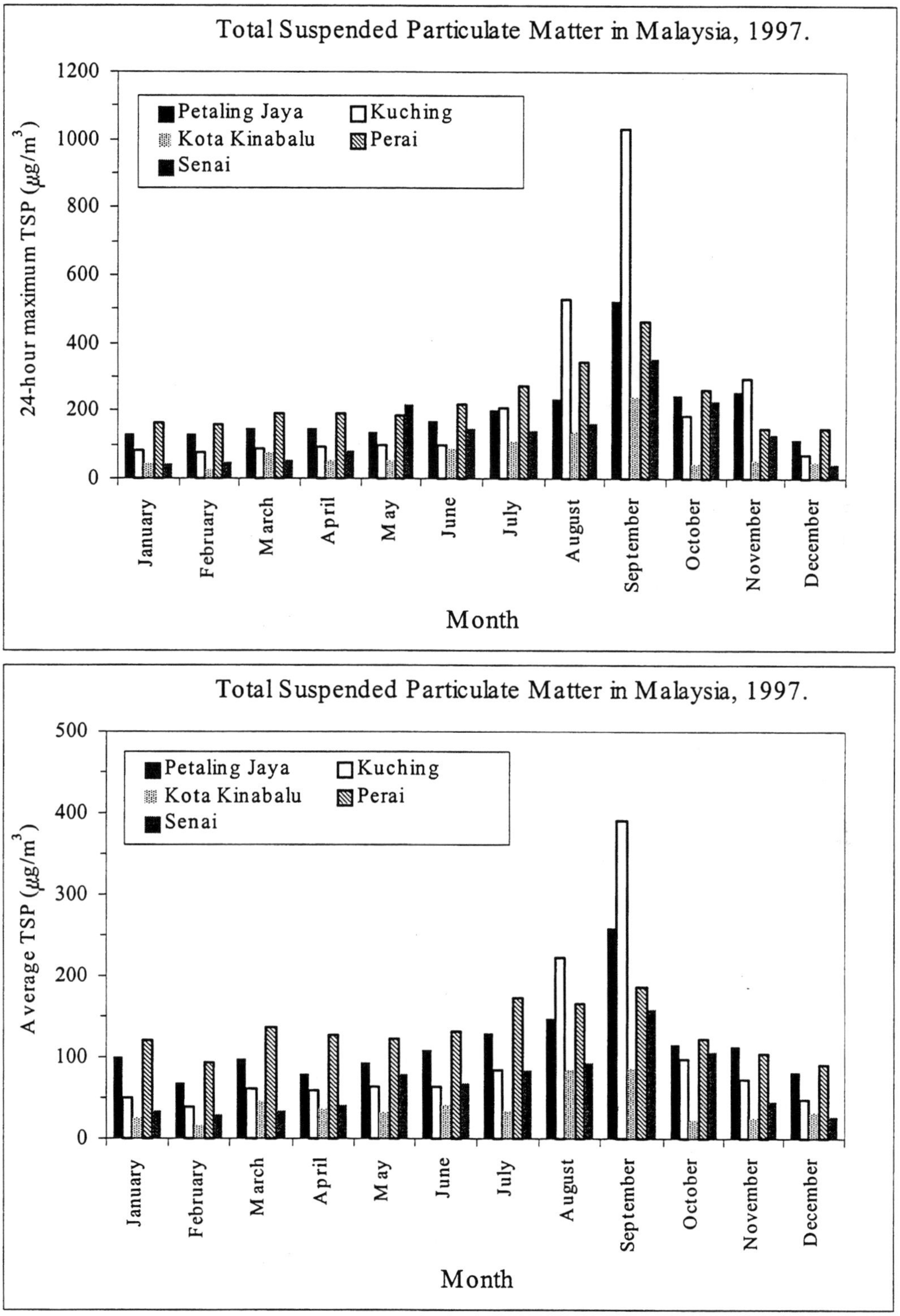

Figure 1 Monthly values of Total Suspended Particulate Matter at five locations in Malaysia for 1997. In the upper panel, the 24-hour maximum values for the months are shown, while in the lower panel, the monthly average values are presented.

In this analysis, we use the convective cloud differential (CCD) method of Ziemke *et al.* (1998) for a long-term analysis of tropospheric column ozone over Southeast Asia with special emphasis on the El Niño years. The aerosol index as derived by Herman *et al.* (1997) is also used to illustrate the amount of absorbing aerosols (in this case black carbon soot emitted from the forest fires) in the column of atmosphere.

Figure 2 exhibits the time series in the CCD tropospheric ozone and the aerosol index over Indonesia for a period of almost twenty years. Some general conclusions can be inferred from the data. In case of tropospheric ozone, a weak seasonal behavior is observed. The tropospheric column ozone has a maximum in the dry seasons of MAM (March, April and May) and SON (September, October and November), whereas the minima occurs in JJA (June, July and August). However, during the El Niño years of 1982-83 and 1997-98, a pronounced increase in tropospheric column ozone is observed. Column ozone amounts exceeded 35 DU during the 1982-83 El Niño, while they were greater than 45 DU during fall 1997. In contrast to these remarkably elevated ozone amounts, the tropospheric ozone column during the normal years is 20-30 DU (Komala *et al.*, 1996).

Enhanced tropospheric ozone over Indonesia during the El Niño years is a result of its production from precursors such as CO and NOx, which are emitted in large quantities during the forest fires. Together with these trace gases, particulate matter in the form of carbon soot is also released to the atmosphere during the incomplete combustion process. Since carbon soot is optically absorbing, a positive signal is obtained in the aerosol index calculation from the satellite retrieval. Superimposed on the tropospheric column ozone in Figure 2 is the time series of the aerosol index for the same period. As in the time series of tropospheric ozone, a weak seasonal cycle is also observed in case of aerosol index, with values ranging between 0.05 and 0.4. However, during the two El Niño's of 1982-83 and 1997-98, sharp increases in the aerosol index (greater than 0.8) are observed. It should be noted that the peak in the aerosol index in 1992 was due to the eruption of Mt. Pinatubo in the Philippines.

Ozonesonde Measurements

Ozonesondes have been launched at least twice each month from Petaling Jaya (PJ), Malaysia (3.1°N, 101.65°E) and Watukosek (W), Indonesia (7.5°S, 112.6°E) since 1993. The vertical profile of ozone from these measurements is used in the following analysis to investigate the changes in the column amounts of ozone during an El Niño year in comparison with a normal non-El Niño year. The tropospheric column amounts at five altitudes viz., 1, 2, 5, 10 and 16 km at the two sites are presented in Figure 3. The column ozone from the surface to 16 km is the tropospheric column ozone. As can be seen in the figure, there is a seasonal behavior in the column ozone over both sites, with spring and fall maxima. On average, the contributions of the various layers to the tropospheric column ozone are ~2, 5, 10, 18 and 25 % respectively. In addition, NOx from lightning may also be an important precursor to ozone formation in the tropical upper troposphere.

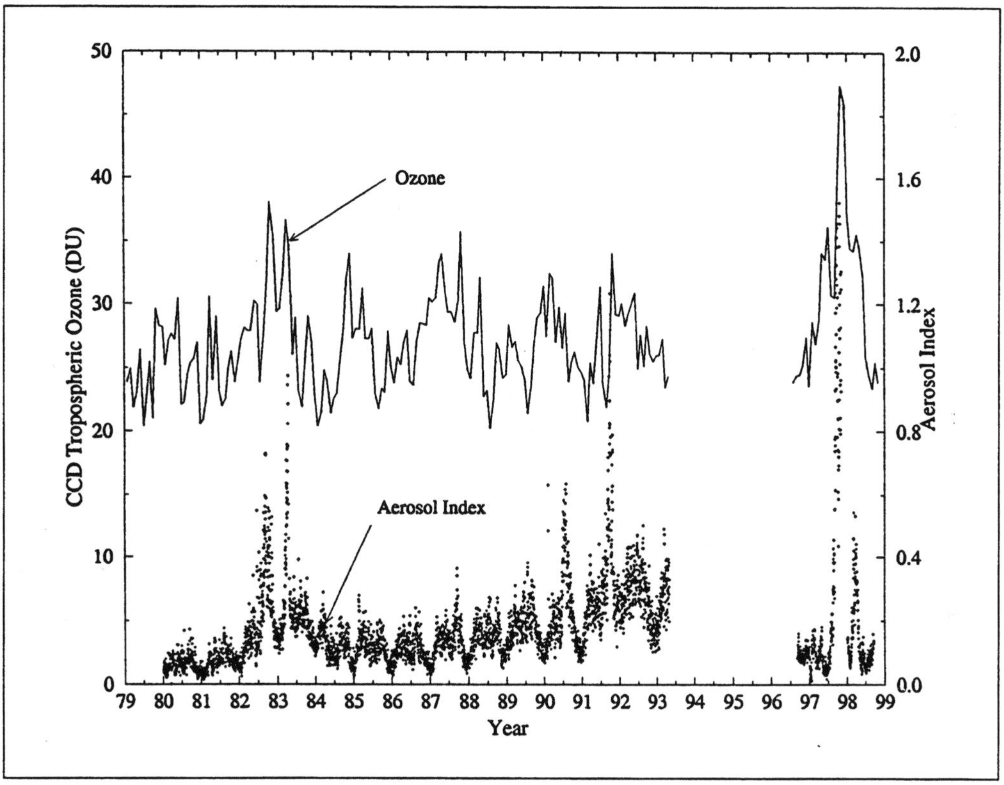

Figure 2 Annual trends in CCD Tropospheric Ozone and Aerosol Index over Indonesia and the central Pacific Ocean between 1979 and 1999. (note: data were not available from 1993 to 1996)

Hence the larger column amounts of ozone in the upper troposphere are probably due to the chemical production of ozone. In contrast, near the surface, in the presence of strong actinic flux and water vapor, ozone is mostly destroyed. Therefore the near surface low ozone contribute minimally to the tropospheric column ozone in the tropics. In addition to these general features, large tropospheric column amounts are often observed during the annual dry seasons (MAM or SON). For example, during October 1994, the tropospheric column ozone exceeded 35 DU. This was a period of large-scale biomass burning over Indonesia (Nichol, 1996, Fujiwara *et al.*, 1999). In 1997-98, the tropospheric column ozone was more than 40 DU, and a significant portion of this amount (~38%) was in the 10-16 km layer. This is large fraction in the upper troposphere during the El Niño season compared with a non-El Niño year, when it is ~25%. As convection is normally subdued during El Niño, this large fraction of ozone in the upper troposphere may have stratospheric origins (Chandra *et al.*, 1998). We shall investigate this effect later in the modeling section of this chapter.

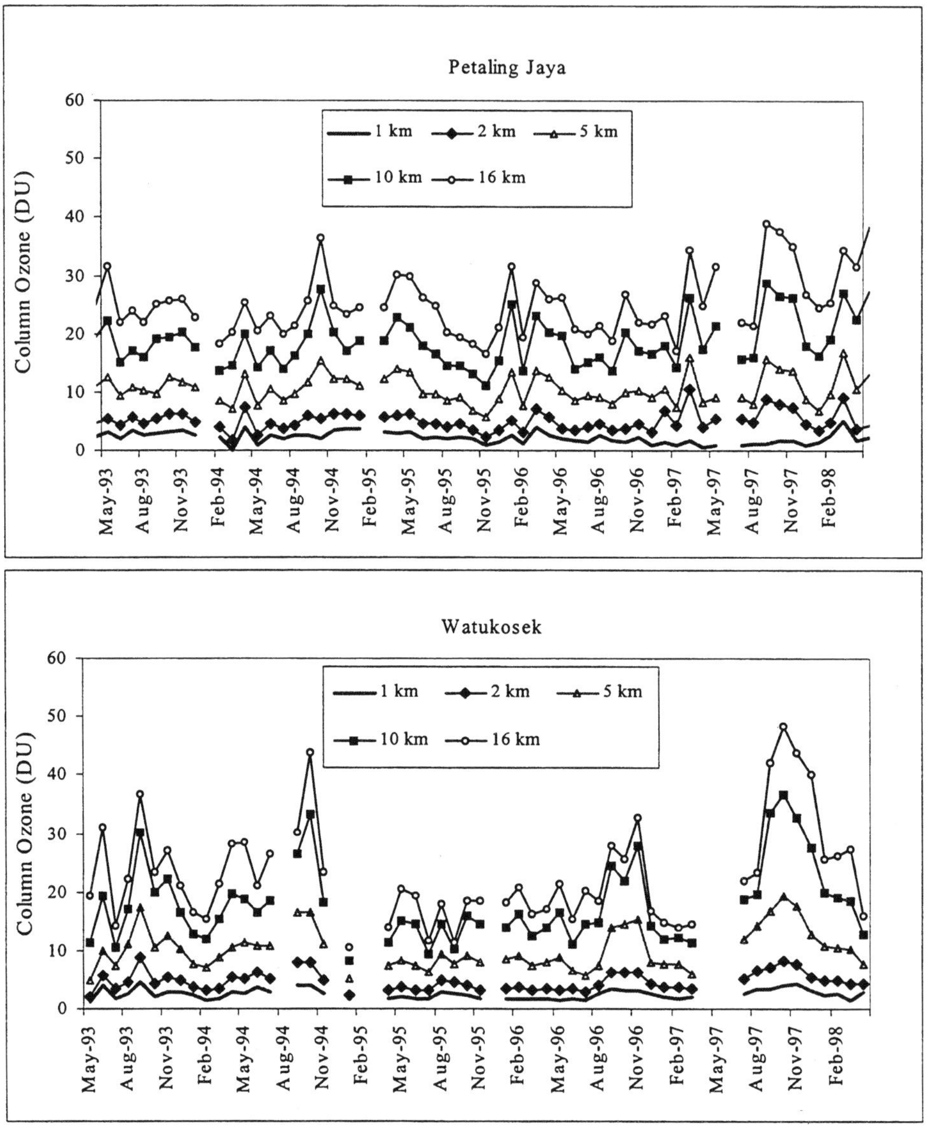

Figure 3 Temporal trends in column ozone over (a) Petaling Jaya, Malaysia, and (b) Watukosek, Indonesia from 1993 to 1998 as calculated from ozonesonde measurements.

Ozone Analysis for the 1996 and 1997 Dry Seasons

So far we discussed the long-term features of tropospheric column ozone and particulate matter over Indonesia. In this section we closely examine the distribution of ozone and particulate matter during the 1997 El Niño dry season of the fall of 1997. Also, for comparison purposes, we will consider the corresponding season during a non-El Niño year (in this case we have chosen 1996). This comparison can aid in delineating the effects of this particular El Niño event on the regional distribution of air pollutants in the atmosphere.

Large-scale ozone features over Southeast Asia and the western Pacific Ocean during October 1996 and 1997 are presented in Figure 4. The figure shows the spatial distribution of tropospheric column ozone (TCO) and the TOMS total ozone (TTO). The TOMS total ozone includes both the tropospheric and the stratospheric components of ozone in the atmosphere. As seen in the figure, the TCO values over Southeast Asia during October 1996 were within the normal range of those observed during a normal non-El Niño dry season. Over much of Indonesia, the tropospheric column ozone was between 20 and 25 DU. The situation was vastly different during 1997. A dramatic increase in the TCO levels was observed in 1997. TCO amounts exceeded 45 DU in October 1997. This is in sharp contrast to the normal observed levels of TCO. An anomaly of the 1997 TCO over the 1996 TCO reflects the effects of El Niño on the tropospheric ozone. Over Sumatra, Indonesia, more than 21 DU anomaly was observed. In the western Pacific Ocean, the anomaly gradually decreased to zero and was in fact negative towards the central Pacific Ocean. Similar results were also obtained by Chandra *et al.* (1998) and Thompson and Hudson (1999).

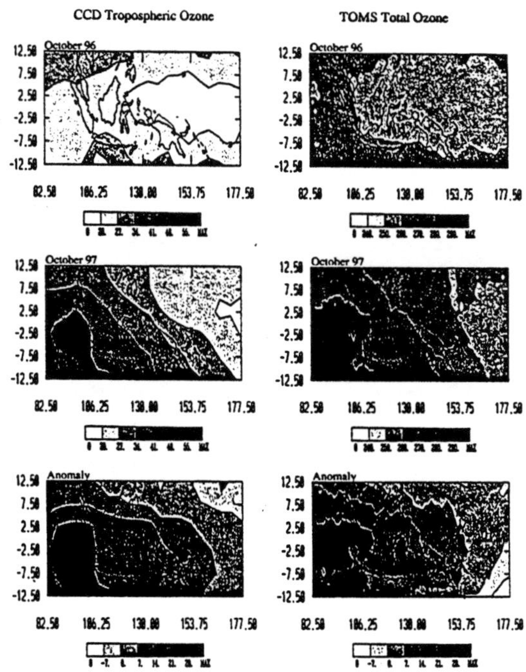

Figure 4 Satellite data analysis for ozone over Southeast Asia. Convective Cloud Differential (CCD) Tropospheric Ozone and Anomaly for September and October 1996 and 1997 (left panel). TOMS Total Ozone and Anomaly for September and October 1996 and 1997 (right panel).

A similar comparison with the TOMS total ozone also exhibits large-scale features in the ozone distribution. Local maxima are observed over Indonesia in 1997, with values exceeding 280 DU against normal TTO values of ~260 DU. The distribution in the anomaly of the TTO also shows the same general features as those observed in that due to TCO. The anomaly due to TTO is more than 24 DU over Sumatra and west of Indonesia.

The fact that the time profile of the difference by both TTO and TCO methods show similarities indicates that the variability in the total column ozone over these tropical regions arises mainly due to changes in the tropospheric component of the total ozone. The contribution due to the stratospheric column is relatively zonally invariant (Ziemke *et al.*, 1996).

Monthly snapshots of the tropospheric column ozone during the 1997 El Niño season over Indonesia are presented in Figure 5. In August, the TCO was mostly between 20 and 30 DU over most of Southeast Asia. As the El Niño strengthened, the tropospheric column ozone started exhibiting a gradual increase. For example, over much of western Indonesia, the TCO increased to more than 40 DU in September and more than 45 DU in October. After October, a decreasing trend in the column ozone was observed. In summary, the tropospheric column amounts in the fall of 1997 were significantly large compared with the normal column amounts of 20-30 DU.

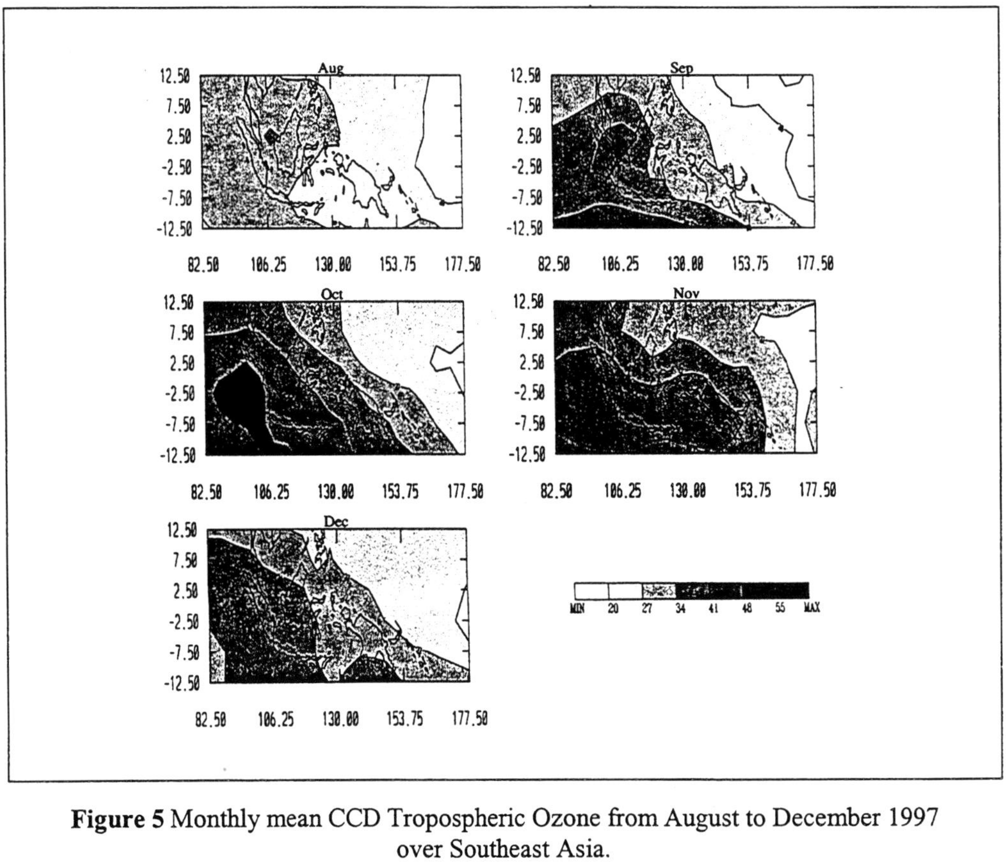

Figure 5 Monthly mean CCD Tropospheric Ozone from August to December 1997 over Southeast Asia.

Aerosol Analysis for the 1997 Dry Season

As mentioned previously, the widespread forest fires over Indonesia during 1997 not only led to emissions of trace gases and ozone precursors to the atmosphere, but also to emissions of large amounts of particulate matter. These excessive levels of particulate matter in the atmosphere near the surface were responsible for reduced visibilities and the resultant severe air pollution episodes were commonly referred to as smoke haze. The particulate matter emitted to the atmosphere was largely carbon soot from the burning of the forest biota. From a radiation standpoint, carbon soot absorbs incoming solar radiation. The reduced visibilities even during daytime were due to the inability of the solar radiation to penetrate the thick aerosol cloud.

Large-scale features of the extent of the spread of the particulate matter over the Southeast Asian region during fall 1997 are presented in Figure 6. Spatial distribution of the monthly mean aerosol index from August through December 1997 is shown in the figure. September and October stand out as the most severely affected months in terms of particulate air pollution. Additionally, two distinct regions of elevated levels of aerosol index are observed; one over Kalimantan and the other over Sumatra. An interesting feature apparent from the plots is that the fires were severe over Kalimantan in September. On the other hand, in October, Sumatra seemed to be more affected. In addition to these two major sources, a small but significant source could also be seen over Irian Jaya in the eastern part of Indonesia.

In our discussion thus far about the forest fires during 1997, we used a combination of satellite and surface based measurements to analyze the data and explain the peculiarities in the observations. In particular, the ozone analysis was useful in understanding the large-scale and vertical extent of the changes in ozone that occurred in 1997. The particulate matter analysis gave more information about the source regions of the forest fires. We shall make use of these results in the next section to investigate the effects of the forest fires on the regional tropospheric trace gas chemistry of Southeast Asia.

Numerical Investigation of the Tropospheric Chemistry of Southeast Asia during the Fall of 1997

To better understand the behavior of trace gases and particulate matter in time and space over a wider region, we have used a mathematical model to simulate the tropospheric trace gas chemistry over Southeast Asia. This can permit us to deduce whether the observed behavior of a trace species was just a local effect close to the observation station or more widespread.

In this section we investigate the regional scale impacts of the gaseous and particulate matter emissions from the forest fire using a numerical model. The transport and distribution of tropospheric trace gases and particulate matter in Southeast Asia is investigated using the STEM-III regional-scale model. The model domain for the

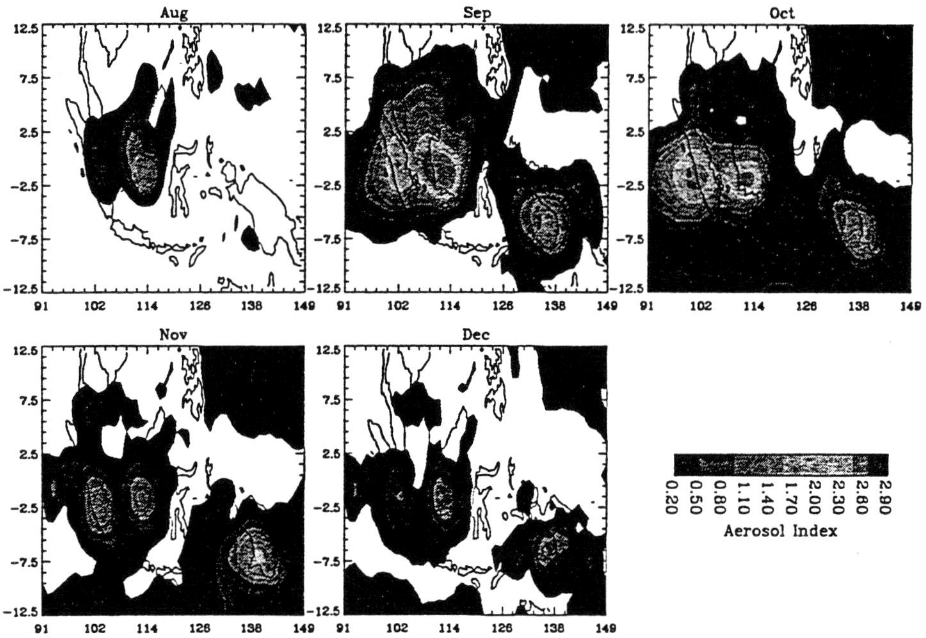

Figure 6 Monthly mean aerosol index from August to December 1997 over Southeast Asia.

numerical simulations is shown in Figure 7. STEM-III is a three-dimensional, Eulerian numerical model, which accounts for the transport, chemical transformation and deposition of atmospheric pollutants in gas, cloud, rain, snow and aerosol. The current 3-D version of the model is based on the condensed chemical mechanism of Lurmann *et al.* (1986) and Atkinson *et al.* (1989), and modified to include low NOx conditions and explicit treatment of isoprene (Carmichael *et al.*, 1991). The numerical integration of the chemistry is done by a fully implicit integrator based on the Runge-Kutta-Rosenbrock type of solvers (Sandu *et al.*, 1997).

Emissions of Trace Gases in Southeast Asia

Anthropogenic NOx emissions used in the model simulations were taken from Van Ardenne *et al.* (1999) and Benkovitz *et al.* (1996). The regions of high NOx emissions (> 0.01 Tg N/yr) in Southeast Asia are mostly centered around urban areas such are Bangkok, Singapore and Jakarta. The NOx emissions considered in our model study

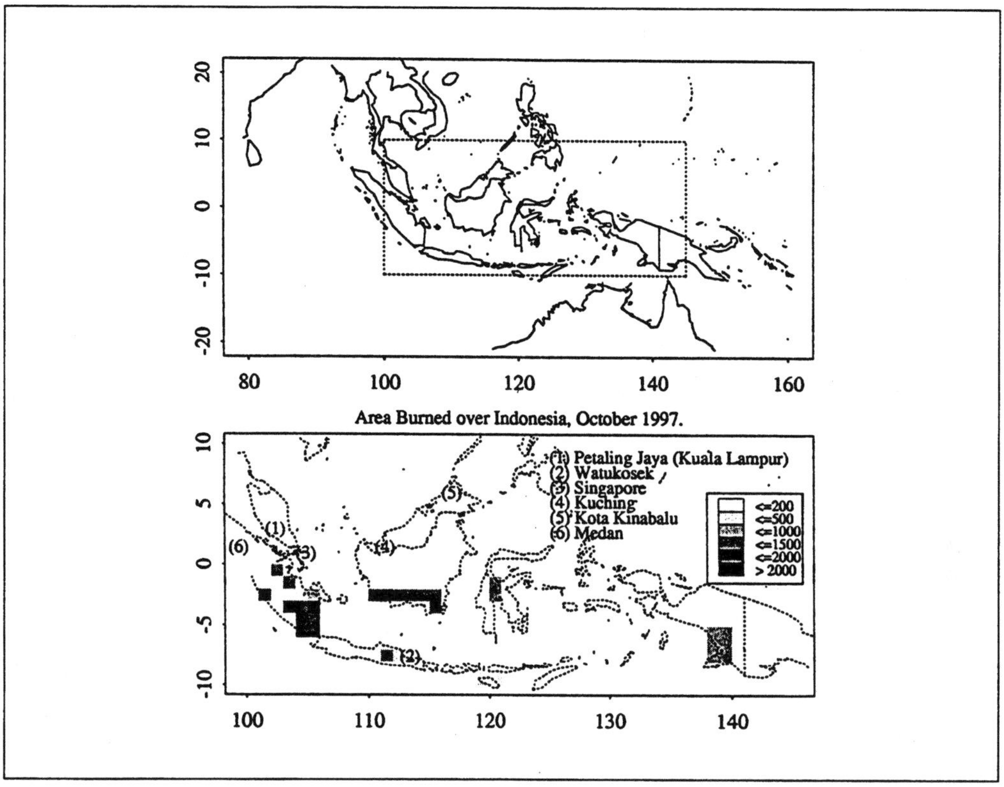

Figure 7 Map of the model domain (top) and the burning areas considered in the model simulations (bottom). Also shown are the locations over which the ozone comparisons are made.

Table 1 Anthropogenic and forest fire emissions of trace gases in Southeast Asia.

	Malaysia	Indonesia	Thailand	Philippines	Forest Fire[a]
NO_x (Tg N/yr)	0.091	0.243	0.100	0.103	0.97
SO_2 (Tg S/yr)	0.075	0.455	0.228	0.185	0.05
CO (Tg C/yr)	1.257	4.233	0.748	0.660	51.60
NMHC (Tg C/yr)	0.744	2.425	0.282	0.259	5.10

[a]forest fire emissions are for one month

include emissions from fossil fuel, bio-fuels as well as soil NOx (data taken for October from Yienger and Levy, 1995). The countrywide emissions of NOx are given in Table 1 (note that the emissions used in the model are for the year 1990).

Both carbon monoxide and non-methane hydrocarbon emissions were taken from the EDGAR database (Olivier *et al.*, 1996). As in the NOx emissions, CO and NMHCs exhibit higher emission intensity over the major urban centers of Southeast Asia. For CO, the emission flux is > 0.1 Tg C/yr over the highly populated regions in the model domain, while in case of NMHCs it is > 0.02 Tg C/yr. The aggregated emissions of CO and NMHCs for the various countries in Southeast Asia are also presented in Table 1.

As seen from Table 1, emissions of major ozone precursors from Indonesia are much larger than those from the other countries in Southeast Asia. Total NOx emissions from Indonesia are ~0.25 Tg N/yr whereas CO emissions are ~4.2 Tg C/yr, which constitute 45-60% of the total NOx and CO emissions from Southeast Asia. Natural biogenic emissions of isoprene were taken for the month of October from Guenther *et al.* (1995).

Emissions of Trace Gases and Particulate Matter from Forest Fires in Indonesia

The estimate of emissions of trace gases and particulate matter was complicated by the nature of the fires, which involved a combination of agricultural areas, forest fires and peat bogs. Peat is partially decomposed organic matter formed in the acid water of bogs. Indonesia and Malaysia have more than 20 million hectares of peat or 60% of all tropical peatlands (Levine *et al.*, 1998).

Calculation for the emissions from the fires in Indonesia between August and December 1997 has been done by Levine (1999a, b). Estimate of land area burned during the forest fires based on SPOT (Satellite Pour l'Observation de la Terra) image analysis was ~45600 km^2, more than twice the earlier estimated of ~17100 km^2 (Levine, 1998). From the distribution of fire hot spots in the analysis, Levine (1999a, b) concluded that the actual area burned was: agricultural and plantation, 50%; forests and bushes, 30%; and peat swamp forests, 20%.

Figure 7 exhibits the extent of the area burned during the forest fires in 1997. Most of the area burned was located on Sumatra and Kalimantan. From this figure, it is possible to examine the diversity in the area burned. More than 2000 km^2 of area was burned in Sumatra and Kalimantan during October 1997, while over Java it was only ~350 km^2. These maps of the area destroyed by the fires provide a means to spatially distribute the emissions from the fires, which we shall discuss subsequently.

The estimates of trace gases and particulate matter from the calculations of Levine (1999a, b) are also given in Table 1. A comparison of the anthropogenic and forest fire emission estimates indicates a disproportionately large amount of emissions from the fires. For example, the sum of the carbon monoxide emissions from the four major emitters in Southeast Asia is ~7 Tg C/yr, while those from the forest fires were ~51 Tg C for October. In case of NOx, the total anthropogenic annual flux of ~0.55 Tg N was exceeded during the one month (0.97 Tg N emitted from the fires).

Meteorological Patterns over Southeast Asia during October 1997

In contrast with the normal convective patterns over the western Pacific Ocean and Indonesia that create surface low pressure conditions, during an El Niño, the region is dominated by the presence of long stagnant high pressure system. As a result, the dry season is prolonged and thus adds to the intensity with which fires burn over Indonesia. A combination of the dominant high pressure and upper level lows may in part be responsible for the downward transport of ozone along the boundary of the two systems. The abnormal increases in column ozone as reported in many observations may be a result of this phenomenon.

Simulation Procedure

Numerical model simulations were conducted for the month of October 1997. This period was dominated by mostly clear skies and hence any convective activity as a result of the presence of clouds was assumed to be suppressed. In our model simulations, no deep convections (which normally occur in the tropics) were considered. The simulations were started on October 1, 0 Z (7 a.m. local west Indonesian time). Initial conditions for most of the species were representative of their mixing ratios in the tropical regions. For ozone, a variable initial condition was employed. Observations from ozonesonde measurements indicated an increase in ozone from surface to the tropopause. Hence ozone mixing ratios were scaled from 20 ppb at the surface to 120 ppb at 16 km. Three kinds of simulations were performed to assess the atmospheric impact of El Niño as well as the emissions from the forest fires. These simulations are described briefly in Table 2. In short, one simulation without the forest fire was carried out. In the second simulation, emissions from the forest fires were also included along with the anthropogenic emissions. In the last simulation, emissions from the forest fires were included along with a decrease in the insolent solar radiation below the aerosol cloud. One should keep in mind that the visibility was greatly reduced during large haze episodes. This visibility reduction was mostly due to the presence of huge amounts of aerosols in the lower troposphere. Hence we included a simulation to investigate the impacts of reduced solar radiation on the photochemical oxidant cycle of the region.

Table 2 Description of various cases considered in the model simulations.

Simulation Case	Description
Case 1	Chemistry included – effect of anthropogenic emissions on tropical trace gas levels
Case 2	Forest fires included – effect of the emissions from the forest fires on the trace gas levels
Case 3	Haze effect on the photochemistry included – reduction in solar intensity below 3 km

RESULTS AND DISCUSSION

In this section we discuss the model results for calculations of the perturbation of the photochemical oxidant cycle because of the influence of the forest fires as well as the distribution of particulate matter from the fires.

Large-scale features from case 1 simulation are presented in Figure 8. Ozone mixing ratios at three altitudes (1, 5 and 12 km) are shown in the figure. Ozone mixing ratios north of the equator are lower than those south of the equator at all levels with a gradual increase from the surface to the upper troposphere. At 5 km, similar features as those observed at lower altitudes are observed. At 12 km, large values are mostly calculated in the south and east regions of the model domain.

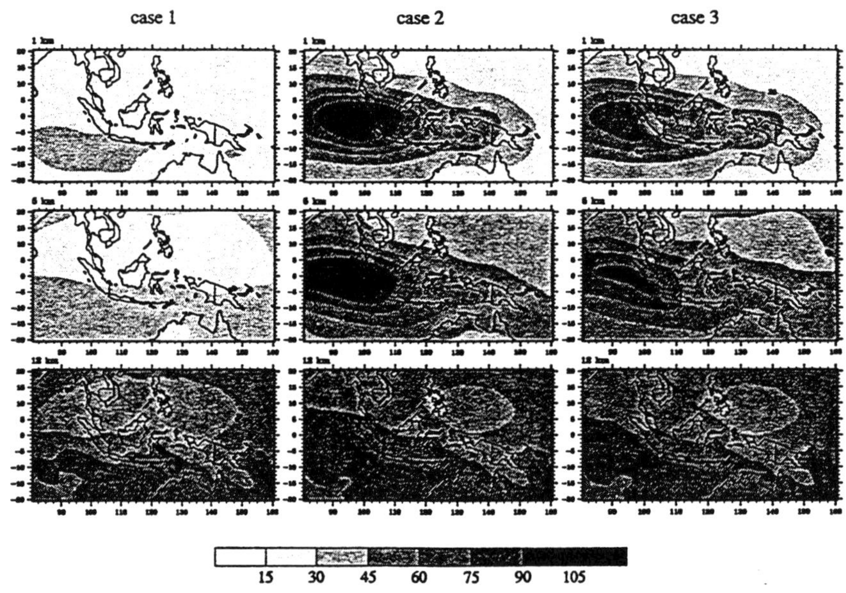

Figure 8 Model calculated monthly mean distribution of ozone at 1, 5 and 12 km. case 1 (left), case 2 (center), case 3 (right). (units = ppb)

These large-scale ozone features are explained by Phadnis (1999) with the help of the monthly mean spatial distribution of the vertical velocity to delineate the role of the tropospheric dynamics. It was found that enhancements in the upper tropospheric ozone corresponded closely with strong downward vertical velocities and subsidences over the model domain south of the equator.

The model was further simulated for two more cases (case 2 and case 3) as described in Table 2. For case 2, the emissions from the forest fires were injected into the grid boxes shown in Figure 7. In case 3, the effect of a decrease in the incoming solar radiation due to the haze layer on the ozone photochemistry was investigated. Recently Herman *et al.* (1999) calculated the change in the UV flux at the surface as a result of the presence of aerosols in the atmospheric column. For an episode of the Indonesian forest fires in September 1997, they calculated more than 75% reduction in UV radiation over areas of high aerosol column loading (and hence a large value of the aerosol index). We used this approach to assign reductions in the solar radiation in our model. Furthermore, the height of the aerosol cloud was assumed at ~3 km and hence the radiative influence below this level was taken into consideration.

The results from these cases on the ozone levels over Southeast Asia are also presented in Figure 8. In general, it can be seen that due to the prevailing meteorological conditions, the region affected by the fire emissions was mostly west of Indonesia into the Indian Ocean. A gradual reduction in the impact of the emissions is seen from the surface to the upper troposphere. For example, in case 2, more than 90 ppb increase in boundary layer ozone is predicted by the model over Sumatra and west of Kalimantan. Enhancements of ozone of more than 20 ppb reach as far as the southern portions of India. At 5 km, the increase is modest and the shift is more toward northwest. Over the Indian Ocean, ozone levels increase by 60 ppb over case 1 values.

If the radiative impact of the haze layer is considered (as shown on the right side of Figure 8) the model calculates a decrease in ozone production and hence the ambient levels of ozone. Over the same regions where ozone levels were more than 90 ppb in case 2, in this case, these values are just ~75 ppb. Moreover, the impact of this reduction in radiation is not just below the aerosol cloud, but our calculations indicate that small reductions in ozone can occur at upper elevations also.

Ozonesondes were launched over Malaysia (twice) and Indonesia (five times) during October 1997. Here we briefly discuss the model calculated vertical profile of ozone in relation to the observed profiles. Vertical profiles of ozone from model calculations along with observed ozone from ozonesonde over Watukosek, Indonesia are presented in Figure 9. Only four of the five profiles are used here for comparison since one ozonesonde was launched on October 1, the start time of our model simulations.

In general, the model captures the vertical distribution of ozone from the ozonesonde profiles favorably on all four days. In both the model calculations as well as observations, ozone increases from the surface to the tropopause. For example, on October 9, the model predicts ozone mixing ratios ~90 ppb at the tropopause compared with ~80 ppb observed. Similar behavior is also observed on other days. Some interesting features in the observations are noticeable. Ozone mixing ratios close to the surface are close to 60 ppb on three of the four days shown in the figure. The observation site is located close to areas of high population density. Hence anthropogenic influences on the surface ozone at this location may be quite significant. In addition, in-situ ozone production from precursors emitted during the fires may be in part responsible for the observed high surface ozone levels. Apart from the large ozone mixing ratios at the surface, higher values are also observed in the mid-troposphere

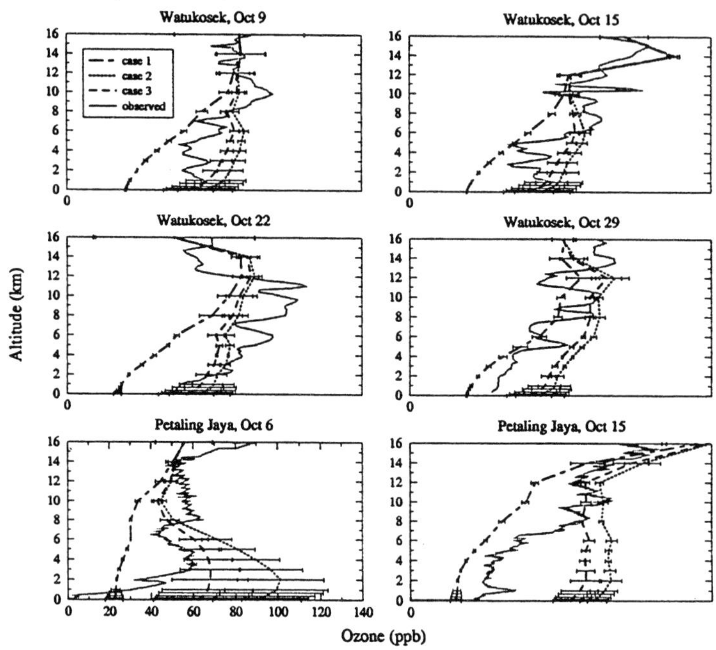

Figure 9 Vertical profiles of ozone at Petaling Jaya, Malaysia and Watukosek, Indonesia during October 1997.

One explanation is that this ozone may be a result of the in-situ production from precursors emitted from the forest fires and advection of ozone-rich air mass from distant sources, although the possibility of downward transport from the upper troposphere cannot be excluded.

Also shown in the figure are the ozone profiles for case 2 and case 3. As seen in the figure, surface ozone increases by ~40 ppb as a result of photochemical production from precursors. This enhancement in ozone is calculated into the middle troposphere to ~10 km, after which the effect of in situ ozone production is reduced. In addition to these general features, it can be seen that in some cases (October 15 and 22), the model calculated the surface ozone quite close to the observations. While this surface feature is well captured, the middle tropospheric ozone profile is misrepresented. On other occasions, such as October 9 and 15, the middle tropospheric ozone is captured well by the model.

Over Petaling Jaya, Malaysia, the effect of the forest fire on the vertical structure of ozone is quite significant (Figure 9). On October 6, a sharp increase in surface and lower tropospheric ozone is predicted by the model for case 2 (ozone greater than 100 ppb). The model calculated ozone is close to the observed ozone between 3 and 7 km. Between 7 and 12 km, the model underpredicts the observed ozone (~20 ppb). This could be due to distant sources contributing to the ozone levels over Petaling Jaya. On October 15, a much closer agreement between the model predictions and observations is seen at higher elevations. The effect of haze on ozone is quite significant at this location. Surface level ozone is found to decrease by ~20 ppb due to reduction in the solar radiation below the smoke cloud.

Distribution of Carbon Monoxide

Carbon monoxide, along with NOx emitted from the fires, is responsible for the production of tropospheric ozone in the vicinity of the emissions. Carbon monoxide levels in the troposphere can reach very high values as a result of emissions from biomass burning and forest fires. Monthly mean carbon monoxide levels calculated by our model are presented in Figure 10. For the three cases presented, the data is shown at 1 km (boundary layer), 5 km (lower troposphere) and 12 km (upper troposphere). The model predicts extremely high levels of carbon monoxide over the source regions of the forest fire. For example, over Sumatra and Kalimantan, CO levels more than 500 ppb are calculated by the model. At 5 km, the mean CO values reduced to ~300 ppb. Additionally, there is a positive effect of the radiative impact of the aerosols on the CO levels. In case 3, more CO is predicted in comparison with that in case 2.

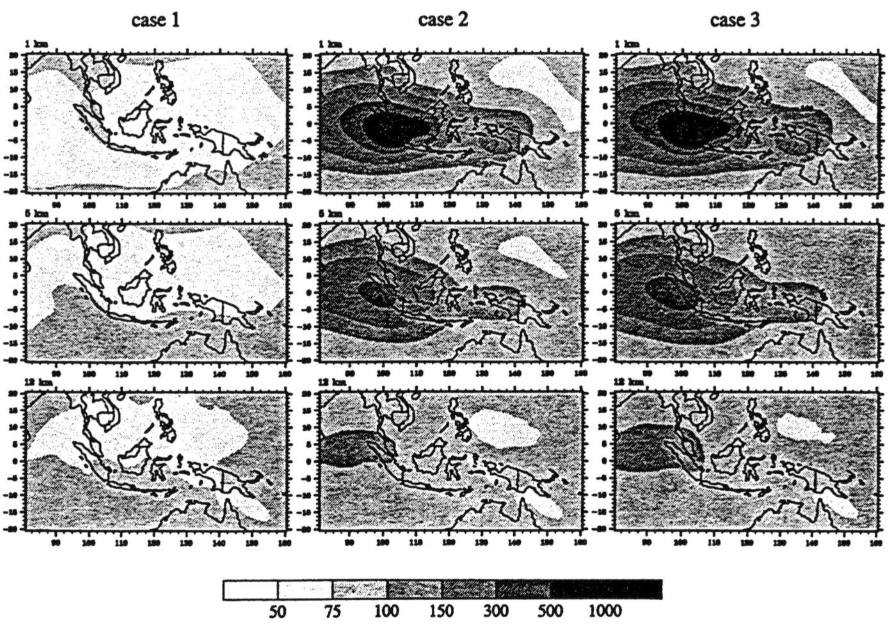

Figure 10 Model calculated monthly mean distribution of carbon monoxide for case 1 (left), case 2 (center) and case 3 (right) at 1 km (top) and 5 km (bottom). (units = ppb).

When our results are compared with aircraft data from Sawa *et al.* (1999) over Kalimantan, the model calculations are lower than the observations. The aircraft observations by Sawa *et al.* (1999) indicate CO mixing ratios ~8000 ppb at 1.3 km. These values are significantly higher that those calculated in our model. The CO in the troposphere as observed from the aircraft data remains stratified up to about 2.5 km where the levels are ~4000 ppb. Above this altitude there is a sudden drop in the CO levels to ~600 ppb at 4 km. The mixing of the emitted CO in the boundary layer below the haze layer may be responsible for this behavior in CO. The sinking air from the high pressure over Indonesia probably prevented the CO from getting vertically mixed throughout the troposphere. Additionally, the radical chemistry was suppressed below the haze layer to cause a photochemical degradation of CO, though this might be less effective given the long lifetime of CO.

Distribution of Particulate Matter

Along with the emissions of trace gases into the atmosphere from the forest fires, particulate matter is also emitted in large amounts. This particulate matter comprises mostly of carbon soot particles, which are produced as a result of incomplete combustion of the forest biota. Unlike carbon monoxide, which is emitted from the fires as well as produced in-situ by the oxidation of non-methane hydrocarbons, carbon soot has a surface-only source. Hence carbon soot represents a good indicator of the surface emissions from forest fires.

To assess the extent of the spread of pollutants from forest fire, spatial distribution of the particulate matter (PM) concentrations is presented in Figure 11. Monthly mean PM levels at four altitudes are shown in the figure. In general, the PM levels decrease from the surface to higher elevations. For example, at 0.2 km, PM concentrations exceed 100 $\mu g/m^3$ near the source region of the fire on the islands of Sumatra and Kalimantan. PM levels of 5-10 $\mu g/m^3$ are also seen over Java. Between 5 and 10 μg/m3 PM is seen affecting the central portions of Malaysia and extending into the Indian Ocean. At upper elevations, the core of the high concentration of PM has moved westward due to the prevailing easterly winds aloft.

Surface-based data for particulate matter were available daily at a number of locations in Malaysia. High volume sampling was employed to obtain the total concentration of particulate matter. Here we show comparisons with data from two stations. Time profiles of calculated and observed PM levels at Petaling Jaya and Kuching are presented in Figure 12. Model calculations compare favorably with observations. The synoptic features in the observed PM levels are captured well by the model. For example, the observed peaks of ~160 μg/m3) on day 9 and 10 at Kuching can also be seen in the model predictions. Furthermore, the variability in the observations during the middle of October is also reproduced by the model. From these calculations, we can assume that the forest fire emissions used in our model are reasonable.

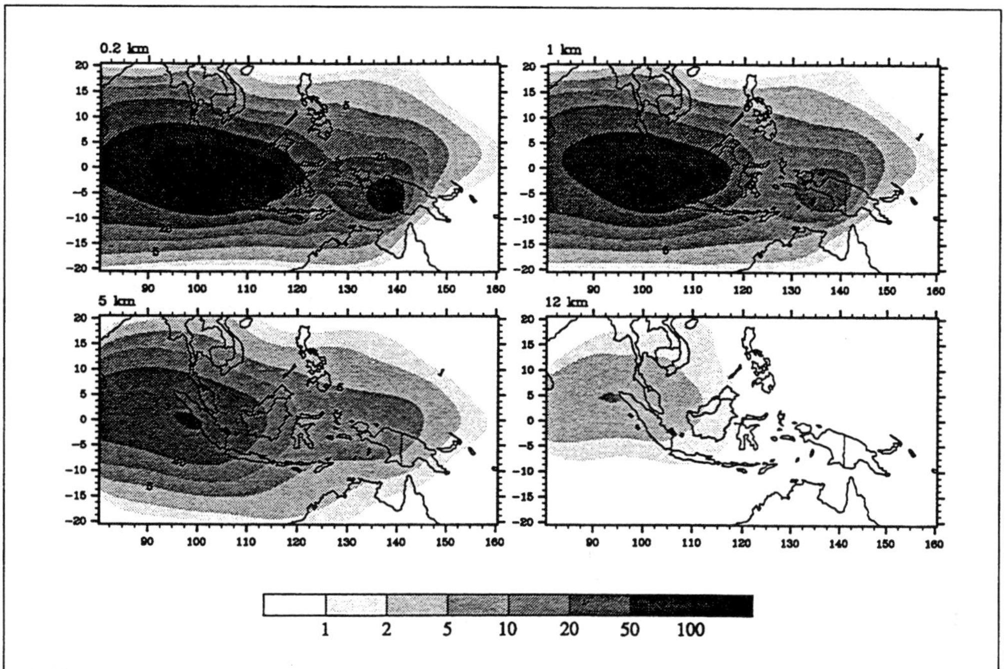

Figure 11 Spatial distribution of monthly mean particulate matter at four levels over Southeast Asia. (units = $\mu g/m^3$)

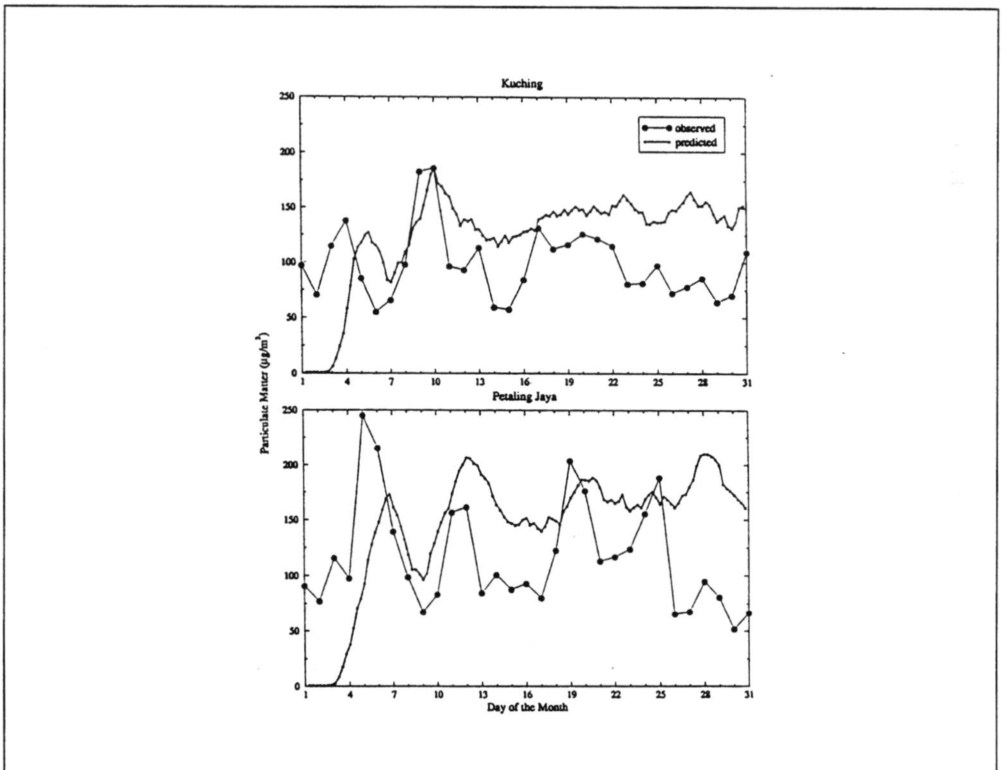

Figure 12 Time profile of the model predicted and observed particulate matter concentration at Petaling Jaya and Kuching in Malaysia during October 1997.

Closing Statements

In this chapter, we employed a combined approach of analysis of observed data and numerical model simulations to quantify the atmospheric impacts of the forest fires over Indonesia during the fall of 1997. Measurements taken both on a near and longer-term scale exhibited significant enhancements in the levels of tropospheric pollutants above their normal levels during the El Niño compared with a non El Niño year. Model simulations were carried out to delineate the importance of in-situ chemical production versus the role of tropospheric dynamics in the observed high amounts of ozone in the tropospheric column over most of Southeast Asia. Our results indicate that the column ozone amounts were a result of an inter-play of both processes. While tropospheric dynamics favored the downward transport of ozone-rich air from the lower stratosphere into the upper troposphere, chemical production of ozone from its precursors was responsible for the observed ozone increases in the middle and lower troposphere. Our results indicate reduction in ozone production in the boundary layer close to the source regions and over areas severely affected by the haze episodes. This was a result of a decrease in the solar radiation below the aerosol cloud. Model calculations for particulate matter showed a favorable agreement in the synoptic variability of the observations.

Acknowledgments

We wish to thank the Malaysian Meteorological Service and the Earth Observation Research Center, National Space Development Agency of Japan for providing us the ozonesonde data over Malaysia and Indonesia respectively.

REFERENCES

1. Atkinson, R., D. L. Baulch, R. A. Cox, R. F. Hampson Jr., J. A. Kerr, and J. Troe, 1989. Evaluated Kinetic and Photochemical Data for Atmospheric Chemistry: Supplement III. *International Journal of Chemical Kinetics* **21**: 115-150.
2. Benkovitz, C.M., M. T. Scholtz, J. Pacyna, L. Tarrasón, J. Dignon, E. C. Voldner, P. A. Spiro, J. A. Logan and T. E. Graedel. 1996. Global gridded inventories of anthropogenic emissions of sulfur and nitrogen. *Journal of Geophysical Research* **101**: 29239-29263.
3. Brauer, M. and J. Hisham-Hashim, 1998. Fires in Indonesia: Crisis and Reaction. *Environmental Science and Technology* **32**: 404-407.
4. Carmichael, G. R., L. K. Peters, and R. D. Saylor, 1991. The STEM-II Regional Scale Acid Deposition and Photochemical Oxidant Model: An Overview of Model Development and Applications. *Atmospheric Environment* **25**: 2077-2090.

5. Chandra, S., J. Ziemke, W. Min and W. G. Read, 1998. Effects of 1997-1998 El Niño on tropospheric ozone and water vapor. *Geophysical Research Letters* **25**: 3867-3870.
6. Fishman, J., C. E. Watson, J. C. Larsen and J. A. Logan, 1990. The distribution of tropospheric ozone determined from satellite data. *Journal of Geophysical Research* **95**: 3599-3617.
7. Fujiwara, M., K. Kita, S. Kawakami, T. Ogawa, N. Komala, S. Saraspriya and A. Suprito, 1999. Tropospheric ozone enhancements during the Indonesian forest fire events in 1994 and in 1997 as revealed by ground-based observations. *Geophysical Research Letters* **26**: 2417-2420.
8. Geernaert, G. L. and P. W. Waahlin, 1998. Source Apportionment Measurements, Receptor Analysis and Meteorological Influences during the 1997 Haze Episode in Southeast Asia. *Proceedings of the American Geophysical Union*, San Francisco.
9. Guenther, A., C. N. Hewitt, D. Erickson, R. Fall, C. Geron, T. Graedel, P. Harley, L. Klinger, M. Lerdau, W. A. McKay, T. Pierce, B. Scholes, R. Steinbrecher, R. Tallamraju, J. Taylor and Pat Zimmerman, 1995. A global model of natural volatile organic compound emissions. *Journal of Geophysical Research* **100**: 8873-8892.
10. Hao, W. M. and M. H. Liu, 1994. Spatial and temporal distribution of tropical biomass burning. *Global Biogeochemical Cycles* **8**: 495-503.
11. Herman, J. R., P. K. Bhartia, O. Torres, C. Hsu, C. Seftor, and E. Celarier, 1997. Global distribution of UV-absorbing aerosols from Nimbus 7/TOMS data. *Journal of Geophysical Research* **102**: 16911-16922.
12. Herman, J. R., N. Krotkov, E. Celarier, D. Larko and G. Labow, 1999. Distribution of UV radiation at the Earth's surface from TOMS-measured UV-backscattered radiances. *Journal of Geophysical Research* **104**: 12059-12076.
13. Hudson, R. D. and A. M. Thompson, 1998. Tropical tropospheric ozone (TTO) from TOMS by a modified-residual method. *Journal of Geophysical Research* **103**: 22129-22145.
14. Komala, N., S. Saraspriya, K. Kita and T. Ogawa, 1996. Tropospheric Ozone Behavior Observed in Indonesia. *Atmospheric Environment* **30**: 1851-8156.
15. Levine, J. S. 1999a. The 1997 fires in Kalimantan and Sumatra, Indonesia: Gaseous and particulate emissions. *Geophysical Research Letters* **26**: 815-818.
16. Levine, J. S. 1999b. Correction to the The 1997 fires in Kalimantan and Sumatra, Indonesia: Gaseous and particulate emissions. *Geophysical Research Letters* **26**: 2407.
17. Levine, J. S., T. D. Edwards, T. E. McReynolds and C. W. Dull, 1998. Gaseous and Particulate Emissions Released to the Atmosphere During the Fires in Kalimantan, Sumatra and Java, Indonesia, September 1997 - March 1998. Paper presented at the *World Meteorological Organization Workshop on Regional Transboundary Smoke and Haze in Southeast Asia*, Singapore.
18. Lewan, T. 1997 Ecological disaster likely in dangerously dry Amazon. *Daily Iowan*, 4 December 1997, Iowa City, Iowa.
19. Lim, S. F., C. P. Leong, H. Tsurita and S. Yonemura, 1998. The increase of tropospheric ozone over Peninsular Malaysia by the 1997 biomass burning in

tropical east Asia. *Proceedings of the Joint International Symposium on Global Atmospheric Chemistry*, Seattle.

20. Lurmann, F. W., A. C. Lloyd, and R. Atkinson, 1986, A Chemical Mechanism for use in Long-Range Transport/Acid Deposition Computer Modeling. *Journal of Geophysical Research* **91**: 10905-10936.
21. Nichol, J. 1997. Bioclimatic impacts of the1994 smoke haze event in Southeast Asia. *Atmospheric Environment* **31**: 1209-1219.
22. Olivier, J. G. J., A. F. Bouwman, C. W. M. Van der Mass, J. J. M. Berdowski, C. Veldt, J. P. J. Bloos, A. J. H. Visschedijk, P. Y. J. Zandveld and J. L. Haverlag, 1996. *Description of EDGAR Version 2.0. A set of global emission inventories of greenhouse gases and ozone-depleting substances for all anthropogenic and most natural sources on a per country basis and on 1oX1o grid.* RIVM/TNO report, December 1996. RIVM, Bilthoven. RIVM report no. 7771060 002. [TNO MEP report no. R96/119]
23. Phadnis, M. J. 1999. Tropospheric Air Pollution Modeling on a Regional Scale: Case Studies for East and Southeast Asia. Ph.D. dissertation, The University of Iowa, Iowa City, Iowa.
24. Sandu, A., J. G. Verwer, J. G. Blom, E. J. Spee, G. R. Carmichael, and F. A. Potra, 1997. Benchmarking Stiff ODE Solvers for Atmospheric Chemistry Problems II: Rosenbrock Solvers. *Atmospheric Environment* **31**: 3459-3472.
25. Sawa, Y., H. Matsueda, Y. Tsutsumi, J. B. Jensen, H. Y. Inoue and Y. Makino, 1999. Tropospheric carbon monoxide and hydrogen measurements over Kalimantan in Indonesia and northern Australia during October, 1997. *Geophysical Research Letters* **26**: 1389-1392.
26. Simons, L. M. 1998. Indonesia's Plague of Fire. *National Geographic* **194**: 100-119.
27. Thompson, A. M. and R. D. Hudson, 1999. Tropical tropospheric ozone (TTO) maps from Nimbus 7 and Earth Probe TOMS by the modified-residual method: Evaluation with sondes, ENSO signals, and trends from Atlantic regional time series. *Journal of Geophysical Research* **104**: 26291-26975.
28. Tsutsumi, Y., Y. Sawa, Y. Makino, J. B. Jensen, J. L. Gras, B. F. Ryan, S. Diharto and H. Harjanto, 1999. Aircraft measurements of ozone, NOx, CO and aerosol concentrations in biomass burning smoke over Indonesia and Australia during October 1997: Depleted ozone layer at low altitude over Indonesia. *Geophysical Research Letters* **26**: 595-598.
29. Van Ardenne, J. A., G. R. Carmichael, H. Levy II, D. Streets, and L. Hordijk, 1999. Anthropogenic NO_x Emissions in Asia in the Period 1990 to 2020. *Atmospheric Environment* **33**: 633-646.
30. Yienger, J. J. and H. Levy II, 1995. Empirical model of global soil-biogenic NO_x emissions. *Journal of Geophysical Research* **100**: 11447-11464.
31. Zaizen Y., M. Ikegami, I. Okada, Y. Makino, J. B. Jensen and J. L. Gras, 1998. Aircraft observation of aerosol size distribution over forest fires in Kalimantan and northern Australia. Submitted to *Geophysical Research Letters*.

32. Ziemke, J. R., S. Chandra, and P. K. Bhartia, 1998. Two new methods for deriving tropospheric column ozone from TOMS measurements: The assimilated UARS MLS/HALOE and convective-cloud differential techniques. *Journal of Geophysical Research* **103**: 22115-22127.
33. Ziemke, J. R., S. Chandra, A. M. Thompson and D. P. McNamara, 1996. Zonal asymmetries in southern hemisphere column ozone: Implications of biomass burning. *Journal of Geophysical Research* **101**: 14421-14427.

Chapter 7

IMPACT OF BIOMASS BURNING EMISSIONS ON REGIONAL ATMOSPHERIC CHEMISTRY

R. Balasubramanian
Department of Chemical and Environmental Engineering, Engineering Drive 4, National University of Singapore, 117576 Singapore

INTRODUCTION

Biomass burning, the burning of living and dead vegetation for land-clearing and land-use change, is a persistent activity in many tropical countries such as Brazil, Indonesia, Nigeria, and Mexico. Intense forest fires can also ignite subsurface organic soil components (e.g. peat), which can continue to smoulder long after the original surface fires are put out. Combustion of vegetation and peat has only recently been recognised as a major source of atmospheric pollution. The immediate effect of burning is the production and release into the atmosphere of gases and particles, in quantities that in some cases can be significant not only locally and regionally but also on a global scale. Vegetation fires account for roughly half of the atmospheric constituents of hydrocarbon, carbon monoxide, nitrogen oxides, and all precursors of tropospheric ozone (Crutzen and Andreae, 1990) and may be the second largest source of anthropogenic aerosols after the production of sulphates from SO_2 (Andreae *et al.*, 1998). Clearly, smoke from large vegetation fires is one of the most conspicuous impacts of human activity on the atmospheric environment. As a result, there has been much interest in studying the contribution of forest fires to enhance the production of air pollutants (e.g. Andreae *et al.*, 1988; Kirchhoff and Rasmussen, 1990; Delany *et al.*, 1985; Kauffman *et al.*, 1992; Kirchhoff *et al.*, 1992; Kirchhoff and Marinho, 1994; Dixon and Krankina, 1993; Auclair and Carter, 1993; Lacaux *et al.*, 1992; Ward *et al.*, 1992; Komala *et al.*, 1996; Balasubramanian *et al.*, 1999, Radojevic and Hassan, 1999).

In recent years, Southeast Asian countries have been affected by periodic episodes of smoke haze from forest fires in Indonesia. In 1997 and 1998, the region was again shrouded in a haze worsened by prolonged drought linked to the El Niño weather phenomenon. Satellite pictures confirmed that the transboundary pollutants have been carried by southwesterly monsoon winds from fires covering an estimated 800,000 hectares of southern Kalimantan, Sumatra, and Java across neighbouring Malaysia and Singapore to as far as Thailand and the Philippines. The impact of these extensive fires on the health of the residents and ecology in the affected areas was substantial, costly, and possibly long lasting.

OVERVIEW OF BIOMASS BURNING EMISSIONS AND THEIR IMPACTS ON THE GLOBAL ENVIRONMENT

Biomass burning is a distinct biogeochemical process with links to the biosphere, atmosphere, and geosphere. In recent years, the scientific research community has recognised the need to gain a better understanding of the extent of global biomass burning activities to assess their effects on the global environment. In regions such as South Africa, Brazil, and Southeast Asia, where many different types of fires burn simultaneously, airborne studies have been successful in quantifying some of the important atmospheric impacts (Andreae *et al.*, 1988, 1994; Kaufman *et al.*, 1992; Talbot *et al.*, 1996; Tsutsumi *et al.*, 1999). The work of Ward *et al.* (1992), along with others, has helped explain how different types of fuels and burning conditions, associated with specific land-use practices, affect the initial emissions of particles and specific compounds (CO_2, CO, and hydrocarbons). Laboratory experiments have also contributed to our overall understanding of the emissions from biomass burning (Lobert *et al.*, 1991; Yokelson 1996, 1997; Muraleedharan *et al.*, 2000b).

To assess the impact of biomass fires, it is important to consider the treatment of emissions, transport, chemical transformations, and deposition processes in an integrated and coupled manner. An overview of the relevant physical and chemical processes is shown in Figure 1.

Fires go through several different phases of burning, with emissions of different compounds tending to be associated with each phase. Most commonly, two phases, a flaming stage followed by a smouldering stage, have been considered. High-temperature chemical transformations proceed rapidly during flaming combustion with a relatively high release rate of products to the atmosphere. During smouldering combustion, low-temperature products are produced with a low rate of release although the duration may be relatively long. It has been recently suggested that, based on their laboratory measurements of emissions from the burning of biomass materials, a third stage with pyrolysis products from heating of the fuel, should also be considered (Yokelson *et al.*, 1996).

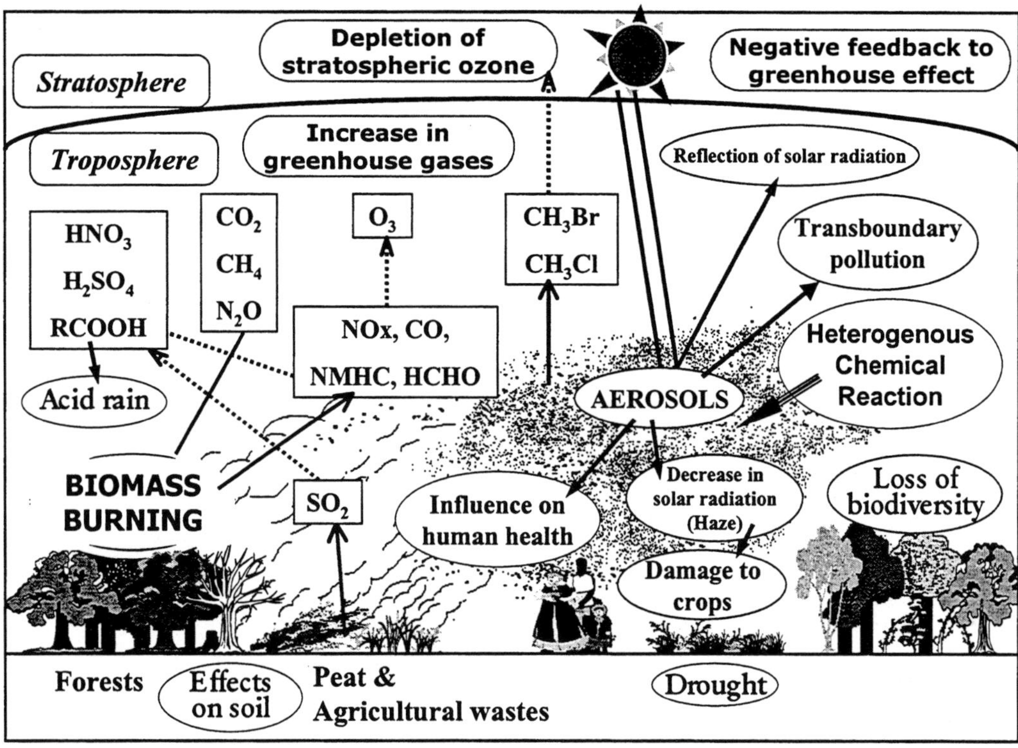

Figure 1 Influence of biomass burning on the global environment

Attempts to estimate the emissions from biomass burning have proven to be very complicated since the chemical products of the transformations of carbon-based biomass fuels by fire are dependent upon a large number of variables such as fuel size, moisture, ratio of living/dead fuel, burn rate, wind speed, ambient temperature, etc. Sorting the total amounts of products that are produced (the chemistry of the products is very much dependent upon the type of combustion) and subsequently released into the atmosphere during the different stages of combustion is difficult (Cofer *et al.*, 1993).

The initial, hot, flaming stage of the fire is associated with emissions of CO_2, NO_x, N_2O, SO_2, and black/elemental carbon. During the smouldering phase, CO, CH_4, NH_3, HCN, CH_3CN, some light hydrocarbons such as methane and acetylene, polynuclear aromatic hydrocarbons (PAH), and aerosols with higher organic carbon (and lower black carbon) contents are emitted (Finlayson-Pitts and Pitts, 2000). Thus the gaseous pollutants and particle formation mechanisms for purely flaming and purely smouldering combustion are quite different. Emissions of individual compounds are usually expressed as enhancement factors relative to CO_2 or CO. Pyrogenic products are attributed to flaming or smouldering combustion based on their temporal correlation with CO_2 or CO, respectively. However, some compounds have been measured in significant quantities that do not correlate temporally with either CO_2 or CO. These compounds are known biomass pyrolysis products contained in the white smoke produced by heating of the primary fuel (Yokelson *et al.*, 1997). *In situ* measurements using FTIR and remote infrared spectroscopy have been used in field studies to detect such compounds such as HCHO, formic acid, methanol, ethene, and ammonia (Yokelson *et al.*, 1996; 1997). In addition, remote sensing from satellites has been used to detect NO_2 and HCHO (Thomas *et al.,* 1998).

Many physical and chemical processes can influence the gases and particles in smoke as it ages. After being emitted, the smoke particles rise to a few kilometres altitude and disperse rapidly into the atmosphere. Smoke from hundreds of fires mixes with biogenic emissions from forests, suspended soil particles, and anthropogenic pollutants (Reid *et al.*, 1998). As the smoke disperses, it undergoes photochemical transformations, gas-to-particle conversion, and particle coagulation. Furthermore, smoke can be entrained into clouds in the boundary layer, thereby increasing the rates of some heterogeneous chemical reactions. These various processes result in a thick photochemical "soup" up to the height of the inversion layer that can have physical, chemical, and optical properties quite different from those of young plumes (Reid *et al.*, 1998).

The environmental consequences of biomass fires are diverse. Denitrification during biomass burning, where as much as 50% of the fuel nitrogen is released as N_2 (Khulbusch *et al.*, 1991), may cause considerable loss of fixed nitrogen from tropical ecosystems (Crutzen and Andreae, 1990). Burning emissions of some C, N, and S compounds leads to production of acidic gases, and these subsequently cause acid rainfall on a regional scale. On the other hand, losses of biomass and nutrients in burning areas, arising from aerosol transport, trace gas losses, erosion, and leaching following fire, may be severe and the resulting biogeochemical consequences may be considerable. Release of smoke particles from biomass fires may provide copious amounts of cloud condensation nuclei, and therefore play a role in cloud formation and precipitation in the tropics (Crutzen and Andreae, 1990). In addition, biomass fire emissions may lead to widespread photochemical production of O_3, a greenhouse gas, coupled with its coincident transport. This phenomenon has been observed in the tropics on regional (Andreae *et al.*, 1992; Kirchhoff *et al.*, 1990, 1994; Burrows *et al.*, 1999) to hemispheric scales (Fishman *et al.*, 1991). Biomass burning is also a source of trace gases such as CH_3Cl, CH_3Br, CH_4, and COS which are stable enough to reach the stratosphere and thus, to interfere with the chemistry of the stratosphere.

In the smoke plume, the aerosol particles may interact with water vapour and numerous other gaseous organic and mineral species emitted concurrently during the combustion of the vegetation. From these interactions, the carbonaceous particles may acquire a coating of condensation products leading rapidly to an internally mixed aerosol phase. Additional transformations by the accretion of particles and photochemical reactions are also likely to occur. Young aerosols emitted from vegetation burning undergo both physical and chemical evolution while being transported by prevailing winds. For example, the coagulation of carbonaceous particles alters their size distributions and the organic component of aerosols may disappear due to evaporation or oxidation processes. SO_2 may undergo oxidation conversion to sulfate in aged particles. Such modifications of the particles significantly influence their size distribution, composition, chemical reactivity, and optical properties, especially the light absorption capacity, solubility, and spatial distribution (Reid *et al.*, 1998).

Submicron pyrogenic particles with sulphate and partially oxidised organic materials are of particular interest as they are both efficient scatterers and absorbers of sunlight. They can influence climate by the "direct" effect, through the scattering and absorption of incident sunlight. Smoke aerosols are also efficient cloud condensation nuclei (CCN) and therefore interact with clouds, leading to the "indirect" effect through the modification of cloud properties, since both the inorganic and organic components of biomass smoke are water soluble (Novakov and Corrigen, 1996). Furthermore, smoke aerosols with high levels of acidic species and carcinogenic compounds are prone to affect human health. Thus the effect of smoke on radiation, cloud microstructure, and air pollution are of major importance on local and regional scales.

MEASUREMENTS OF BIOMASS BURNING EMISSIONS AND THE RESULTING HAZE IN SE ASIA

Methods ranging from satellite remote sensing images to very small controlled combustion experiments can be used to measure emissions released from vegetation fires. The AVHRR (Advanced Very High Resolution Radiometer) images are being increasingly used to estimate the extent of the burning activity and to assess the associated atmospheric effects at one end of the scale. Controlled combustion laboratory experiments are very effective in confirming hypotheses concerning processes affecting the release of emissions. For example, emissions released as a function of flaming and smouldering combustion can be easily studied using combustion laboratory facilities. For modeling atmospheric chemical reactions, it is important to know the temporal and spatial scales of release of emissions from biomass burning, which are usually obtained from ground-based and airborne measurements.

In airborne studies, instrumented aircraft fly through forest fire plumes making *in situ* measurements and/or taking samples for chemical analysis, while in ground-based studies emissions are measured using either fixed monitoring stations or instrumented motor vehicles. Both types of measurements have their own advantages and disadvantages. For example, airborne measurements can provide information on both horizontal and vertical distributions of biomass burning emissions and their products in young and aged plumes. However, these "snapshot" measurements do not provide temporal distributions of ecologically and radiatively important trace gases and particles, which can only be obtained from ground-based measurements. In addition, ground-based measurements are of use in assessing the impact of gaseous pollutants and particles on human health especially in SE Asia where pyrogenic emissions are mixed with biogenic and anthropogenic pollutants during the long-range transport. In order to evaluate the composition of biomass burning emissions and to assess their environmental impact, it is important to integrate laboratory, ground-based, airborne, and spaceborne studies.

Such extensive and expensive measurements have been employed in North America, South America, and Central Africa to estimate pollutant fluxes from biomass burning and to predict global impacts of this pollution. Several studies have been reported on the emission factors and chemical compositions of biomass burning aerosols (e.g. Ward *et al.*, 1992; Artaxo *et al.*, 1988) and precipitation (e.g. Hendry *et al.*, 1984; Andreae *et al.*, 1990; Lacaux *et al.*, 1987, 1988, 1992). However, although regional haze episodes in SE Asia have been recorded repeatedly since 1982 (Radojevic and Hassan, 1999), there have been only few detailed chemical studies of the haze in SE Asia. Forest fires in SE Asia account for a significant fraction of worldwide biomass burning emissions. In addition, relatively little is known about the influence of different vegetation types on the mixture of chemical species emitted from biomass fires in SE Asia.

Comprehensive forest fire studies in SE Asia have been conducted only during the 1997 and 1998 episodes. The few reported studies of earlier haze events have been of a general or qualitative nature (Nichol, 1997; Tussin, 1995; Rindam, 1995; Sani *et al.*, 1991) and/or focused on a limited number of chemical components. Concentrations of criteria pollutants (PM_{10}, NO_2, CO, O_3, SO_2) are available for Kuala Lumpur and Singapore for the 1994, 1997 and 1998 haze episodes, and for Brunei for the 1998 episode. For earlier haze events, such as the 1991 episode, only suspended particulate matter (SPM) and CO measurements were made at some stations in SE Asia. For many sites, especially those in Indonesia, no chemical measurements of the haze are available and only visibility data have been reported. Particle concentrations and meteorological conditions during the earlier haze episodes in Malaysia (1990, 1991 and 1994) have recently been reviewed by Soleiman *et al.* (2000). Multiphase investigations on chemical transformations of fire emissions are discussed in the following sections in order to explain the enrichment in the concentration of several secondary pollutants in the regional atmosphere during the burning periods. The contribution of vegetation and peat fires to atmospheric pollution is compared with that of other natural and anthropogenic sources.

EMISSIONS OF TRACE GASES AND PARTICLES FROM VEGETATION BURNING

Since Crutzen *et al.* (1979) first suggested that biomass fires could be a significant contributor to the budgets of several important atmospheric trace gases and particulate matter, extensive research has been conducted to understand trace gas and aerosol production from vegetation fires and their subsequent interaction within the environment. The release of trace gases by large-scale biomass burning contributes to the increase of greenhouse gases (CO_2, CH_4, N_2O, CH_3Cl) in the atmosphere, and of reactive compounds such as CO, NMHCs, NO_x, which are, as methane, photochemical ozone precursors. In addition, biomass burning has been reported to be an important source of atmospheric acids such as nitric (HNO_3), formic (HCOOH), and acetic (CH_3COOH) acids (Talbott *et al.*, 1988; Helas *et al.*, 1992; Lefer *et al.*, 1994). These atmospheric species contribute significantly to rainwater acidification in tropical regions, especially in Africa (Lacaux *et al.*, 1991, 1992). Rondon and Sanhueza (1989) reported determinations of nitrous acid (HONO) during the day and night in the Venezuelan savanna region during the vegetation burning season. A range of values from 1.07 ppbv (nighttime) to 0.05 ppbv (daytime) was observed.

The smoke particulate fraction is composed primarily of carbonaceous particles for which the carbon content has been estimated to be of the order of 66% (Crutzen and Andreae, 1990; Muraleedharan *et al.*, 2000b). In these particles, two main components may be distinguished: the black carbon and the organic carbon. Accompanying the carbonaceous component of the particles are numerous water-soluble inorganic ions such as sulphates (SO_4^{2-}), nitrates (NO_3^-), phosphates (PO_4^{3-}), and chlorides (Cl^-). This combination of carbon and inorganic material can produce hydrophilic particles that can act as cloud condensation nuclei and ultimately contribute to the formation of acidic rain. Water soluble cations (+ve ions) typically associated with the inorganic constituents consist primarily of potassium (K^+), sodium (Na^+), ammonium (NH_4^+), and calcium (Ca^{2+}). It should be noted that these inorganic compounds are the primary soluble inorganic particulate emissions, and that other trace minerals also exist in vegetation fire emissions. The large enrichment of potassium in vegetation fire produced aerosols has permitted its use as a tracer for these aerosols (Andreae, 1983).

Most of the research on the environmental aspects of biomass burning has been directed towards the tropics, specifically to the tropical rain forests of Brazil and to the savannas of Africa and South America. However, the results of these studies are of little use in assessing the environmental impacts of biomass fires in Southeast Asia as fires in diverse ecosystems are very different in terms of the production of gaseous and particulate emissions. Emissions depend on the type of ecosystem; the moisture content of the vegetation; and the nature, behavior, and characteristics of the fire (Levine *et al.*, 1995).

Most of the information on the smoke pollution in Southeast Asia has been obtained from air quality monitoring stations in the region. Concentrations of criteria pollutants (PM_{10}, SO_2, NO_2, CO, and O_3) are available from Malaysia and Singapore for the 1994, 1997, and 1998 haze episodes, and for Brunei for the 1998 episode. In Indonesia, only total suspended particulates (TSP) were measured during the recent haze episodes. The full environmental impact assessment of the haze in Southeast Asia requires much more detailed chemical characterisation. Comprehensive field studies of biomass burning impacted air masses in Southeast Asia have been conducted recently in Singapore (Balasubramanian *et al.*, 1998, 1999) during the 1997 and 1998 haze episodes and in Brunei Darussalam during the 1998 haze episode (Radojevic and Hassan, 1999; Muraleedharan *et al.*, 2000b). While the haze in Singapore during 1997 was due to fires in Kalimantan and Sumatra, that in Brunei was mainly influenced by local fires in Brunei and the neighbouring Malaysian states of Sarawak and Sabah, rather than from the long–range transport of pollutants from forest fires in Indonesia (Radojevic and Hassan, 1999). The results emerging from ground-based measurements of acidic gases and particles are discussed in this section.

Observational Data of Acidic Gases and Particles

The major acidic gases in the atmosphere are SO_2, HNO_3, HNO_2 (or HONO), HCOOH and CH_3COOH. While SO_2, HCOOH, and CH_3COOH are due to both direct emissions from a number of sources and atmospheric production, both HONO and HNO_3 are mainly produced in the atmosphere as secondary pollutants through reactions between NO_x and the OH radicals. As part of an urban air quality study in Singapore, SO_2, HONO, HNO_3, and inhalable fine particles ($PM_{2.5}$) were sampled simultaneously using an annular denuder system (ADS) and subsequently analysed in the laboratory from June1997 to February 1998 on a daily basis (Balasubramanian *et al.*, 1998, 1999). During the course of this study, biomass burning impacted air masses were received in Singapore from Sumatra and Kalimantan between August and November 1997 as evident from satellite images (Liew *et al.*, 1998) and a considerable increase in the daily PSI values (Balasubramanian *et al.*, 1999). In order to assess the impact of the haze on the budget of acidic gases and particles, the results are classified into three periods, pre-haze, during haze and post-haze in Table 1.

Table 1 Mean concentrations ($\mu g\ m^{-3}$) of air pollutants before, during and after the haze of 1997.

	Pre-haze [a]	During haze [b]	Post-haze [c]
Pollutant	**Mean ± σ**	**Mean ± σ**	**Mean ± σ**
HONO	2.3 ± 1.4	4.9 ± 2.0	2.4 ± 1.3
HNO_3	0.3 ± 0.2	1.0 ± 0.6	0.5 ± 0.4
SO_2	16.2 ± 7.1	21.6 ± 11.4	16.4 ± 2.2
SO_4^{2-}	4.6 ± 1.7	12.8 ± 6.2	2.8 ± 1.8
NO_3^-	0.5 ± 0.2	2.5 ± 1.7	0.6 ± 0.2
PSI[d]	51 ± 6	80 ± 23	33 ± 8

[a]June10 to August 22 , 1997
[b]August. 23 to November 12, 1997
[c]November 14, 1997 to February 9, 1998
[d]Pollutant Standards Index

All the pollutants collected by the ADS were elevated during the haze period as compared to both pre-haze (dry) and post-haze (wet) periods, with the exception of SO_2, which only increased slightly. The average concentrations of HONO and HNO_3 were at least twice their average "background" values. On the other hand, the acidic components of airborne particles were greater by a factor of more than 3. Thus the smoke pollution from Indonesia appears to be responsible for a substantial increase in the concentration of atmospheric acidic pollutants which are of both health and environmental concern. The following major factors are thought to be responsible for the enhancement in the concentration of the gaseous and particulate pollutants.

(1) Direct Emission

Certain amounts of HONO, HNO_3, SO_2, and particulate NO_3^- and SO_4^{2-} may have been emitted directly from the forest fires and subsequently transported towards Singapore. High atmospheric HONO concentrations during vegetation burning in the tropical savanna have been reported (Rondon and Sanhueza, 1989) in fresh plumes. Lefer *et al.* (1994) suggested that HNO_3 could be directly emitted from biomass combustion based on aircraft measurements in both fresh and aged biomass burning plumes. SO_2 was probably emitted by combustion of a number of coal deposits beneath the surface in East Kalimantan upon exposure to direct sunlight (Nichol, 1997) and/or by combustion of peat bogs in sub-surface soil (Nakajima and Higurashi, 1998; Gras *et al.*, 1999).

(2) Atmospheric Production

During the forest fires, NO and NO_2, the precursors of HONO and HNO_3 are emitted from burning of the N containing substances; NO is the major component of NO_x emitted from vegetation burning. The NO thus emitted can be photochemically oxidised to NO_2 and then to HNO_3 through either the homogeneous reaction of NO_2 with the OH radical, reaction of NO_3 radical with aldehydes (RCHO) and other hydrocarbons (RH), or hydrolysis of N_2O_5 as shown below.

$$NO + HO_2 \text{ (or } RO_2) \rightarrow NO_2 + OH \text{ (or RO)} \quad (1)$$

$$NO_2 + OH + M \rightarrow HNO_3 + M \quad (2)$$

$$NO_2 + O_3 \rightarrow NO_3 + O_2 \quad (3)$$

$$NO_3 + RCHO \rightarrow HNO_3 + RCO \quad (4)$$

$$NO_3 + RH \rightarrow HNO_3 + R \quad (5)$$

$$NO_2 + NO_3 + M \leftrightarrow N_2O_5 + M \quad (6)$$

$$N_2O_5 + H_2O \rightarrow 2HNO_3 \quad (7)$$

where M in equations (2) and (6) is a third body reactant. The reaction of NO_2 with the OH radical is important during daytime, while at night, N_2O_5 hydrolysis is believed to become an important source of HNO_3. The presence of aerosols in the atmosphere can alter the above reactions in a variety of ways. For instance, nitric acid can be deposited onto the aerosol surface via the overall reaction:

$$HNO_3 + \text{aerosol surface} \rightarrow \text{particulate nitrate.} \quad (8)$$

In addition, NO_3 can react on aerosol surfaces to form particulate nitrate:

$NO_3 + H_2O$ + aerosol surface $\rightarrow$ particulate nitrate. (9)

Furthermore, the conversion of N_2O_5 to HNO_3 can also take place on wetted aerosol surfaces. Each of these reactions represents a sink for gaseous NO_x. The rate of production of particulate nitrate is a function of the surface area of the aerosols and the ambient concentrations of these species (Zhang *et al.*, 1998). Atmospheric particulate loading is likely to influence both of these factors. As the loading increases, the surface area of particles increases. Atmospheric particles may also alter the concentrations of the OH and HO_2 radicals. Additionally, HNO_3 can also be removed from the gas phase through wet and dry deposition. Due to its high water solubility, wet deposition of HNO_3 is usually dominant, leading to increased acidity of hydrometeors (cloud, rain, and fog) as will be seen from the next section on rain chemistry. It should be noted that during the haze sunlight is much less intense and consequently the photochemical formation of HNO_3 would be expected to be less favorable. However, its concentration during the haze increased substantially, indicating that HNO_3 and/or its precursors were enriched in biomass burning impacted air masses.

Previous studies clearly show that HONO is present in the urban atmosphere (Calvert *et al.*, 1994) and also in biomass combustion plumes (Rondon and Sanhueza, 1989), and that the maximum concentrations occur at night. As the sun rises, photodecomposition of HONO becomes important, and its concentration decreases to much lower daytime values. HONO absorbs sunlight strongly in the UV region (<390 nm). Its dissociation in sunlight occurs by the following reaction with a quantum yield near unity:

$HONO + h\nu \rightarrow OH + NO$ (10)

Since the gas phase photolysis of HONO can significantly enhance photooxidation processes in the early morning due to the rapid production of OH radicals, HONO is of particular importance in atmospheric chemistry. Wet precipitation makes a small contribution for the removal of atmospheric HONO as its Henry's Law Coefficient at 298K is 49 (mol L^{-1} atm^{-1}) compared with the value of 2.1x 10^5(mol L^{-1} atm^{-1}) for nitric acid.

The production of atmospheric HONO in biomass combustion plumes is less clear as compared to HNO_3. HONO can be produced by the following homogeneous gas-phase reactions:

$NO + OH + M \rightarrow HONO + M$ (11)

$NO + NO_2 + H_2O \rightarrow 2HONO$ (12)

$2NO_2 + H_2O \rightarrow HONO + HNO_3$ (13)

While reaction (11) may lead to significant production of HONO during the day, the other two reactions (12 and 13) may be responsible for the formation of HONO at night. As HONO builds up only at night, reaction (11) is unimportant. Additionally, reactions (12) and (13) are much too slow to explain the observed night-time HONO production rates (Junkermann and Ibusuki, 1992). Based on the reported smog chamber studies and laboratory kinetic studies (Sakamaki *et al.*, 1988; Svensson *et al.*, 1987; Weisen *et al.*, 1995), it can be concluded that the heterogeneous equivalent of reactions (12) and (13) is not important in atmospheric production of HONO. The rationale for this conclusion is that the rate law for HONO formation is reported to be first order in NO_2 and H_2O with no dependence on NO concentration. Both atmospheric observations as well as laboratory studies point towards the importance of one or several heterogeneous HONO formation mechanisms whose rate laws are not consistent with either reaction (12) or (13) as the rate limiting process (Gerecke *at al.*, 1998). One such pathway is a heterogeneous reaction of NO_x on biomass combustion aerosols, which typically have a significant quantity of elemental and organic carbon (carbonaceous aerosols). Recent laboratory and simulation studies have indicated that the interaction between NO_2 and soot (carbonaceous aerosols) leads to HONO production, which supports the above hypothesis (Gerecke *at al.*, 1998; Amman *et al.*, 1998; Aumont *et al.*, 1999). More field studies, especially near forest fires, are needed to gain a better understanding of how HONO is produced from vegetation combustion and its role in photooxidation processes.

Although SO_2 increased only marginally during the haze, the total S concentration (i.e SO_2 + particulate-sulfate) was enriched as compared to the non-haze period. This observation suggests that part of the SO_2 emitted from combustion of coal deposits and peat bogs was oxidised to particulate SO_4^{2-}. The oxidation of SO_2 may have taken place in the gas-phase by the OH radicals and/or on combustion aerosols by oxidants such as O_3. However, the gas-phase oxidation of SO_2 is known to be relatively slow. In view of the fact that the particulate size distribution during haze is dominated by fine particles (Muraleedharan and Radojevic, 2000) with large surface area, SO_2 is likely to undergo efficient interfacial mass transfer. Additionally, since haze aerosols are hygroscopic (Gras *et al.*, 1999), the heterogeneous oxidation of SO_2 is quite likely, provided there are sufficient oxidants. Particulate sulphate is of concern because of its ability to interact with incoming solar radiation and its potential contribution to acid rain. It should be noted that the overall acidity of airborne particles is dependent on the concentration of NH_4^+, and other ions (e.g. Ca^{2+} and Mg^{2+}) which tend to neutralise the acidity introduced by both sulphate and nitrate. In addition to elemental carbon, organic carbon, and water-soluble ions, haze aerosols in SE Asia also contain toxic metals (Cd, Ni, V, Hg, and As) as reported by Muraleedharan *et al.* (2000b).

IMPACT OF BIOMASS BURNING ON RAINWATER CHEMISTRY

Wet deposition is known to be the major sink for gases and particles. It also plays an important role in controlling the concentrations of these species in the troposphere. The chemical content of rainwater integrates numerous physical and chemical mechanisms such as advection, convection, gas to particle conversion, cloud formation, in-cloud reactions, and below-cloud scavenging of atmospheric particles, etc. Therefore, determination of the composition of rainwater helps to resolve the relative importance of the different sources of gases and particulate matter. Moreover, wet and dry deposition play an important role as a source of nutrients for biological systems. Because of these concerns, precipitation chemistry has been the subject of intense research in biomass burning impacted areas in Africa (Lacaux *et al.,* 1987, 1991-1993), South America (Sanhueza *et al.*, 1989, 1992), Southeast Asia (Balasubramanian *et al.*, 1999; Radojevic and Tan, 2000). The rainwater data from Singapore collected during haze and non-haze periods in 1997 are discussed to elucidate the impact of forest fires on rain chemistry (Balasubramanian *et al.*, 1999).

Major inorganic and organic ions were determined in 104 wet-only samples collected from July through December (54 rain events). The pH was measured immediately after a rain event with a calibrated pH electrode and meter. pH values ranged from 3.79 to 6.20 with a volume-weighted mean of 4.35. Approximately 85% of the samples had pH values less than 5.0; 15% of the samples had pH values between 5.0 and 6.2. Generally, the pH of precipitation is the result of the acid-base balance relationship. For the purpose of estimating the impact of biomass burning on rainwater acidity and composition, rain events were categorised into two types: haze (August–October) and non-haze (July, November-December) and the concentrations of the measured 11 ions are given in Table 2. On several occasions during September and October, the PSI increased above 100 (unhealthy levels) when the region was shrouded with visible smoke haze. After the end of October, the onset of the northeast monsoon brought a sudden return to PSI levels below 50 and good air quality, which supports the above classification.

Rainwater samples collected during the haze period had greater concentrations of Cl^-, NO_3^-, SO_4^{2-}, $HCOO^-$, CH_3COO^-, Ca^{2+}, NH_4^+, and K^+ when compared to those collected during the non-haze period ($p<0.05$). The enrichment in the concentration of these species in rainwater was most likely due to below-cloud scavenging of gases and particles in the haze layer; gray to black particles were noted in rainwater collection buckets and on filter substrates following sample filtration. Concentrations of most of the ionic components in rainwater declined rapidly during the wet season. Decreases in ion concentrations were presumably due to a combination of decreased input of acidic gases and particles and frequent washout of atmospheric gases and particles. High concentration events for each ion are identified and included in Table 2 along with their corresponding enrichment factors (ratio of haze to non-haze concentrations). For most of the ions exhibiting enrichment during the dry season, a large number of exceedances were observed when the biomass burning impacted air masses were received at the sampling site. These exceedances represent those concentrations greater than overall volume-weighted mean plus two standard deviations for each ion measured. In terms of exceedances, there was a significantly large number of rain events that had elevated concentrations of K^+, NH_4^+, and inorganic and organic acids during the burning period. Similar observations were made in airborne particles as well (Balasubramanian, unpublished results).

Table 2 Summary of the analytical results of rainwater during burning and non-burning periods.

	Concentration (meq L^{-1})											
	pH	H^+	Ca^{2+}	Mg^{2+}	Na^+	K^+	NH_4^+	nss SO_4^{2-}	NO_3^-	Cl^-	Acetate	Formate
Haze (n=32)												
Min	3.5	1.4	3.1	2.1	13.5	0.4	1.78	23.7	7.6	10.3	1.9	1.4
Max	5.9	323.6	338.8	53.3	136.0	27.9	201.5	541.6	200.3	230.3	58.9	108.6
Mean	4.4	56.1	31.1	6.2	45.4	7.8	45.9	95.6	22.8	44.2	18.4	20.5
S.D.	1.2	21.4	11.6	1.9	13.5	3.1	12.3	28.4	5.6	13.0	6.4	5.8
No. of exceedences*		7	12	9	9	21	22	17	11	19	12	10
Non-haze (n=72)												
Min	4.0	0.5	6.0	1.9	4.1	2.0	1.2	9.1	3.3	2.4	0.3	1.7
Max	6.3	95.5	274.0	36.9	185.5	48.1	82.2	212.4	71.2	195.5	11.9	38.2
Mean	4.5	42.1	27.7	8.5	45.4	5.3	16.6	56.4	17.5	20.5	3.8	5.4
S.D.	0.8	8.9	4.0	1.9	8.7	0.7	3.2	10.3	3.7	3.2	0.9	0.9
No. of exceedences*		5	5	4	13	4	4	5	5	7	0	1
Haze/Non-haze ratio	1.0	1.3	1.1	0.7	1.0	1.5	2.8	1.7	1.3	2.2	4.8	3.8

*Rain events with concentrations greater than mean plus 2 standard deviations.

The relatively high concentrations of SO_4^{2-}, NO_3^-, and NH_4^+ in rainwater suggest that emissions of SO_2, NO_x, NH_3, and particles containing SO_4^{2-}, NO_3^-, and NH_4^+ were considerable upwind from Singapore during the burning period. This is supported by the fact that SO_4^{2-} had a strong correlation with NO_3^- ($R^2 = 0.77$; $p<0.05$) and significant correlation was found between NH_4^+ and SO_4^{2-} ($R^2 = 0.56$; $p <0.05$), and NH_4^+ and NO_3^- ($R^2 = 0.67$; $p < 0.05$). Additionally, the transport time of biomass burning impacted air masses between the source(s) and sampling site was of the order 3 or 4 days under slowly moving wind conditions experienced in 1997 (~3 m s^{-1}). This long atmospheric residence time of air masses would result in progressive gas to particle conversion of biomass burning emission components. This process presumably led to a steadily increasing proportion of ammonium-containing particles.

Both NO and NH_3 are known to be emitted from vegetation burning during flaming and smouldering combustion, respectively (Hegg *et al.*, 1988; Yokelson *et al.* 1997). In addition, NO and NH_3 may be emitted due to interactive effects of fertiliser use and soil moisture (Clark *et al.*, 1998). The concentrations of both formate and acetate increased tremendously during the burning period, which is consistent with earlier reports on biomass burning generated organic acids (Sanhueza *et al.*, 1989; Helas *et al.*, 1992). Elevated formate and acetate concentrations in precipitation could indicate the existence of direct vegetation emission and indirect production of organic acids from a chain of oxidative steps of hydrocarbons released by vegetation burning. Increase in the concentrations of chloride and sulfate is rather intriguing as such a dramatic change has not been reported from other biomass burning impacted areas such as Brazil. Therefore changes in the concentrations of these species merit further discussion.

The Cl/Na ratio is close to 1 during the haze period and about 0.45 during the non-haze period. It appears that the chloride concentration was depleted during the non-haze period relative to the ratio of 1 as the Na^+ concentration remained unchanged. As the ratios were calculated based on volume-weighted mean concentrations of Na^+ and Cl^-, the decrease in the Cl^- content is not due to the dilution effect resulting from increased rainfall during the non-haze period. The decrease is due mainly to degassing of HCl produced from the heterogeneous reaction between marine NaCl aerosols and gaseous HNO_3 (Clegg and Brimblecombe, 1985). This heterogeneous reaction was probably suppressed during the haze period as the atmosphere was loaded with carbonaceous aerosols released from vegetation and peat burning. As for the enrichment of the sulphate concentration in rainwater, it may be due to below-cloud scavenging of sulfate-containing secondary aerosols produced through gas to particle conversion process of SO_2 released from combustion of coal deposits and/or peat bogs. As H_2O_2 is reported to be directly emitted in air impacted by biomass burning (Lee and Heikes, 1997), the possibility of in-cloud oxidation of SO_2 by H_2O_2 should also be considered. In short, the enrichment of sulphate in rainwater during the haze period is a result of nucleation and below-cloud scavenging of sulphate aerosols and in-cloud oxidation of SO_2 by H_2O_2.

Neutralisation of Acidity

It is clear from Table 2 that the concentrations of both inorganic and organic acids during the haze period have increased tremendously as compared to the non-haze period. However, the decrease in pH in response to the increased concentrations of acids is not significant, implying that high concentrations of inorganic and organic acids are not always associated with low pH. Further insight into rainwater composition may be gained by examination of fractional acidity, $[H^+]/([NO_3^-] + [SO_4^{2-}])$. This ratio would be unity if the acidity contributed by these strong acids is not neutralised at all. However, the calculated ratio turns out to be about 0.5. This simple calculation clearly indicates that a substantial fraction of the acidity in precipitation samples was neutralised. Such neutralisation is frequently attributed to NH_3 and/or $CaCO_3$ (Vong, 1990). If $CaCO_3$ and NH_3 play a major role in neutralising the acidity of precipitation, then the sum of the concentrations of the inorganic and organic acids should be linearly related to the sum of the concentrations of H^+, Ca^{2+}, and NH_4^+. The rainwater data are indeed consistent with this hypothesis. Similar observations on rainwater acidity were made in Brunei (Radojevic and Tan, 2000) and in Venezuela (Sanhueza *et al.*, 1989) during vegetation burning. Neutralization factors (NF) were calculated to evaluate the relative neutralisation of rainwater by the soil-derived species (divalent cations) and NH_4^+ using the following formula:

$$NF_x = x/(NO_3^- + SO_4^{2-}) \qquad (14)$$

where x represents the concentration of the cations (μeq L^{-1}).

It was observed that NH_4^+ was the main neutralising species from August to October as it had the highest NF among the three ions. During all the other months Ca^{2+} played an important role in neutralisation. In general, neutralization by Mg^{2+} was negligible. These results pertaining to neutralisation of acidity during the burning period contrast with the results obtained in the equatorial forest of Africa (Lacaux *et al.*, 1987), where an increase in rain acidity was observed during upwind savannah fires. This difference in rainwater acidity between different locations implies that the acidity of rain during vegetation burning is dependent on a number of factors including the type of vegetation being burned, combustion efficiency, residence time of gases and particles released into the atmosphere, the location of the sampling site relative to biomass burning areas, etc.

PHOTOCHEMICAL FORMATION OF OZONE FROM BIOMASS BURNING EMISSIONS

Forest fires in the tropics are considered to significantly affect the regional and global distributions of tropospheric ozone because they emit a large amount of ozone precursor gases such CO, hydrocarbons and their oxidation products, and oxides of nitrogen (Crutzen *et al.*, 1979). It is estimated that roughly 15% of the global ozone formation in the tropics is due to the impact of natural and anthropogenic biomass fires. By using the TOMS (Total Ozone Mapping Spectrometer), Fishman *et al.* (1990) suggested that tropospheric ozone had increased over Borneo Island in March-April 1983 when extensive biomass burning had been reported there. Recent studies have revealed that significant enhancements in tropospheric ozone above Indonesia were observed in September and October 1993, 1994, and 1997 based on the TOMS data, aircraft measurements, satellite measurements, and ground-based Brewer total ozone measurements when biomass burning occurred (Fujiwara *et al.*, 1996; Komala *et al.*, 1996; Fujiwara *et al.*, 1999; Burrows *et al.*, 1999; Tsutsumi *et al.*, 1999). A build-up in the concentration of tropospheric ozone is of much concern as it is a greenhouse gas whose radiative forcing is largest in the upper troposphere where temperatures are coldest. The injection of Southeast Asian biomass burning emissions into the upper atmosphere by deep convection can therefore enhance the climatic effects of biomass burning.

Although biomass burning does occur in Southeast Asia, the extent to which these fires influence the photochemistry of the surrounding region is not well characterised. Chatfield and Delany (1990) have discussed the influence of transport on the formation of ozone. They distinguished between two cases: fresh emissions within the plume are first chemically processed and then mixed ("cook-then-mix"), and the other limiting case where during the early stages dilution takes place and most of the chemistry occurs later on ("mix-then-cook"). The latter case is particularly important in Southeast Asia where biomass burning emissions tend to be mixed with anthropogenic emissions from urban areas during the long-range transport of plumes emerging from biomass burning sites. Thus, atmospheric mixing processes downwind from the fire can influence the amount of ozone produced. Systematic ground-based measurements of O_3 and its precursor gases made at the National University of Singapore elucidate the processes involved in photochemical production of ozone in biomass burning plumes (Balasubramanian, unpublished data).

The intermittent biomass burning in Indonesia in 1998 led to several photochemical air pollution episodes in Singapore, with elevated levels of ozone (>70 ppbv). A particularly interesting episode was observed on 8 October, which is shown in Figure 2a. On 7 October, there were "hot spots" in Sumatra according to the on-line satellite images and surface weather maps from the Singapore Meteorological Service (Figure 3a). In addition, the meteorological data obtained at the sampling site indicated that biomass burning impacted air masses originating from Sumatra were transported towards Singapore by the prevailing winds. There was a considerable increase in the concentration of airborne particles (PM_{10}) in Singapore, which provided further support regarding the advection of biomass burning emissions from Sumatra into Singapore (Ministry of the Environment, Singapore, personal communication). This is a typical example of a photochemical smog episode or photochemical air pollution induced by biomass burning emissions, which were presumably mixed with local anthropogenic emissions. The concentration of NO was very high for a prolonged period in the morning, with the maximum being 260 ppbv. The maximum value of NO on this day was much higher than the values observed due to local traffic emissions. Moreover, the temporal variation of NO, with concentrations exceeding 70 ppbv for a few hours, has never been observed at this site. The concentration of SO_2, the other primary pollutant, was also elevated during this episode. The level of HCHO, which is an indicator of pollution by hydrocarbons, was exceptionally high (42 ppbv at 12:30 p.m.) during this particular event. The overall reaction involved in the photochemical production of O_3 is as follows:

$$NO + HCs + O_2 + h\nu \rightarrow NO_2 + O_3 + \text{other pollutants} \qquad (15)$$

Ozone is formed chemically by the photolysis of NO_2 and subsequent reaction of the emerging ground state oxygen atom with molecular oxygen.

$$NO_2 + h\nu \rightarrow NO + O \tag{16}$$

$$O + O_2 + M \rightarrow O_3 + M \tag{17}$$

On the other hand, ozone reacts with NO:

$$NO + O_3 \rightarrow NO_2 + O_2 \tag{18}$$

Considering only reactions (16) to (18), the sum of $[O_3]$ and $[NO_2]$ is constant. Thus, the above reaction chain is a zero cycle for O_3. Efficient ozone formation in the atmosphere can only be achieved in reactions that oxidise NO to NO_2 without consuming simultaneously O_3. The reactions of NO with HO_2:

$$HO_2 + NO \rightarrow NO_2 + OH \tag{19}$$

and the organic peroxy radicals, RO_2:

$$RO_2 + NO \rightarrow NO_2 + RO \tag{20}$$

are of this kind. HO_2 and RO_2 are in turn formed by reaction of CO and hydrocarbons with the OH radicals:

$$CO + OH \rightarrow CO_2 + H \tag{21}$$

$$H + O_2 + M \rightarrow HO_2 + M \tag{22}$$

$$HCs + OH \rightarrow R + H_2O \tag{23}$$

$$R + O_2 + M \rightarrow RO_2 + M \tag{24}$$

During the conversion of NO to NO_2 by the RO_2 radicals, RO radicals are formed leading to the production of aldehydes (e.g. HCHO). HCHO in turn produces hydroperoxy radicals (HO_2) and carbon monoxide upon photolysis as shown below:

$$HCHO + h\nu \rightarrow H + HCO \tag{25}$$
$$HCHO + h\nu \rightarrow H_2 + CO \tag{26}$$

$$H + O_2 + M \rightarrow HO_2 + M \tag{27}$$

$$HCO + O_2 \rightarrow HO_2 + CO \tag{28}$$

Therefore, HCHO can enhance the accumulation of O_3 and CO in the atmosphere during a photochemical air pollution episode. Since the formation of the two secondary pollutants, HCHO and O_3, is dependent on photodegradation of hydrocarbons, there appears to be a similarity in the concentration profiles of HCHO and O_3 during daytime (Figure 2).

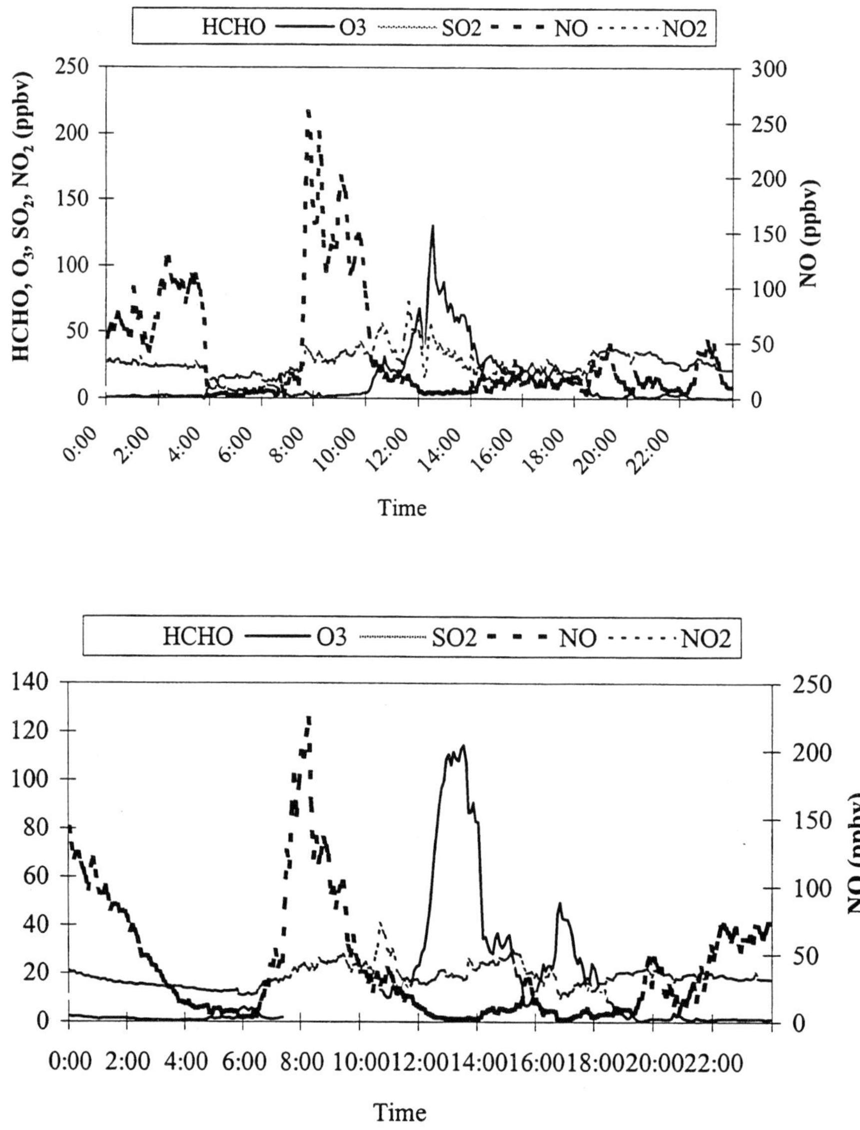

Figure 2 (a) A typical photochemical air pollution episode on 8 October 1998, (b) Concentrations of gaseous pollutants on 9 October 1998

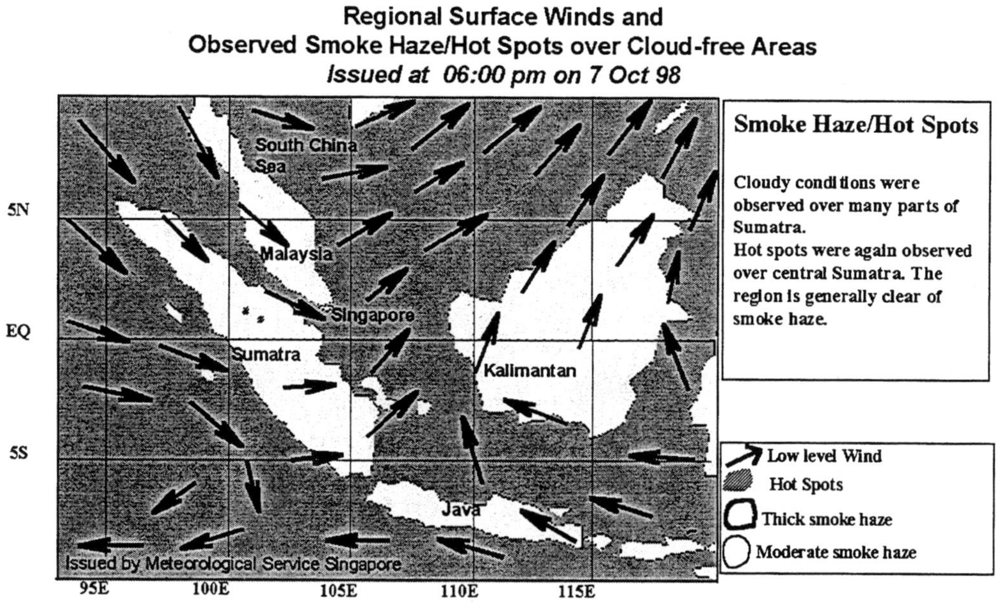

Figure 3 (a) Haze map showing the dominant wind direction and the hot spots in Sumatra on 7 October 1998

During the burning of vegetation in Sumatra, the emissions of HCs and NO appear to be accompanied by elevated levels of SO_2 which may come from combustion of peat bogs and/or coal deposits. While NO results from flaming combustion, both HCs and SO_2 are thought to be products of low-temperature smouldering combustion of materials in sub-surface soil like peat (Gras *et al.*, 1999). Therefore, SO_2 can probably be used as an indicator of HCs emissions from combustion of subsurface organic soil components. As HCs are the source of aldehydes, the concentration profiles of SO_2, HCHO and O_3 show a similar trend. The O_3 and HCHO concentrations decreased to low levels after 18:00. This decrease may be attributed to both the lack of photochemical reactions and the depletion of precursors such as NO_x and HCs from the burning sites. The hot spots were not visible on 8 October, 6 p.m. (Figure 3b). Furthermore, the concentrations of the gaseous pollutants were relatively low on October 9 (Figure 2b). These field observations suggest that the sporadic increase in O_3 concentration in Singapore was mainly due to photochemical production from precursor gases emitted by the fires.

It should be noted that the amount of O_3 in biomass burning plumes is not only controlled by the concentration of its precursors but also by the atmospheric loading of particles. Tsutsumi *et al.* (1999) have reported that O_3 had a maximum concentration of 80.5 ppbv in the middle layer of the smoke haze and very low concentrations of ~20 ppbv in the lower smoke over Kalimantan based on aircraft measurements.

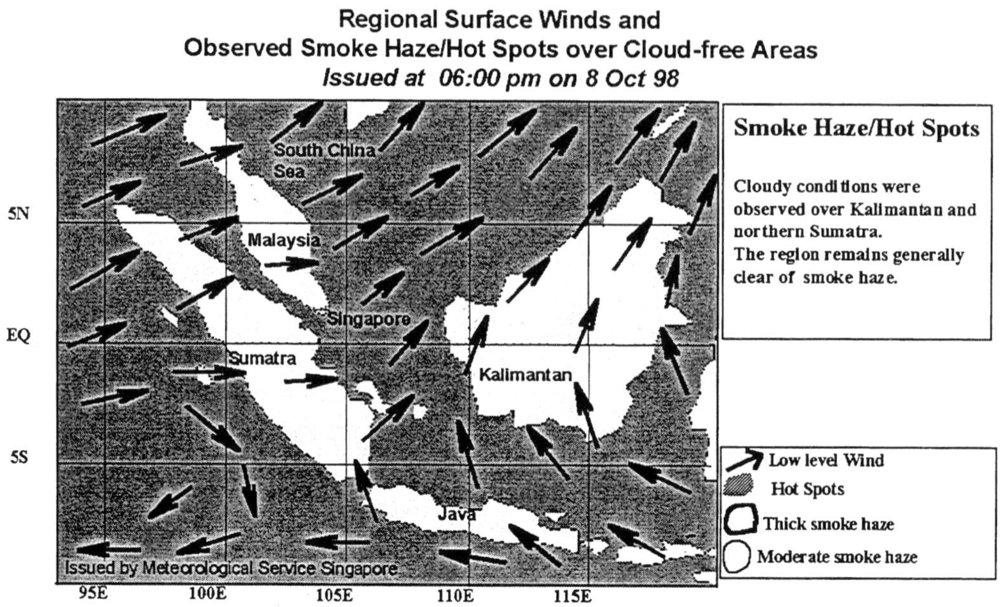

Figure 3 (b) Haze map showing the dominant wind direction and the hot spots in Sumatra on 8 October 1998

They proposed the following hypotheses for the low O_3 concentration at low altitudes over Kalimantan:

(1) advection of O_3 depleted tropical maritime air mass;
(2) reduction of O_3 production efficiency due to reduction of solar radiation intensity;
(3) influence of O_3 deposition at the earth's surface due to active convection in the smoke plume;
(4) O_3 destruction by heterogeneous reactions on the aerosol surfaces.

In light of this interesting aircraft observation, more ground-based measurements of O_3 and its precursors are needed in "young" and "aged" plumes emerging from biomass burning sites so that regional atmospheric photochemistry can be characterised more comprehensively.

Acknowledgments

I thank Miroslav Radojevic for his constant encouragement, helpful comments on the chapter, and for providing preprints of his papers. Technical assistance provided by T. Victor and Qin Zhen is greatly appreciated. This research was supported by the National University of Singapore through Grant-RP980896.

REFERENCES

1. Andreae, M.O., 1983. Soot carbon and excess fine potassium: Long-term transport of combustion derived aerosols. *Science* **220**: 1148-1151.
2. Andreae, M. O., E.V. Browell, M. Garstang, G.L. Gregory, R.C. Harris, G.F. Hill, D.J. Jacob, M.C. Pereira, G.W. Sachse, A.W. Setzer, P.L. Dias Silva, R.W. Talbot, A.L. Torres and S.C. Wofsy, 1988. Biomass burning emissions and associated haze layers over Amazonia. *Journal of Geophysical Research* **93**: 1509-1527.
3. Andreae, M.O., A. Chapius, B. Cros, J. Fontan, G. Helas, C. Justice, Y.J. Kaufman, A. Minga and D. Nganga, 1992. Ozone and Aitken nuclei over equatorial Africa: Airborne observations during DECAFE 88. *Journal of Geophysical Research* **97**: 6137-6148.
4. Andreae. M.O., T. W. Andreae, H. Annegarn, J. Beer, H. Cachier, P. le Canut, W. Elbert, W. Maenhaut, I. Salma, F. G. Wienhold, and T. Zenker, 1998. Airborne studies of aerosol emissions from savanna fires in southern Africa: 2. Aerosol chemical composition, *Journal of Geophysical Research*, **103**: 32,119-32,128.
5. Amman, M., M. Kalberer, D. T. Jost, L. Tobler, E. Rossler, D. Piguet, H. W. Gaggeler, and U. Baltensperger, 1998. Heterogeneous production of nitrous acid on soot in polluted air masses. *Nature* **395**: 157-160.
6. Artaxo, P., H. Storms, F. Bruynseels, R. Van Gricken, and W. Maenhaut, 1988. Composition and sources of aerosols from the Amazon Basin. *Journal of Geophysical Research* **93**: 1605-1615.
7. Auclair, A. N. D. and T. B. Carter, 1993. Forest wildfires as a recent source of CO_2 at northern latitude, *Canadian Journal of Forest Research* **23**, 1528-1537.
8. Aumont, B., S. Madronich, M. Amman, M. Kalberer, U. Baltensperger, D. Hauglustaine, and F. Brocheton, 1999. On the NO_2 + soot reaction in the atmosphere. *Journal of Geophysical Research* **104**: 1729-1736.
9. Balasubramanian, R. Z. Qin, Z. Zhong, and T. Victor, 1998. A field study of biomass burning impacted air masses in Southeast Asia, paper presented at the *Joint International Symposium on Global Atmospheric Chemistry*, Seattle, Washington, USA.
10. Balasubramanian, R., T. Victor, and R. Begum, 1999. Impact of biomass burning on rainwater acidity and composition in Singapore. *Journal of Geophysical Research* **104**: 26,881-26,890.

11. Burrows, J. P., M. Weber, M. Buchwitz, V. Rozanov, A. Ladstatter-Weibenmyaer, A. Richter, R. DeBeek, R. Hoogen, K. Bramstedt, K-U Eichmann, and M. Eisinger, 1999. The global ozone monitoring experiment (GOME): Mission concept and first scientific results. *Journal of Atmospheric Science* **56**, 151-175.
12. Cachier, H., P. Buat-Menrad, M. Fontugne, and R. Chesslet, 1985. Source terms and source strengths of the carbonaceous aerosols in the tropics. *Journal of Atmospheric Chemistry* **3**: 469-489.
13. Cachier, H., P. Buat-Menrad, M. P. Bremond, and P. Buat-Menrad, 1992. Carbonaceous aerosols from different tropical biomass burning sources. *Nature* **359**: 812-815.
14. Calvert, J. G., G. Yarwood and A. M. Dunker, 1994. An evaluation of the mechanism of nitrous acid formation in the urban atmosphere. *Research in Chemical Intermediates* **20**: 463-502.
15. Chatfield, R. B., and A. C. Delany, 1990. Convection links biomass burning to increased tropical ozone: however, models will tend to overpredict O_3. *Journal of Geophysical Research* **95**: 18473-18488.
16. Chatfield, R.B., J. A. Vastano, H. B. Singh, and G. Sachse, 1996. A general model of how fire emissions and chemistry produce African/oceanic plumes (O_3, CO, PAN, smoke) in Trace-A. *Journal of Geophysical Research* **101**: 24,279-24,306.
17. Clark, K. L., N. M. Nadkarni, D. Schaefer, and H. L. Gholz, 1998. Cloudwater and precipitation chemistry in a tropical Montane forest, Monteverde, Costa Rica. *Atmospheric Environment* **32**: 1595-1603.
18. Clegg, S. L., and P. Brimblecombe, 1985. Potential degassing of hydrogen chloride from acidified sodium chloride droplets, *Atmospheric Environment* **20**: 1717-1725.
19. Cofer III, W. R., J. S. Levine, E. L. Winstead, B. J. Stocks, D. R. Cahoon, and J. P. Pinto, 1993. Trace gas emissions from tropical biomass fires: Yucatan Peninsula, Mexico. *Atmospheric Environment* **27A**: 1903-1907.
20. Crutzen, P.J., L.E. Heidt, J. P. Krasnec, W.H. Pollock, and W. Seiler, 1979. Biomass burning as a source of atmospheric gases CO, H_2, N_2O, NO, CH_3Cl, and COS. *Nature* **282**: 253-256.
21. Crutzen, P. J., and M. O. Andreae, 1990. Biomass burning in the tropics: Impact on atmospheric chemistry and biogeochemical cycles. *Science* **250:** 1669-1678.
22. Dixon, R. A., and O. N. Krankina, 1993. Forest fires in Russia: Carbon dioxide emissions to the atmosphere. *Canadian Journal of Forest Research* **23**: 700-706.
23. Delany, A.C., P. Haagensen, S. Walters, and A. F. Wartburg, 1985. Photochemically produced ozone in the emission from large-scale tropical vegetation fires. *Journal of Geophysical Research* **90**: 2425-2429.
24. Ferek, R. J., J. S. Reid, P. V. Hobbs, D. R. Blake, and C. Liousse, 1998. Emission factors of hydrocarbons, halocarbons, trace gases, and particles from biomass burning in Brazil. *Journal of Geophysical Research* **103**: 32,107-32,118.
25. Finlayson-Pitts, B. J., and J. N. Pitts, Jr., 2000. *Chemistry of the Upper and Lower Atmosphere*, Academic Press.

26. Fishman, J., C. E. Watson, J. C. Larsen, and J. A. Logan, 1990. Distribution of tropospheric ozone determined from satellite. *Journal of Geophysical Research* **95**: 3599-3617.
27. Fishman, J., K. Fakhruzzaman, B. Cros, and D. Ngganga, 1991. Identification of widespread pollution in the southern hemisphere deduced from satellite analyses. *Science* **252**: 1693-1696.
28. Fujiwara, M., K. Kita, T. Ogawa, N. Komala, S. Saraspriya, A. Suripto, and T. Sano, 1996. Total ozone enhancement in September and October 1994 in Indonesia, paper presented at the *XVIII Quadrennial Ozone Symposium*, International Ozone Commission, L'Aquila, Italy.
29. Fujiwara, M., M. Kita, S. Kawakami, T. Ogawa, N. Komala, S. Saraspriya, and A. Suripto, 1999. Tropospheric ozone enhancements during the Indonesian forest fire events in 1994 and in 1997 as revealed by ground-based observations. *Geophysical Research Letters* **26**: 2417-2420.
30. Garstang, M., J. Sacala, S. Greco, R. Harris, S. Beck, E. Bowell, G. Sachse, G. Gregory, G. Hill, J. Simpson, W-K. Teo, and A. Torres, 1988. Trace gases exchanges and convective transports over the Amazonian rain forest. *Journal of Geophysical Research* **93**: 1528-1550.
31. Gerecke, A., A. Thielmann, L. Gutzwiller, and M. J. Rossi, 1998. The chemical kinetics of HONO formation resulting from heterogeneous interaction of NO_2 with flame soot. *Geophysical Research Letters* **25**: 2453-2456.
32. Gras, J. L., J. B. Jensen, K. Okada, M. Ikegami, Y. Zaizen, and Y. Makino, 1999. Some optical properties of smoke aerosol in Indonesia and Tropical Australia. *Geophysical Research Letters* **26**: 1393-1396.
33. Hegg, D. A., L. F. Radke, P. V. Hobbs, and P. J. Riggan, 1988. Ammonia emissions from biomass burning. *Geophysical Research Letters* **15**: 335-337.
34. Helas, G., H. Bingemer, and M. O. Andreae, 1992. Organic acids over equatorial Africa: Results from DECAFE 88. *Journal of Geophysical Research* **97**: 6187-6193.
35. Junkermann, W., and T. Ibusuki, 1992. FTIR spectroscopic measurements of surface products of nitrogen oxides on aerosol surfaces-implications for heterogeneous HNO_2 production. *Atmospheric Environment* **26**: 3099-3103.
36. Kaufman, Y. J., A. Setzer, D. Ward, D. Tanre, B. N. Holben, P. Menzel, M. C. Pereira, and R. Rasmussen, 1992. Biomass burning airborne and spaceborne experiment in the Amazonas (Base-A). *Journal of Geophysical Research* **97**: 14,581-14,599.
37. Kaufman, Y. J., and T. Nakajima, 1993. Effect of Amazon smoke on cloud microphysics and albedo-analyses from satellite imagery. *Journal of Applied Meteorology* **32**: 729-744.
38. Kirchhoff, V. W. J. H., Y. Nakamura, E. V. A. Marinho, and M. M. Mariano, 1992. Excess ozone production in Amazonia from large scale burning. *Journal of Atmopheric and Terrestrial Physics* **54**: 583-588.

39. Kirchhoff, V. W. J. H., and E. V. A. Marinho, 1994. Layer enhancements of tropospheric ozone in regions of biomass burning. *Atmospheric Environment* **28**: 69-74.
40. Kirchhoff, V. W. J. H., and R. A. Rasmussen, 1990. Time variations of CO and O_3 concentrations in a region subject to biomass burning. *Journal of Geophysical Research* **95**: 7521-7532.
41. Komala, N., S. Saraspriya, K. Kita, and T. Ogawa, 1996. Tropospheric ozone behavior observed in Indonesia. *Atmospheric Environment* **30**: 1851-1856.
42. Koppmann, R., A. Khedim, J. Rudolph, D. Poppe, M.O. Andreae, G. Helas, M. Welling, and T. Zenker, 1997. Emissions of organic trace gases from savanna fires in southern Africa during the 1992 Southern African Fire Atmosphere Research Initiative and their impact on the formation of tropospheric ozone. *Journal of Geophysical Research* **102**: 18,879-18,888.
43. Kuhlbusch, T. A., J. M. Lobert, P. J. Crutzen, and P. Warneck, 1991. Molecular nitrogen emissions from denitrification during biomass burning. *Nature* 351, 135-137.
44. Lacaux, J. P., J. Servant, and J. G. R. Baudet, 1987. Acid rain in the tropical forests of the Ivory Coast. *Atmospheric Environment* **21**: 2643-2647.
45. Lacaux, J. P., J. Servant, M. L. Huertas, B. Cros, and R. Delmas, 1988. Precipitation chemistry from remote sites in the African equatorial forest. *EOS* **69**: p. 1069.
46. Lacaux, J. P., J. Loemba-Ndembi, B. Lefeivre, B. Cros, and R. Delmas, 1992. Biogenic emissions and biomass burning influences on the chemistry of the chemistry of the fogwater and stratiform precipitations in the African equatorial forest. *Atmospheric Environment* **26A**: 541-551.
47. Lacaux, J.P., H. Cachier, and R. Delmas, 1993. Biomass burning in Africa: An overview of its impact on atmospheric chemistry. In: *Fire in the Environment: The ecological, atmospheric, and climatic importance of vegetation fires,* (P.J. Crutzen and J. G. Goldammer, Eds.) John Wiley and Sons, Chichester, pp. 159-191.
48. Lee, M., and B. G. Heokes, 1997. Hydrogen peroxide, methyl hydroperoxide, and formaldehyde as primary pollutants from biomass burning. *Journal of Geophysical Research* **102**: 1301-1309.
49. Lefer, B. L. et al., 1994. Enhancement of acidic gases in biomass burning impacted air masses over Canada. *Journal of Geophysical Research* **99**: 1721-1737.
50. Leslie, A.C.D., 1981. Aerosol emissions from forest and grassland burning in the southern Amazon Basin and central Brazil. *Nuclear Instruments and Methods in Physics Research, Section B* **181**: 345-351.
51. Liew, S.C., O. K. Lim, L. K. Kwoh, and H. Lim, 1998. A study of the 1997 fires in Southeast Asia using SPOT quicklook mosaics. Paper presented at *The 1998 International Geoscience and Remote Sensing Symposium*, July 6-10, 1998, Seattle, Washington, 3 pages.
52. Lobert, J. M., and J. Warnatz, 1993. Emissions from the combustion process in vegetation. In: *Fire in the Environment: The Ecological, Atmospheric, and Climatic*

Importance of Vegetation Fires, (P.J. Crutzen and P. J. Goldammer, Eds.), John Wiley, Chichester, pp. 15-37.

53. Lobert, J. M., D. H. Scharffe, W.-M. Hao, T. A. Kuhlbusch, R. Seuwen, P. Warneck, and P. J. Crutzen, 1991. Experimental evaluation of biomass burning emissions: Nitrogen and carbon containing compounds. In: *Global Biomass Burning: Atmospheric, Climatic, and Biospheric Implications* (J. S. Levine, Ed.), MIT Press, Cambridge, Mass. pp. 289-304.
54. Muraleedharan T.R. and M. Radojevic, 2000. Personal particle exposure monitoring using nephelometry during haze in Brunei. *Atmospheric Environment* **34**: 2733-2738.
55. Muraleedharan T.R., M. Radojevic, A. Waugh, and A. Caruana, 2000a. Emissions from the combustion of peat: an experimental study. *Atmospheric Environment* **34**: 3033-3035.
56. Muraleedharan T.R., M. Radojevic, A. Waugh. and A. Caruana, 2000b. Chemical characterisation of the haze in Brunei Darussalam during the 1998 episode. *Atmospheric Environment* **34**: 2725-2732.
57. Nakajima, T., and A. Higurashi, 1998. Satellite and ground-based study of optical propeties of 1997 Indonesian forest fire aerosols. In *An Environmental study of the Indonesian Fire Smoke Disaster,* Center for the Climate Systems Research, The University of Tokyo, Japan, p17.
58. Nichol J. 1997. Bioclimatic impacts of the 1994 smoke haze event in Southeast Asia. *Atmospheric Environment* **31**: 1209-1219.
59. Novakov, T., and C. E. Corrigen, 1996. Thermal characterisation of biomass smoke particles. *Mikrochemica Acta*: **119**: 157-166
60. Olson, J., M. Prather, T. Berntsen, G. Carmichael, R. Chatfield, P. Connell, R. Derwent, L. Horowitz, S. Jin, M. Kanakidou, P. Kasibhatla, R. Kotamarthi, M. Kuhn, K. Law, J. Penner, L. Perlikski, S. Sillman, F. Stordal, A. Thompson, and O. Wild, 1997. Results from the Intergovernmental Panel on Climate Change Photochemical Model Intercomparison (PhotoComp). *Journal of Geophysical Research* **101**, 5979-5991.
61. Radke, L. F., 1989. Airborne observations of cloud microphysics modified by anthropogenic forcing. In: *Symposium on Atmospheric Chemistry and Global Climate*, American Meteorological Society, Boston, Mass.
62. Radojevic M. and K.S. Tan, 2000. Impacts of biomass burning and regional haze on the pH of rainwater in Brunei Darussalam. *Atmospheric Environment* **34**: 2739-2744.
63. Reid, J. S., P. V. Hobbs, R. J. Ferek, D. R. Blake, J. V. Martins, M. R. Dunlap, and C. Liousse, 1998. Physical, chemical, and optical properties of regional hazes dominated by smoke in Brazil. *Journal of Geophysical Research* **103**: 32,059-32,080.
64. Rindam M., 1995. The haze phenomenon in Malaysia: problems and challenges. In: *Proceedings of the International Symposium on Climate and Life in the Asia Pacific*, (K.U. Sirinanda, Ed.) 10-13 April 1995, University of Brunei Darussalam, Bandar Seri Begawan, Brunei Darussalam, pp. 74-84.

65. Rondon, A., and E. Sanhueza, 1989. High HONO atmospheric concentration during vegetation burning in the tropical savanna. *Tellus* **41B**: 474-477.
66. Rudolph, J. A., A. Khedim, and B. Bonsang, 1992. Light hydrocarbons in the tropospheric boundary layer over tropical Africa. *Journal of Geophysical Research* **97**: 6181-6186.
67. Sakamaki, F. and H. Akimoto, 1988. HONO formation as unknown radical source in photochemical smog chamber. *International Journal of Chemical Kinetics* **20**: 11-116.
68. Sanhueza, E., W. Elbert, A. Rondon, M. Corina Arias, and M. Hermoso, 1989. Organic and inorganic acids in rain from a remote site of the Venezuelan savannah. *Tellus* **41B**, 170-176.
69. Sanhueza, E., Arias, M.C., Donoso, L., Graterol, N., Hermoso, M., Marti, I., Romero, J., Rondon, A., and Santana, M. 1992. Chemical composition of acid rains in the Venezuelan savannah region. *Tellus* **44B**: 54-62.
70. Sani S., C. B. Khan, L. C. Peng, and L. S. Fook, 1991. *The August 1990 Haze in Malaysia with Special Reference to the Klang Valley Region, Technical Note No. 49*, Malaysian Meteorological Service, Petaling Jaya, Malaysia.
71. Soleiman A., M. Othman, A. A. Shamah, N. M. Suleiman, and M. Radojevic, 2000. The occurrence of haze in Malaysia: a case study in an urban industrial area, Klang Valley. Submitted to *Pure and Applied Geophysics*.
72. Svensson, R., E. Ljungstrom, and O. Lindquist, 1987. Kinetics of reaction between nitrogen dioxide and water vapour. *Atmospheric Environment* **21**: 1529-1539.
73. Talbot, R. W., M. O. Andreae, T. W. Andreae, and R. C. Harris, 1988. Regional aerosol chemistry of the Amazon basin during the dry season. *Journal of Geophysical Research* **93**: 1499-1508.
74. Talbot, R. W., J. E. Dibb, K. I. Klemm, J. D. Bradshaw, S. T. Sandholm, D. R. Blake, G. W. Sachse, J. Collins, B. G. Heikes, G. L. Gregory, B. E. Anderson, H. B. Singh, D. C. Thornton, and J. T. Merril, 1996. Chemical characteristics of continental outflow over the tropical South Atlantic Ocean from Brazil and Africa. *Journal of Geophysical Research* **101**: 24,187-24,202.
75. Thomas, W., E. Hegels, S. Slijkhuis, R. Spurr, and K. Chance, 1998. Detection of biomass burning combustion products in Southeast Asia backscatter data taken by the GOME spectrometer. *Geophysical Research Letters* **25**: 1317-1320.
76. Tsutsumi, Y., Y. Sawa, Y. Makino, J.B. Jensen, J.L. Grass, B.F. Ryan, S. Diharto, and H. Harjanto, 1999. Aircraft measurements of ozone, NO_x, CO, and aerosol concentrations in biomass burning smoke over Indonesia and Australia in October 1997: Depleted ozone layer at low altitude over Indonesia. *Geophysical Research Letters* **26**: 595-598.
77. Tussin A.M. 1995. Thick haze in Malaysia in the 90's. In: *Proceedings of the International Symposium on Climate and Life in the Asia Pacific*, (K.U.Sirinanda, Ed.) 10-13 April 1995, University of Brunei Darussalam, Bandar Seri Begawan, Brunei Darussalam, pp. 56-73.
78. Vong, R. J., 1990. Mid-latitude Northern Hemisphere background sulfate concentration in rainwater. *Atmopheric Environment* **24A**: 1007-1018.

79. Ward D.E., R.A. Susott., J.B. Kaufmann, R.E. Babbitt, D.L. Cummings, B. Dias, B.N. Holben, Y.J. Kaufman, R.A. Rasmussen, and A.W. Setzer, 1992. Smoke and fire characteristics for Cerrado and deforestation burns in Brazil-BASE-B Experiment. *Journal of Geophysical Research* **97**: 14,601-14,619.

80. Weisen, P., J. Kleffmann, R. Kurtenbach, and K. H. Becker, 1995. Mechanistic study of the heterogeneous conversion of NO_2 into HONO and N_2O on acid surfaces.*Faraday Discussions* **100**: 121-127.

81. Yokelson, R. J., D. W. T. Griffith, J. B. Burkholder, and D. E. Ward, 1996. Open-path Fourier transform infrared studies of large-scale laboratory biomass fires. *Journal of Geophysical Research* **101**: 21,067-21,080.

82. Yokelson, R. J., R. Susott, D. E Ward, J. Reardon, and D. W. T. Griffith, 1997. Emissions from smouldering combustion of biomass measured by open-path Fourier transform infrared spectroscopy. *Journal of Geophysical Research* **101**: 18,865-18,877.

Chapter 8

FIRES AND SMOKE: EFFECTS ON TROPICAL RAIN FORESTS IN SOUTHEAST ASIA

Stuart J. Davies

Institute of Biodiversity and Environmental Conservation, Universiti Malaysia Sarawak, 94300 Kota Samarahan, Sarawak, Malaysia

and

Center for International Development, Harvard University, Cambridge, MA 02138, USA

INTRODUCTION

Deforestation in the tropics has the potential to significantly affect atmospheric chemical composition, biogeochemical cycles, global climate, and consequently ecosystem diversity and functioning (Myers, 1988; Bazzaz, 1990; Kauffman *et al.*, 1993). Tropical deforestation continues at the unprecedented worldwide rate of about 15.4 x 10^6 ha yr^{-1} (Dixon *et al.,* 1994; Houghton, 1995). For Southeast Asia[1] recent estimates of deforestation rates range from 0.95 to 2.30 x 10^6 ha yr^{-1} (Myers, 1991; Houghton and Hackler, 1994). In the past century, the area of forest and wetland in Southeast Asia[2] declined by 30%, a reduction of 69 x 10^6 ha (Richards and Flint, 1994). Most of this deforestation was due to agricultural expansion (Palm *et al.* 1986, Richards and Flint, 1994). Associated with increases in agricultural area, biomass burning has been widely practiced to release nutrients for shifting cultivation cycles, and to clear biomass for permanent agriculture and development activities (Crutzen and Andreae, 1990; Tinker *et al.*, 1996).

In recent decades, large-scale human-induced tropical forest fires have become common in Amazonia, Africa and Southeast Asia (Malingreau *et al.,* 1985; Setzer and

---

[1] Area for this estimate includes Indonesia, Malaysia, Philippines and Papua New Guinea.

[2] Excluding New Guinea.

Pereira, 1991; Isichei *et al.,* 1995). The environmental consequences of tropical biomass burning are immense. Deforestation and habitat fragmentation result in the loss of biodiversity and forest resources, the erosion of soil nutrients, the pollution of waterways, and changes in the ability of the ecosystems to supply humanity with services such as climate regulation, catchment protection, and radiation absorption and reflection (Lovejoy, 1995). Furthermore tropical biomass burning contributes about 22% of annual global CO_2 emissions (Dale, 1997), as well as substantial quantities of other radiatively active greenhouse gases (e.g., CH_4), and atmospheric pollutants (e.g. SO_2) (Crutzen and Andreae, 1990). The impact of these emissions may result in climate change that will most likely alter the area and distribution of tropical rain forest.

This chapter analyzes the effects of fire and the resulting smoke-haze on tropical rain forests in Southeast Asia, focussing on aseasonal forests of the Sunda Shelf. Initially, the impact of fire on tropical forests is considered, including an assessment of the history of fire in wet tropical areas, the effect of fire on plant survival, and the factors controlling forest regeneration after fire. The second part of the chapter focuses on the more indirect effects of fire on remaining undisturbed tropical forests. The impacts of fire on intact forests are considered to occur as a consequence of deforestation and fragmentation, and through the release of gaseous and particulate emissions. In many cases research and information do not exist that is specifically relevant to understanding how tropical forests respond to biomass burning, so studies from temperate ecosystems were used to assess potential effects on tropical forests. This is in many ways inadequate and emphasizes the urgent need for detailed research on many of the issues discussed in this chapter. Several recent studies have assessed the role of fire in seasonal and naturally fire-prone tropical forests (e.g. Stott *et al.*, 1990), and they are not considered here. The impacts of fire on forest animal communities are also not discussed, but the limited available evidence suggests that animal communities are seriously affected by forest fire and smoke emissions (Mackinnon *et al.,* 1996, O'Brien *et al.,* 1998).

EFFECTS OF FIRE ON TROPICAL RAINFORESTS

Fire in Tropical Rainforest

Charcoal remains in soil cores from east Kalimantan (Goldammer and Seibert, 1990) and Amazonia (Sanford *et al.,* 1985; Fearnside, 1990) indicate that fires have occurred in what are now wet tropical areas, over at least the past 20,000 years. The extent to which these forest fires were natural (e.g. from lightning strikes) or due to human activities is less well known. However, it is clear that the fires had very long return intervals, as the ecology of the forests and the autecology of the species indicate that fire is not an integral component of these ecosystems, but rather an extreme disturbance event (Kauffman and Uhl, 1990). Nevertheless large forest fires have occurred in Borneo throughout this century (Goldammer and Seibert, 1990), and more recently have come to have an enormous impact on regional ecological, atmospheric and biogeochemical processes. The area burned in recent fires in Southeast Asia is immense (see also Setzer and Pereira, 1991 for Amazonia). In 1982-3, approximately 3.5 million ha burned in east Kalimantan (Malingreau *et al.,* 1985; Mackinnon *et al.,*1996); in 1994 a similarly large area burned in Borneo and Sumatra (Yanhong *et al.,* 1996; Nichol, 1997); and most recently, in September and October 1997 fires in Kalimantan and Sumatra burned an estimated 0.3 - 1.7 million ha (WWF and FAO estimated figures).

Although high rainfall and humidity typically exclude fire from wet tropical rain forests, extreme climatic oscillations coupled with changes in land-use patterns and habitat fragmentation have greatly increased the vulnerability of forests to fire (Uhl and Kaufmann, 1990; Whitmore and Burslem, 1998). El Niño Southern Oscillation (ENSO) events result in dramatically reduced rainfall in Southeast Asia leading to drought stress for forest plants (Becker, 1992; Becker and Wong, 1993; Walsh, 1996). Drought may lead to increased litterfall from water-limited plants and unusual levels of drying of the forest floor litter (Mueller-Dombois and Goldammer, 1990). This effect may be most evident in forests on extreme soils, such as tropical heath (*Kerangas*) forests, with naturally deep leaf litter, less humid conditions, and low decomposer activity (Mueller-Dombois and Goldammer, 1990; Becker and Wong, 1992). More importantly however, degradation and fragmentation of forest resources has been the most significant catalyst for increased forest fire ignitions. Selectively logged forests have significantly greater fuel mass on the forest floor which, coupled with higher temperatures, lower relative humidity, and higher vapor pressure deficits and fuel drying rates, leads to greatly increased chances of forest fire ignition (Kauffman and Uhl, 1990; Uhl and Kaufmann, 1990). Furthermore, the fragmentation of once continuous forest areas creates a mosaic of primary forest interspersed with fire-prone degraded and cleared forests (Skole and Tucker, 1993), greatly increasing the susceptibility of primary forest fragments to fire and other damage (Laurance *et al.,* 1998). It was estimated that only 16% of the burned area in the 1982-3 Borneo fires was primary forest (Woods, 1989; Goldammer and Seibert, 1990). Extensive habitat fragmentation and degradation, coupled with unusually dry weather associated with ENSO events, has been the major cause of the huge forest fires in Borneo and Sumatra in recent years.

Fire Effects on Tropical Tree Survival

Tropical rain forests are dynamic ecosystems that experience spatially and temporally unpredictable disturbances which range in size from branch-falls that create small canopy openings, to pathogen outbreaks or hurricanes that open up large tracts of forest (Denslow, 1987; Bazzaz and Wayne, 1994; Zimmerman *et al.*, 1994). Rain forest species and communities have evolved in the context of these disturbances and respond through successional processes to reorganize and re-establish following disturbance (Holling *et al.,* 1995). Early successional species have a suite of ecophysiological and life-history characteristics (e.g. small, well-dispersed seeds, light induced seed germination, and high photosynthetic and growth rates) that facilitate access to the flux of resources released following natural disturbance (Bazzaz, 1991; Davies, 1998). Similarly, plants in normally fire-prone ecosystems have a suite of traits (e.g., fire-tolerant stems, epicormic sprouting, and fire-induced seed germination) that enable them to tolerate or avoid fire (Stott *et al.*, 1990; Whelan, 1995). However, plant species from aseasonal or ever-wet tropical rain forests do not possess traits to deal with fires and consequently their effects on community structure and composition may be substantial (Kauffman and Uhl, 1990).

The combined effects of the ENSO-induced drought and fire in east Kalimantan in 1982-3 were devastating to rain forest plant survival, particularly in degraded forests (Woods, 1989; Goldammer and Seibert, 1990). Several studies found extremely high fire-induced mortality rates in tree seedlings and saplings in drought-affected primary forest; Leighton (1984 cited in Goldammer and Seibert, 1990) reported 99% mortality for saplings <4 cm dbh, Woods (1989) reported >80% mortality for saplings 2.5-10 cm dbh, and Wirawan (1983, see Goldammer and Seibert, 1990) found 75% mortality for saplings <5 cm dbh in primary ironwood (*Eusideroxylon*) forest. For mature trees in primary forest, fire-induced mortality rates were generally lower than for saplings and declined with increasing tree size, but still ranged from 20-35% (Leighton, 1984), 8-15% (Wirawan, 1983; Goldammer and Seibert, 1990), and 19-71% (Woods, 1989). In all cases burned forests had two to three times higher mortality than forests subjected to drought only.

The degree of fire-induced forest damage from the 1982-3 fires was strongly related to land use history, and also varied according to forest type (see also Uhl and Kaufmann, 1990). In contrast to primary forests, damage was much more severe in logged forests (Woods,1989; Goldammer and Seibert, 1990); when fires were intense enough to burn the upper canopy forest composition was greatly changed. Peat forests, which dried out due to the drought, suffered extremely high mortality rates, whereas *Eusideroxylon* forests were apparently less sensitive to drought and readily resprouted after fire. Kauffman and Uhl (1990) suggested that post-fire survival of tropical rain forest trees depends on one of several possible "fortuitous adaptations" such as thick bark, anomalous stem tissue arrangements (e.g., well protected apical meristems, as in palms) or vegetative sprouting, that may have evolved in these species for reasons independent of a history of exposure to fire. Several Southeast Asian rain forest species have been reported to be able to survive fire, including the Bornean ironwood, *Eusideroxylon zwageri*, as well as some species of the genera *Eugenia* and *Diospyros* (Goldammer and Seibert, 1990), however no systematic studies have been undertaken to examine floristic composition of resprouting species in south-east Asian wet forests. Resprouting ability is important as it in part determines forest composition following fire.

Regeneration after Fire

The pattern of plant regeneration in aseasonal tropical rain forests following fire has a huge impact on subsequent forest structure and composition, and hence local and regional levels of biodiversity. Rapid regeneration of native pioneer trees will limit the degree of weed invasion, minimize native species loss (Mueller-Dombois and Goldammer, 1990), limit the extent to which soil processes are affected by erosion, and lessen deforestation impacts on local stream ecosystems (Shukla *et al.,* 1990). The regeneration of tropical rain forests following fire is also important through its impact on atmospheric and ecosystem processes (see discussion below). The vegetation that regenerates on burned land will determine the net contribution of carbon released from the fires to greenhouse gas emissions (Houghton *et al.*, 1991; Fearnside, 1996).

The prospects for forest regeneration following fire depend on a wide range of factors occurring before, during and after the fire, including site disturbance history, fire characteristics, propagule availability, degree of weed invasion, as well as the post-fire climate and environment confronting the regenerating forest (Table 1; see also Finegan, 1996). The disturbance history of a forest has a significant impact on the potential for forest regeneration (Goldammer and Seibert, 1990; Nykvist, 1996). After relatively light selective logging, vegetation may regenerate quite rapidly following low intensity fire, even though forest composition will be significantly modified (Uhl, 1988; Nykvist, 1996). Approximately two years after a forest in East Kalimantan had been burned, logged plots had much lower diversity and density of regenerating tree seedlings and saplings than unlogged plots, whereas grass and fern densities were much higher in logged plots (Woods, 1989). Furthermore, cleared and burned forests, for example following shifting cultivation, have extremely low post-fire plant diversity and very high fern and grass cover (Goldammer and Seibert, 1990; Riswan and Kartawinata, 1991). Cleared and burned forests are also particularly susceptible to invasion by weedy grass species (e.g., *Imperata cylindrica*), or invasive tree species (e.g., *Acacia auriculiformis*, Davies *et al.*, 1996) which may promote repeated fires that permanently exclude native tree species, resulting in a 'deflected' succession of unproductive and low diversity vegetation (D'Antonio and Vitousek, 1992).

Fire characteristics also strongly influence the course of post-fire succession. Intense fires may have a greater effect on post-fire conditions than low intensity fires by greatly changing forest structure, through killing mature trees (Goldammer and Seibert, 1990). Intense fires may combust the soil's humus layer, exposing the soil to erosion and subsequent nutrient loss. Intense fires may also deplete the soil seed bank or soil mycorrhizal populations, however few studies have addressed these issues specifically for Southeast Asian wet tropical rain forests. Successional processes in burned forests differ substantially from unburned forests as much of the seed- and seedling-bank may be removed by fire, and regeneration must therefore rely on dispersal of propagules from unburned areas (Woods, 1989). Even though the soil seed bank may not be completely depleted, the composition of the soil seed bank may be changed in favor of particular species (e.g. hard-seeded species), thereby changing the course of post-fire succession. Post-fire climatic conditions also influence the prospects for forest regeneration by influencing the likelihood of subsequent fires which may be more likely early in succession when temperatures are higher and humidity levels are lower.

Table 1. Summary of factors influencing the regeneration of tropical rain forest species following fire.

• site disturbance history:
- logging damage
- shifting cultivation
- extent of soil erosion and nutrient loss

• fire characteristics:
- intensity and size of fire
- impact on soil nutrient loss
- impact on soil microorganisms (e.g., mycorrhizae)

• availability of propagules:
- proximity to forest edge
- composition of surrounding vegetation
- composition and survival of soil seed bank
- seedling/sapling survival or resprouting ability

• weed invasion:
- degree of invasion by fire-promoting or fire-tolerant weed (e.g. *Imperata cylindrica*) or plantation (*Eucalyptus* spp.) species
- degree of invasion by regeneration-inhibiting species (e.g. fern species, *Dicranopteris linearis*)

• post-fire climate and environment:
- effect on fuel moisture content and further ignition sources

Effects of Fire on Remaining Undisturbed Tropical Rainforests

The effect of fire on remaining unburned tropical rain forests and their plants is mediated through two major environmental changes: (i) deforestation and the resulting fragmentation of once continuous ecosystems; and (ii) the release of large quantities of CO_2, pollutants, aerosols, and other radiatively active greenhouse gases to the atmosphere. Both of these changes have significant and in some cases irreversible impacts on soil hydrology, atmospheric moisture, nutrient cycling, soil nutrient availability, atmospheric composition and climate that may significantly affect the diversity and functioning of remaining tropical forests (Fosberg *et al.*, 1990; Mueller-Dombois and Goldammer, 1990; Hobbie *et al.*, 1994).

Deforestation and Forest Fragmentation Following Forest Fire

Biodiversity Loss

Tropical forest fires often lead to land clearing for agriculture or development projects, creating a mosaic of remaining forest fragments amidst a sea of cleared land (Skole and Tucker, 1993). Absolute reductions in tropical forest area obviously lead to significant species loss (reviewed by Turner, 1996). However these effects are not uniquely due to fragmentation caused by fire but pertain to all modes of ecosystem fragmentation. Raven (1988) estimated that 25,000 plant species will become extinct by the end of this century, largely due to deforestation in tropical areas.

Species persisting in forest fragments often suffer a wide range of adverse effects that may decrease survival rates and increase their probability of extinction. Forest edge effects, reduced population sizes, reduced genetic diversity, exposure to invasion by weed species, absence of large mammals and birds, and vulnerability to further forest degradation may all change the diversity and functioning of remaining communities (Laurance, 1994; Turner, 1996; Laurance *et al.*, 1998).

Climatic Effects of Deforestation

Deforestation and fragmentation can affect remaining forest ecosystems through their impact on climate. Deforestation is projected to affect climatic patterns through the release of greenhouse gases (see below), however the direct effects of forest removal on water and energy budgets may have an even more significant effect on local and regional climatic patterns (Houghton, 1995; Miller *et al.,* 1995). Conversion of tropical forest to grassland leads to reduced evapotranspiration due to a combination of increased albedo and less radiant energy absorbed by the vegetation, greater diffusive resistance of the vegetation (grassland) to water vapor exchange, and shallower rooting depths (Houghton, 1995). Reduced evapotranspiration leads to reduced rainfall and higher temperatures, leading to fewer clouds and more sunshine, reinforcing the heating effect (Shukla *et al.*, 1990; Nobre *et al.*, 1991; Fearnside, 1995; Tinker *et al.*, 1996). If local rainfall is largely the result of locally evaporated water then changes in forest cover will undoubtedly affect rainfall through reduced evapotranspiration. However, Tinker *et al.* (1996) caution that this effect may be weaker in insular Southeast Asia where oceanic influences on water vapor exchange are important. Nevertheless, the potential for significant impact of deforestation on local and regional climates is cause for serious concern.

Climatic changes affect forest ecosystems through changed growth conditions, which may feed-back on climatic conditions through changes in evapotranspiration and disturbance regimes resulting in permanent changes in climate (Tinker *et al.*, 1996). Reduced rainfall, increased seasonality and increased temperatures are projected climate changes for currently aseasonal tropical forest areas (Hulme and Viner, 1998). These changes may strongly affect tropical rain forest structure and composition, in the direction of a more drought tolerant and fire-prone community. Whitmore (1998) suggested that the economically important red-meranti species of *Shorea* in the Dipterocarpaceae would be particularly sensitive to these climate changes through their low drought tolerance, leading to a contraction of their current geographic distribution. Phenological patterns of individual trees are also likely to be strongly affected by such climatic changes; flowering in the mast fruiting Dipterocarpaceae may be triggered more frequently under more seasonal climates (Borchert, 1998). Fire-induced deforestation and concomitant climate changes would likely further decrease the area of everwet tropical forests in which a large fraction of the world's biodiversity resides.

Effect of Forest Fire Emissions on Intact Tropical Rain Forests

Large scale tropical forest fires lead to the release of giant smoke and haze clouds over tropical forest areas (Setzer and Pereira, 1991; Yanhong *et al.*, 1996; Nichol, 1997). These clouds contain gaseous and particulate emissions with the potential to drastically alter regional atmospheric, climatic, hydrological and biogeochemical processes (Crutzen and Andreae, 1990; Houghton *et al.*, 1992), thereby having a significant impact on remaining tropical rain forest ecosystems in the medium- to long-term. In addition, these emissions may have significant short-term impacts on forest functioning through the direct effects of gaseous exposure and deposition on forest gas exchange, water relations and nutrient dynamics.

Carbon Dioxide

Tropical forest fires emit huge quantities of CO_2 which contribute significantly to global warming through the 'greenhouse effect' (Houghton *et al.,* 1992; Dale, 1994). After fossil fuel burning in industrialized countries, biomass burning in the tropics is the largest contributor to global carbon emissions (Figure 1, Dale, 1997). Changed global climates may have a profound affect on the distribution of the world's remaining vegetation (e.g. Fosberg *et al.*, 1990, see above). The total CO_2 released from the recent Southeast Asian forest fires is likely to have been significant (Nichol, 1997; see also Setzer and Pereira, 1991), however the net impact on atmospheric carbon budgets depends on the total area of forest burned, the biomass loadings of the forest, and the vegetation that is replanted or naturally regenerates following the fires. Natural regrowth or plantations typically have lower carbon storage than the original forests (Houghton *et al.,* 1991; Fearnside, 1996), so these fires will most likely have contributed significantly to atmospheric CO_2 levels. The fires will also have resulted in significant net emissions of other important green-house gases, including methane and carbon monoxide (Houghton, 1991); the potential direct effects of these gases on remaining tropical forests are unknown.

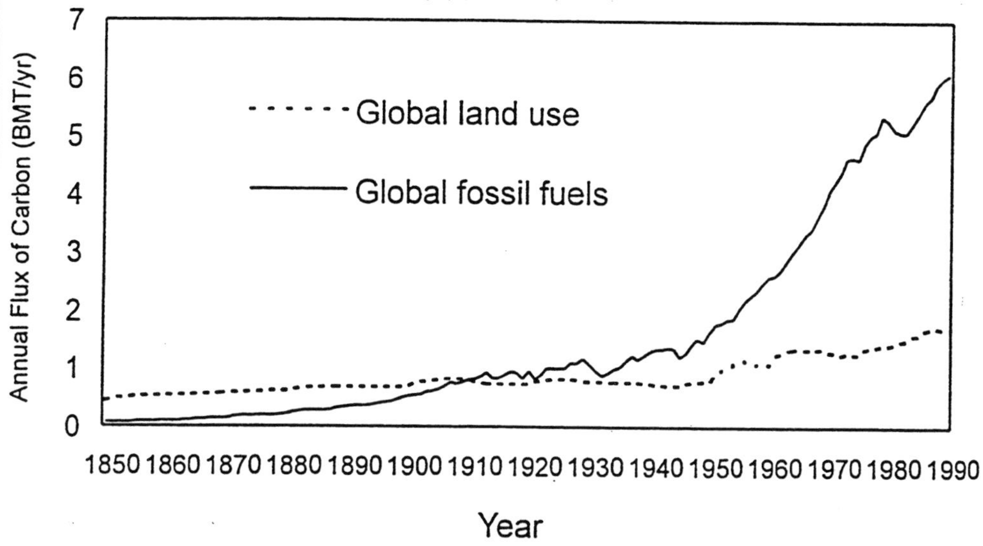

Figure 1 Estimated annual global emissions of carbon in billion metric tons (BMT) due to fossil fuel combustion and changes in land-use. (From Houghton, 1995).

Atmospheric CO_2 concentrations have steadily increased this century to current levels of *ca.* 355 ppm (Figure 2), due to fossil fuel burning and tropical deforestation (Figure 1). As the substrate for photosynthesis, increased atmospheric CO_2 concentrations can significantly alter plant physiology, growth and biomass allocation patterns, and thereby potentially influence community structure and functioning (Bazzaz, 1990; Korner, 1993; Ceulemans and Mousseau, 1994). For tropical plants grown individually at experimentally elevated CO_2 concentrations, several studies have shown that photosynthesis and growth are increased at high CO_2 levels, suggesting that productivity (carbon sequestration) may be higher under high CO_2 conditions (e.g., Ziska *et al.*, 1991; Eamus *et al.*, 1993; although see Oberbauer *et al.*, 1985; Reekie and Bazzaz, 1989). For simple experimental communities of tropical plants, carbon sequestration has been shown to be only slightly higher in elevated CO_2 environments (Korner and Arnone, 1992; Arnone and Korner, 1995). However, these experimental studies have shown that elevated CO_2 significantly changed community composition. Further study of community-level responses to elevated CO_2 in tropical forests is required.

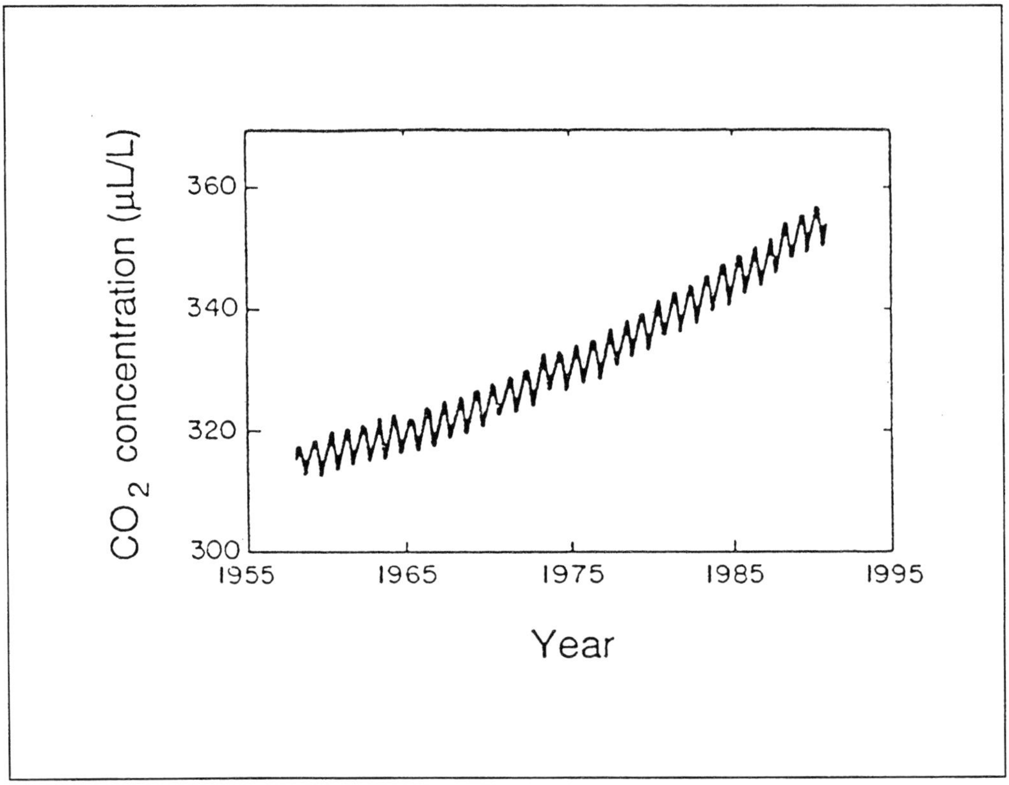

Figure 2 Changes in the atmospheric concentration of carbon dioxide (CO_2) measured in Hawaii from 1960-1990. (From Vitousek, 1994).

In a recent study conducted in Sarawak during the 1997 forest fires in Kalimantan and Sumatra, atmospheric CO_2 levels were elevated by 17-28% by the extensive smoke-haze cloud that covered the region for several weeks (Figure 3; Davies and Unam, 1999). Somewhat unexpectedly, photosynthetic rates in saplings of three economically important timber species were significantly lower on high CO_2 days, than on days without smoke-haze and with normal CO_2 levels. This result could be partly explained by significantly lower levels of photosynthetically active radiation (PAR), higher humidities and lower temperatures on hazy days (Figure 3, see also Yanhong *et al.,* 1996). However, even when light levels were controlled for, photosynthetic rates were still suppressed by the smoke-haze. It was suggested that the abnormally elevated pollutant and aerosol concentrations that also occurred during the smoke-haze, particularly SO_2 levels and particulate concentrations, may have had a significant negative effect on photosynthetic rates (Davies and Unam, 1999; see below). Importantly, the pulse of CO_2 released following the fires did not lead to a short-term increase in carbon fixation, and hence a possible negative feedback with respect to the total CO_2 released due to the fires. Tropical deforestation and biomass burning lead to changes in the atmospheric concentrations of a wide range of gases and pollutants that may affect plant physiology and growth (Crutzen and Andreae, 1990; see below); understanding the effects of interactions between these factors on plants will be important for making predictions on how tropical rain forests will respond to frequent large-scale forest fires.

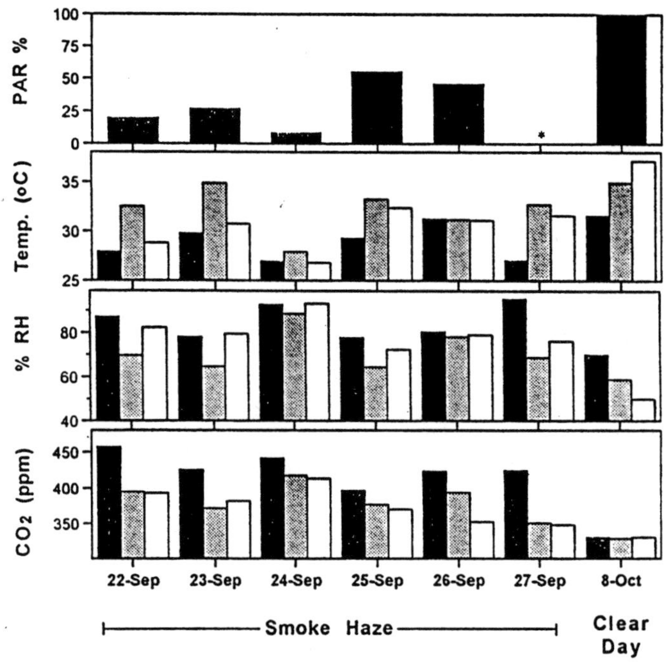

Figure 3 Local changes in climate (photosynthetically active radiation, PAR; ambient temperature; and relative humidity, RH) and atmospheric concentrations of CO_2 measured in Kuching, Sarawak, Malaysia, due to the smoke-haze cloud produced by the 1997 forest fires in Sumatra and Kalimantan. Measurements were taken on six hazy days and one clear day. For temperature, relative humidity and CO_2, measurements are shown for morning (black bars), midday (grey) and afternoon (white) of each day. Modified from Davies and Unam (1998).

Smoke and Aerosols

Large quantities of smoke and aerosols are released by tropical biomass burning (Crutzen and Andreae, 1990). Smoke-haze clouds contain significant quantities of particulate matter, which may be distributed over wide geographical areas (Setzer and Pereira, 1991). Following the 1997 Southeast Asian forest fires, the atmospheric concentration of particulate matter (<10 μm) was elevated to >990 μg m^{-3} in Kuching (*ca.* 300 km from the nearest fires; Davies and Unam, 1999). This is over twenty times higher than normal atmospheric particulate matter levels of 35-50 μg m^{-3} (Figure 4), much higher than levels reported for other Southeast Asian smoke-haze episodes (Nichol,

1997), and is even higher than levels recorded in Riyadh following emissions arising from the 1991 Gulf War (El-Shobokshy and Al-Saedi, 1993).

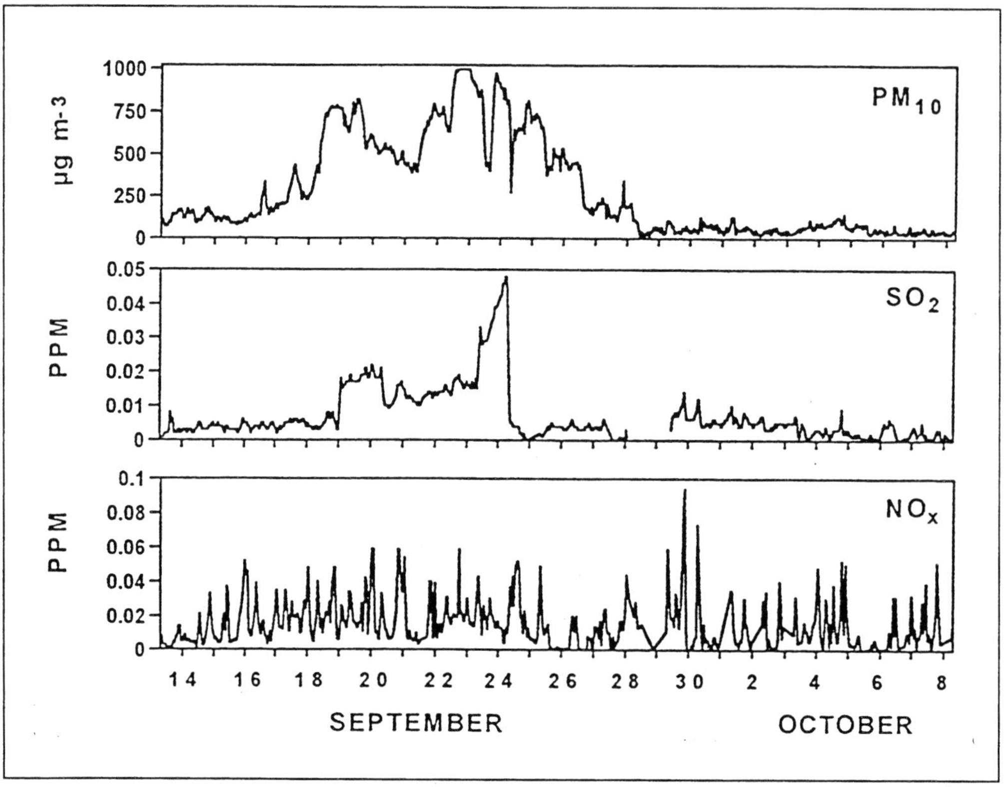

Figure 4 Changes in atmospheric composition of particulate matter (PM_{10}, <10 µg m^{-3}), sulfur dioxide (SO_2) and nitrogen oxides (NO_x) in Kuching, Malaysia, in September and early October 1997 encompassing the period of smoke-haze produced by the 1997 forest fires in Sumatra and Kalimantan. Modified from Davies and Unam (1998).

Aerosols emanating from forest fires may affect climate by changing the earth's radiation balance (Preining, 1993). Aerosols reflect and absorb solar radiation leading to less solar heating of the earth's surface, and significantly changed meteorological conditions (Crutzen and Andreae, 1990; Houghton *et al*., 1992; Tinker *et al*., 1996). Significantly reduced ambient temperatures under smoke-haze clouds were reported following the 1997 forest fires (Davies and Unam, 1999). In addition, smoke particles may lead to changes in rainfall and cloud formation patterns which could significantly alter regional climates (Crutzen and Andreae, 1990). The probable effect of these changes on tropical forest ecosystems is at best speculative; several possible consequences of climate change on tropical forests have been described above. Airborne particles also carry a range of chemical pollutants that may have direct effects on plant tissue functioning.

Other Atmospheric Pollutants

Tropical forest fires also result in significant emissions of nitrous oxide (N_2O), nitrogen oxides (NO_x), sulphur containing gases, and ozone (O_3) (Crutzen and Andreae, 1990). During the 1997 Southeast Asian forest fires, SO_2 levels were significantly elevated due to the smoke-haze (Figure 4), however no increase in NO_x or O_3 concentrations were detected (Davies and Unam, 1999). These gases have a range of complex effects on atmospheric chemistry that influence the radiation balance of the earth, and may therefore have impacts on ecosystems (Barker and Tingey, 1992; Houghton *et al.*, 1992). This section focuses on potential short-term, direct effects of these emissions on tropical forests.

Few studies have investigated the effects of pollution on tropical plants and forests, however studies of temperate plant community responses to atmospheric pollution provide strong evidence that exposure to SO_2, NO_2, O_3, and acid deposition may have significant effects on tropical forests (Binns, 1988). Atmospheric pollution effects on plants may occur through exposure to gases or through particulate deposition on leaf and soil surfaces. Smoke-haze clouds produced by tropical forest fires may have gaseous concentrations sufficient to adversely affect plant functioning in remaining forests. Smith (1992) reports that SO_2 concentrations as low as 20-80 ppb may result in reduced agricultural yields. SO_2 concentrations during the 1997 fires were well above 20 ppb on several days (Figure 4), but the effect on tropical forest species was not studied. Nichol (1997) cites an unpublished report of a 50% reduction in growth of two hybrid rice varieties in Malaysia following the 1994 smoke-haze event; whether this reduction was due to pollution effects or reduced PAR and other climatic changes remains unknown. With respect to the effect of smoke-haze arising from the tropical forest fires, it is also important to note that the effects of different pollutants may interact causing more severe plant toxicity effects. Exposure to combinations of elevated NO_2, SO_2 and O_3 may reduce water-use efficiency, through effects on stomatal function, and thereby increase plant susceptibility to drought (Wolfenden *et al.* 1992); an effect that would be particularly detrimental under changed climatic conditions (see earlier discussion). Furthermore, SO_2 and NO_2 together are more detrimental to plants than either is alone, as SO_2 inhibits the normal path of NO_2 detoxification (Wolfenden *et al.*, 1992).

Atmospheric pollution from forest fires may target particular species, for example, species with typically high rates of stomatal conductance (e.g. early successional trees), or species with well exposed foliage (e.g. canopy trees) (Wolfenden *et al.,* 1992). Atmospheric pollution may also have significant effects on herbivore and pollinator populations that may affect the reproductive biology of tropical plants (Cox, 1992). Any of these changes may have significant effects on tropical forest diversity and functioning, however detecting community-level responses of vegetation to pollutants is methodologically difficult (Armentano and Bennett, 1992).

Deposition of nitrogen and sulphur containing substances through 'acid rain' has the potential to profoundly affect tropical ecosystems, both through effects on soils and on ground water and streams (Brimblecombe, 1993). Several studies in temperate ecosystems have demonstrated that nitrogen deposition can greatly alter community structure (e.g. Aerts and Berendse, 1988). Experimental studies have also demonstrated that soil nitrogen additions can change community composition (Tilman, 1987). These effects are particularly strong in ecosystems that are nitrogen-limited (Vitousek *et al.*, 1997). Species with higher nitrogen-uptake rates will tend to succeed over those with lower nitrogen-uptake rates. Communities on nitrogen-deficient soils may change floristically to communities more similar to those on richer soils, leading to a loss in regional biodiversity. A shift in community composition in favor of particular species, often leads to greater dominance by these species and hence lower community diversity (Tilman, 1987; Vitousek, 1994). For lowland tropical rain forests of the Sunda Shelf, productivity is widely considered to be phosphorus-limited rather than nitrogen-limited, as is common in many temperate forests (Vitousek and Sanford, 1986; Burslem *et al.*, 1994; Sollins, 1998). Therefore, patterns of tropical forest response to increasing acid deposition may differ from temperate forests. The importance of tropical forests for timber and non-timber forest products, as well as a range of ecosystem services (e.g. water catchments) demands that research be conducted to assess the consequences of increasing pollution levels on tropical ecosystem diversity and functioning.

CONCLUSIONS

Large-scale fires are becoming more frequent in tropical rain forests of Southeast Asia due to the coincidence of forest fragmentation and degradation, with ENSO-induced drought events. Such fires have significant effects on atmospheric processes, biogeochemical cycles, climate, and ecosystem diversity and functioning. As most tropical rain forest plant species are not adapted to fire, burned forests suffer significant losses of biodiversity as well as changes in the composition of regenerating ecosystems. Regeneration following fire depends on a range of factors occurring before (e.g. degree of forest degradation), during (e.g. fire intensity effects on the soil seed bank) and after (e.g. climate effects on the probability of subsequent fire) the fire. Current knowledge of the effects on, and responses of, tropical rain forests to fire is inadequate. Active management of forest reserves and protected areas is required to ensure that future forest fires will not further erode the region's biodiversity. Techniques for post-fire land rehabilitation need to be developed to minimize the loss of soil resources, limit native species loss, and restrict the invasion of weed species.

Forest fires also have significant effects on remaining intact forests through the effects of deforestation and forest fragmentation. Loss of forest will lead to inevitable losses of biodiversity, but may also lead to local climatic changes due to altered hydrological cycles. It is projected that reduced rainfall and higher temperatures may lead to a contraction in the area of Southeast Asian tropical rain forests. Methods for the regeneration and restoration of tropical rain forests on cleared and degraded lands need to be developed to ameliorate the effects of forest loss on climatic and hydrological cycles.

Such developments would also help reduce the risks of biodiversity loss through the effects of forest fragmentation.

Large-scale tropical forest fires emit vast quantities of pollutants: aerosols, CO_2, and other radiatively active greenhouse gases to the atmosphere. These emissions have a range of effects on remaining tropical rain forest ecosystems, including indirect effects on climate and cloud formation processes, effects of gaseous exposure on plant physiology, and effects of acid deposition on soil processes. These effects may significantly change tropical forest diversity and functioning. However, there is currently very little information with which to model the likely effects of these emissions on tropical rain forest communities. The release of large quantities of CO_2 and other greenhouse gases is of major concern due to its potential effect on global and regional climates. Forest management strategies that sequester carbon are required in Southeast Asia to mitigate the effects of these fires on carbon emissions (Brown, 1996). A more detailed understanding of the ecological effects of forest fires is crucial to the future sustainable management of the region's invaluable tropical rain forests.

Acknowledgements

The Faculty of Resource Science and Technology, UNIMAS provided technical support for the ecophysiological studies. The Department of the Environment, Sarawak kindly provided access to the air pollution data. Layang ak. Unam assisted with the ecophysiological study, and Khazaki bin Loman greatly assisted with data preparation. James LaFrankie and Indraneil Das provided comments on the manuscript. I thank Peter Wayne, Johann Goldammer and Dawn Delo for sending reference literature.

REFERENCES

1. Aerts, R., and F. Berendse, 1988. "The effect of increased nutrient availability on vegetation dynamics in wet heathlands." *Vegetation* 76: 63-69.
2. Armentano, T. V., and J. P. Bennett. 1992. "Air pollution effects on the diversity and structure of communities". In J. R. Barker and D. T. Tingey, editors. *Air Pollution Effects on Biodiversity*. Van Nostrand Reinhold, New York pp. 159-176.
3. Arnone, J. A., and C. Korner. 1995. "Soil and biomass carbon pools in model communities of tropical plants under elevated CO_2." *Oecologia* **104**: 61-71.
4. Barker, J. R., and D. T. Tingey. 1992. "The effects of air pollution on biodiversity: a synopsis". In J. R. Barker and D. T. Tingey, editors. *Air Pollution Effects on Biodiversity*. Van Nostrand Reinhold, New York, pp.3-9.
5. Bazzaz, F. A. 1990. "The response of natural ecosystems to the rising global CO_2 levels." *Annual Review of Ecology and Systematics* **21**: 167-96.
6. Bazzaz, F. A. 1991. "Regeneration of tropical forests: Physiological responses of pioneer and secondary species" In A. Gomez-Pompa, T. C. Whitmore, and M.

Hadley, editors. *Rain Forest Regeneration and Management.* The Parthenon Publishing Group, Parkridge, New Jersey., pp.91-118.

7. Bazzaz, F. A., and P. M. Wayne. 1994. "Coping with environmental heterogeneity: the physiological ecology of tree seedling regeneration across the gap-understorey continuum." In M. M. Caldwell and R. W. Pearcy, editors. *Exploitation of Environmental Heterogeneity by Plants.* Academic Press, San Diego, pp. 349-390.
8. Becker, P. 1992. "Seasonality of rainfall and drought in Brunei Darussalam." *Brunei Museum Journal* 7: 99-109.
9. Becker, P., and M. Wong. 1992. "Fire damage in tropical heath forest: peculiar features and vulnerability". *Brunei Museum Journal* 7: 110-113.
10. Becker, P., and M. Wong. 1993. "Drought-induced mortality in tropical heath forest."*Journal of Tropical Forest Science* **5**: 416-419.
11. Binns, W. 1988. "Vegetation and soils." In K. Mellanby, editor. *Air Pollution, Acid Rain and the Environment.* Elsevier, London. pp. 13-21.
12. Borchert, R. 1998. "Responses of tropical trees to rainfall seasonality and its longterm changes." *Climate Change* 39: in press.
13. Brimblecombe, P. 1993. "Environmental acidification." In C. N. Hewitt and W. T. Sturges, editors. *Global Atmospheric Chemical Change.* Elsevier Applied Science, London, pp.233-270.
14. Brown, S. 1996. "Mitigation potential of carbon dioxide emissions by management of forests in Asia." *Ambio* **25**: 273-278.
15. Burslem, D. F. R. P., I. M. Turner, and P. J. Grubb. 1994. "Mineral nutrient status of coastal hill dipterocarp forest and adinandra belukar in Singapore: bioassays of nutrient limitation". *Journal of Tropical Ecology* **10**: 579-599.
16. Ceulemans, R., and M. Mousseau. 1994."Effects of elevated atmospheric CO_2 on woody plants." *New Phytologist* **127**: 425-446.
17. Cox, R. M. 1992. "Air pollution effects on plant reproductive processes and possible consequences to their population biology". In J. R. Barker and D. T. Tingey, editors. *Air Pollution Effects on Biodiversity.* Van Nostrand Reinhold, New York, pp. 131-158.
18. Crutzen, P. J., and M. O. Andreae. 1990. "Biomass burning in the tropics: Impact on atmospheric chemistry and biogeochemical cycles". *Science* **250**: 1669-1678.
19. D'Antonio, C. M., and P. M. Vitousek. 1992. "Biological invasions by exotic grasses, the grass-fire cycle, and global change. *Annual Review of Ecology and Systematics* **23**: 63-87.
20. Dale, V. H. 1994 "Terrestrial CO_2 flux: the challenge of interdisciplinary research." In V. H. Dale, editor. *Effects of Land-Use Change on Atmospheric CO_2 Concentrations: South and Southeast Asia as a case study.* Springer-Verlag, New York pp. 1-14.
21. Dale, V. H. 1997. "The relationship between land-use change and climate change."*Ecological Applications* **7**: 753-769.
22. Davies, S. J. 1998. "Photosynthesis of nine pioneer *Macaranga* species from Borneo in relation to life-history." *Ecology* **79**: 2292-2308.

23. Davies, S. J., D. S. Edwards, and W. E. Booth. 1996. "A botanical survey of the proposed Meragang Beach Recreation Park, Brunei Darussalam". *Brunei Museum Journal* in press.
24. Davies, S. J., and L. Unam. 1999. "Smoke-haze from the 1997 Indonesian forest fires: effects on pollution levels, local climate, atmospheric CO_2 concentrations, and tree photosynthesis" *Forest Ecology and Management* **124**(2-3): 137-144.
25. Denslow, J. S. 1987. "Tropical rainforest gaps and tree species diversity." *Annual Review of Ecology and Systematics* **18**: 431-451.
26. Dixon, R. K., S. Brown, R. A. Houghton, A. M. Soloman, M. C. Trexler, and J. Wisniewski. 1994. "Carbon pools and flux of global forest ecosystems." *Science* **263**: 185-190.
27. Eamus, D., C. A. Berryman, and G. A. Duff. 1993. "Assimilation, stomatal conductance, specific leaf area and chlorophyll responses to elevated CO_2 of *Maranthes corymbosa*, a tropical monsoon rain forest species." *Australian Journal of Plant Physiology* **20**: 741-755.
28. El-Shobokshy, M. S., and Y. G. Al-Saedi. 1993. "The impact of the Gulf War on the Arabian environment- I. Particulate pollution and reduction of solar radiation" *Atmospheric Environment* **27A**: 95-108.
29. Fearnside, P. M. 1990. "Fire in the tropical rain forest of the Amazon basin." In J. G. Goldammer, editor. *Fire in the Tropical Biota: Ecosystem Processes and Global Challenges*. Springer-Verlag, Berlin, pp. 106-116
30. Fearnside, P. M. 1995. "Potential impacts of climatic change on natural forests and forestry in Brazilian Amazon." Fores*t Ecology and Management* **78**: 51-70.
31. Fearnside, P. M. 1996. "Amazonian deforestation and global warming: carbon stocks in vegetation replacing Brazil's Amazon forest." *Forest Ecology and Management* **80**: 21-34.
32. Finegan, B. 1996. "Pattern and process in neotropical secondary rain forests: the first 100 years of succession." *Trends in Ecology and Evolution* **11**: 119-124.
33. Fosberg, M. A., J. G. Goldammer, D. Rind, and C. Price. 1990. "Global change: effects on forest ecosystems and wildlife severity." In J. G.Goldammer, editor. *Fire in the Tropical Biota. Ecosystem processes and global challenges*. Springer-Verlag, Berlin, pp.463-486.
34. Goldammer, J. G., and B. Seibert. 1990. "The impact of droughts and forest fires on tropical lowland rain forest of East Kalimantan." In J. G. Goldammer, editor. *Fire in the Tropical Biota. Ecosystem processes and global challenges*. Springer-Verlag, Berlin, pp. 11-31.
35. Hobbie, S. E., D. B. Jensen, and F. S. Chapin. 1994. "Resource supply and disturbance as controls over present and future plant diversity." In E.D. Schulze and H. A. Mooney, editors. *Biodiversity and Ecosystem Function*. Springer, Berlin, pp.385-408.
36. Holling, C. S., D. W. Schindler, B. W. Walker, and J. Roughgarden. 1995. "Biodiversity in the functioning of ecosystems: an ecological synthesis." In C. Perrings, K. Maler, C. Folke, C. S. Holling, and B. Jansson, editors. *Biodiversity*

Loss: Economic and Ecological Issues. Cambridge University Press, Cambridge, UK, pp. 44-83.

37. Houghton, J. T., B. A. Callander, and S. K. Varney. 1992. *Climate Change 1992: The Supplementary Report to the IPCC Scientific Assessment*. Cambridge University Press. Cambridge.
38. Houghton, R. A. 1991. "Biomass burning from the perspective of the global carbon cycle." In J. S. Levine, editor. *Global Biomass Burning. Atmospheric, Climatic, and Biospheric Implications*. The MIT Press, Cambridge, MA pp. 321-325
39. Houghton, R. A. 1995. "Tropical forests and climate." In R. B. Primack and T. E. Lovejoy, editors. *Ecology, Conservation, and Management of Southeast Asian Rainforests*. Yale University Press, New Haven, USA, pp. 263-289.
40. Houghton, R. A., and J. L. Hackler. 1994. "The net flux of carbon from deforestation and degradation in south and south-east Asia." In V. H. Dale, editor. *Effects of Land-Use Change on Atmospheric CO_2 Concentrations: South and Southeast Asia as a Case Study*. Springer-Verlag, New York, pp. 301-327.
41. Houghton, R. A., D. L. Skole, and D. S. Lefkowitz. 1991. "Changes in the landscape of Latin America between 1850 and 1985. II. Net release of CO_2 to the atmosphere."*Forest Ecology and Management* **38**: 173-199.
42. Hulme, M., and D. Viner. 1998. "A climate change scenario for the tropics." *Climate Change* **39**: in press.
43. Isichei, A. O., J. I. Muoghalu, F. A. Akeredolu, and O. A. Afolabi. 1995."Fuel characteristics and emissions from biomass burning and land-use change in Nigeria."*Environmental Management Assessment*. **38**: 279-289.
44. Kauffman, J. B., R. L. Sanford, D. L. Cummings, I. H. Salcedo, and E. V. S. B. Sampaio. 1993. "Biomass and nutrient dynamics associated with slash fires in neotropical dry forests.' *Ecology* **74**: 140-151.
45. Kauffman, J. B., and C. Uhl. 1990. "Interactions of anthropogenic activities, fire, and rain forests in the Amazonian basin." In J. G. Goldammer, editor. *Fire in the Tropical Biota. Ecosystem processes and global challenges*. Springer-Verlag, Berlin, pp.117-134
46. Korner, C. 1993. "CO_2 fertilization: the great uncertainty in future vegetation development." In A. M. Soloman and H. H. Shugart, editors. *Vegetation Dynamics & Global Change*. Chapman & Hall, New York, pp.53-70.
47. Korner, C., and J. A. Arnone. 1992. "Responses to elevated carbon dioxide in artificial tropical ecosystems." *Science* **257**: 1672-1675.
48. Laurance, W. F. 1994. "Rainforest fragmentation and the structure of small mammal communities in tropical Queensland." *Biological Conservation* **69**: 23-32.
49. Laurance, W. F., L. V. Ferreira, J. M. Rankin-de Merona, and S. G. Laurance. 1998. "Rain forest fragmentation and the dynamics of Amazonian tree communities." *Ecology* **79**: 2032-2040.
50. Leighton, M. 1984. The El Niño-Southern Oscillation event in Southeast Asia: Effects of Drought and Fire in Tropical Forest in Eastern Borneo. WWF report, 31 pp.

51. Lovejoy, T. E. 1995. "The quantification of biodiversity: an esoteric quest or a vital component of sustainable development?" In D. L. Hawksworth, editor.*Biodiversity: Measurement and Estimation*. Chapman & Hall, London, U.K, pp. 81-89.
52. Mackinnon, K., G. Hatta, H. Halim, and A. Mangalik. 1996. *The Ecology of Kalimantan*. Periplus Editions, Singapore.
53. Malingreau, J. P., G. Stephens, and L. Fellows. 1985. "Remote sensing of forest fires: Kalimantan and North Borneo in 1982-83." *Ambio* **14**: 314-321.
54. Miller, K., M. H. Allegretti, N. Johnson, and B. Jonsson. 1995. "Measures for conservation of biodiversity and sustainable use of its components." In V. H. Heywood, editor. *Global Biodiversity Assessment*. Cambridge University Press, Cambridge, pp. 915-1061.
55. Mueller-Dombois, D., and J. G. Goldammer. 1990. "Fire in tropical ecosystems and global environmental change: An introduction." In J. G. Goldammer, editor. *Fire in the Tropical Biota. Ecosystem Processes and Global Challenges*. Springer-Verlag, Berlin, pp. 1-10.
56. Myers, N. 1988. "Tropical forests and their species. Going, going ...?" In E. O. Wilson, editor. *Biodiversity*. National Academy Press, Washington DC, pp.28-35.
57. Myers, N. 1991. "Tropical forests: present status and future outlook." *Climatic Change* **19**: 3-32.
58. Nichol, J. 1997. "Bioclimatic impacts of the 1994 haze event in southeast Asia." *Atmospheric Environment* **31**: 1209-1219.
59. Nobre, C. A., P. J. Sellers, and J. Shukla. 1991. "Amazonian deforestation and regional climate change." *Journal of Climatology* **4**: 957-988.
60. Nykvist, N. 1996. "Regrowth of secondary vegetation after the 'Borneo fire' of 1982-1983." *Journal of Tropical Ecology* **12**: 307-312.
61. O'Brien, T., M. F. Kinnaird, Sunarto, A. A. Dwiyahreni, W. M. Rombang, and K. Anggraini. 1998. "Effects of the 1997 fires on the forest and wildlife of the Bukit Barisan Selatan National Park, Sumatra." unpublished report. Wildlife Conservation Society. New York.
62. Oberbauer, S. F., B. R. Strain, and N. Fetcher. 1985. "Effect of CO_2-enrichment on seedling physiology and growth of two tropical tree species." *Physiologia Plantarum* **65**: 352-356.
63. Palm, C. A., R. A. Houghton, J. M. Melillo, and D. L. Skole. 1986."Atmospheric carbon dioxide from deforestation in Southeast Asia." *Biotropica* **18**: 177-188.
64. Preining, O. 1993. "Global climate change due to aerosols." In C. N. Hewitt and W. T. Sturges, editors. *Global Atmospheric Chemical Change*. Elsevier Applied Science, London, pp. 93-22.
65. Raven, P. H. 1988. "Our diminishing tropical forests." In E. O. Wilson, editor. *Biodiversity*. National Academy Press, Washington DC, pp. 119-122.
66. Reekie, E. G., and F. A. Bazzaz. 1989. "Competition and patterns of resource use among seedlings of five tropical trees grown at ambient and elevated CO_2." *Oecologia* **79**: 212-222.
67. Richards, J. F., and E. P. Flint. 1994. "A century of land-use change in south and southeast Asia." In V. H. Dale, editor. *Effects of Land-Use Change on Atmospheric*

CO_2 Concentrations: South and Southeast Asia as a Case Study. Springer-Verlag, New York, pp. 15-66.

68. Riswan, S., and K. Kartawinata. 1991. "Regeneration after disturbance in a lowland mixed dipterocarp forest in East Kalimantan, Indonesia." In A. Gomez-Pompa, T. C. Whitmore, and M. Hadley, editors. *Rain Forest Regeneration and Management*. UNESCO, Man and the Biosphere Series, Paris, pp.295-301.
69. Sanford, R. L., J. Saldarriaga, K. E. Clark, C. Uhl, and R. Herrera. 1985. "Amazonian rain-forest fires." *Science* **227**: 53-55.
70. Setzer, A. W., and M. C. Pereira. 1991. "Amazonia biomass burnings in 1987 and an estimate of their tropospheric emissions."*Ambio* **20**: 19-22.
71. Shukla, J., C. Nobre, and P. Sellers. 1990. "Amazon deforestation and climate change."*Science* **247**: 1322-1325.
72. Skole, D., and C. Tucker. 1993. "Tropical deforestation and habitat fragmentation in the Amazon: satellite data from 1978 to 1988." *Science* **260**: 1905-1910.
73. Smith, W. H. 1992. "Air pollution effects on ecosystem processes." In J.R. Barker and D. T. Tingey, editors. *Air Pollution Effects on Biodiversity*. Van Nostrand Reinhold, New York, pp.234-260.
74. Sollins, P. 1998. "Factors influencing species composition in tropical lowland rain forest: Does soil matter?" *Ecology* **79**: 23-30.
75. Stott, P. A., J. G. Goldammer, and W. L. Werner. 1990. "The role of fire in the tropical lowland deciduous forests of Asia." In J. G. Goldammer, editor. *Fire in the Tropical Biota. Ecosystem Processes and Global Challenges*. Springer-Verlag, Berlin, pp.32-44.
76. Tilman, D. 1987. "Secondary succession and the pattern of plant dominance along experimental nitrogen gradients." *Ecological Monographs* **57**: 189-214.
77. Tinker, P. B., J. S. I. Ingram, and S. Struwe. 1996. Effects of slash-and-burn agriculture and deforestation on climate change. *Agriculture, Ecosystems and Environment*. **58**: 13-22.
78. Turner, I. M. 1996. "Species loss in fragments of tropical rain forest: a review of the evidence." *Journal of Applied Ecology* **33**: 200-209.
79. Uhl, C. 1988. "Restoration of degraded lands in the Amazon basin." In E. O. Wilson, editor. *Biodiversity*. National Academy Press, Washington DC, pp.326-332.
80. Uhl, C., and J. B. Kaufmann. 1990. "Deforestation, fire susceptibility, and potential tree responses to fire in the eastern Amazon." *Ecology* **71**: 437-449.
81. Vitousek, P., and R. L. Sanford. 1986. "Nutrient cycling in moist tropical forests." *Annual Review of Ecology and Systematics* **17**: 137-167.
82. Vitousek, P. M. 1994. "Beyond global warming: Ecology and global change." *Ecology* **75**: 1861-1876.
83. Vitousek, P. M., J. D. Aber, R. W. Howarth, G. E. Likens, P. A. Matson, D. W. Schindler, W. H. Schlesinger, and D. G. Tilman. 1997. "Human alteration of the global nitrogen cycle: sources and consequences." *Ecological Applications* **7**: 737-750.

84. Walsh, R. P. D. 1996. "Drought frequency changes in Sabah and adjacent parts of Borneo since the late nineteenth century and possible implications for rain forest dynamics." *Journal of Tropical Ecology* **12**: 385-407.
85. Whelan, R. J. 1995. *The Ecology of Fire*. Cambridge University Press, Cambridge.
86. Whitmore, T. C. 1998. "Potential impact of climatic change on tropical rain forest seedlings and forest regeneration." *Climatic Change* **39**: 429-438.
87. Whitmore, T. C., and D. F. R. P. Burslem. 1998. "Major disturbances in tropical rainforests." In D. M. Newbery, H. H. T. Prins, and N. D. Brown, editors. *Dynamics of Tropical Communities*. Blackwell Science, Oxford, pp. 549-565.
88. Wirawan, N. 1983. Progress in the Management of Protected Areas in Kalimantan and Consequences of Recent Forest Fires. WWF Project 1687.
89. Wolfenden, J., P. A. Wookey, P. W. Lucas, and T. A. Mansfield. 1992. "Action of pollutants individually and in combination." In J. R. Barker and D. T. Tingey, editors. *Air Pollution Effects on Biodiversity*. Van Nostrand Reinhold, New York, pp. 72-92.
90. Woods, P. 1989. "Effects of logging, drought, and fire on structure and composition of tropical forests in Sabah, Malaysia." *Biotropica* **21**: 290-298.
91. Yanhong, T., K. Naoki, F. Akio, and M. Awang. 1996. "Light reduction by regional haze and its effect on simulated leaf photosynthesis in a tropical forest of Malaysia."*Forest Ecology and Management* **89**: 205-211.
92. Zimmerman, J. K., E. M. Everham, R. B. Waide, D. J. Lodge, C. M. Taylor, and N. V.L. Brokaw. 1994. "Responses of tree species to hurricane winds in subtropical wet forest in Puerto Rico: implications for tropical tree life histories." *Journal of Ecology* **82**: 911-922.
93. Ziska, L. H., K. P. Hogan, A. P. Smith, and B. G. Drake. 1991. "Growth and photosynthetic response of nine tropical species with long-term exposure to elevated carbon dioxide. *Oecologia* **86**: 383-389.

Chapter 9

GLOBAL IMPACTS AND CLIMATE CHANGE

Joel S. Levine

Atmospheric Sciences Research, NASA Langley Research Center, Hampton, Virginia 23681-0001, USA

BIOMASS BURNING: A GLOBAL PHENOMENON

During the 1997-98 period, there were a series of extensive and widespread fires throughout the world, including Southeast Asia, South America, Central America, Africa, Russia, China, Canada, and the United States (Florida). Extensive and widespread tropical forest and peat fires swept throughout Kalimantan and Sumatra, Indonesia, between August and December 1997. The fires resulted from burning for land clearing and land use change. However, the severe drought conditions resulting from El Niño caused small land-clearing fires to become large uncontrolled wildfires. Based on satellite imagery, it has been estimated that a total of 45,600 km^2 burned in Kalimantan and Sumatra between August and December 1997.

The fires in Southeast Asia were widespread and extensive. The gaseous and particulate emissions produced in these fires and released into the atmosphere reduced atmospheric visibility, impacted the composition and chemistry of the atmosphere, and affected human health. Some of the consequences of the fires in Southeast Asia were:

- More than 200 million people were exposed to high levels of air pollution and particulates produced during the fires.
- More than 20 million smoke-related health problems were recorded.
- Fire-related damage cost in excess of $ 4 billion.
- On September 26, 1997, a commercial airliner (Garuda Airlines Airbus 300-B4) crashed on landing in Sumatra due to very poor visibility as a result of smoke from the fires with 234 passengers killed.

- On September 27, 1997, two ships collided at sea due to poor visibility in the Strait of Malacca, off the coast of Malaysia, with 29 crew members killed.

International concern about the environmental and health impacts of these fires was great. Three different agencies of the Unite Nations organized workshops and reports on the environmental and health impacts of these fires: The World Meteorological Organization (WMO) Workshop on Regional Transboundary Smoke and Haze in Southeast Asia, Singapore, June 2-5, 1998, The World Health Organization (WHO) Health Guidelines for Forest Fires Episodic Events, Lima, Peru, October 6-9, 1998, and the United Nations Environmental Program (UNEP) *Report on Wildland Fires and the Environment: A Global Synthesis*, published in February, 1999 (Levine *et al.,* 1999). The Indonesian fires formed the basis of an article in *National Geographic* magazine, entitled, "Indonesia's Plague of Fire." (Simons, 1998).

Biomass burning is the customary practice for land clearing and land-use change in Southeast Asia and the rest of the tropics. Biomass burning is the burning of living and dead vegetation and has been identified as a significant source of gases and particulates to the regional and global atmosphere (Crutzen *et al.*, 1979, Seiler and Crutzen, 1980, Crutzen and Andreae, 1990, Levine *et al.*, 1995). The bulk of the world's biomass burning occurs in the tropics - in the tropical forests of South America and Southeast Asia and in the savannas of Africa and South America. The majority of biomass burning (perhaps as much as 90%) is believed to be human-initiated, with natural fires triggered by atmospheric lightning only accounting for on the order of about 10% of all fires (Andreae, 1991). Biomass burning is truly a multidisciplinary subject, encompassing the following areas: fire ecology, fire measurements and modeling, fire combustion, remote sensing of fires, gaseous and particulate emissions from fires, the atmospheric transport of these emissions and the chemical and climatic impacts of these emissions. Over the last few years, a series of books have documented much of our current understanding of biomass burning. These volumes include: Goldammer (1990), Levine (1991), Crutzen and Goldammer (1993), Goldammer and Furyaev (1996), Levine (1996a), Levine (1996b), and van Wilgen *et al.* (1997).

On an annual global basis, biomass burning is a significant source of gases and particulates to the atmosphere. The gaseous and particulate emissions produced during biomass burning are dependent on the nature of the biomass matter, which is a function of the ecosystem and the temperature of the fire, which is also ecosystem dependent. In general, biomass is composed mostly of carbon (about 45% by weight) and hydrogen and oxygen (about 55% by weight), with trace amounts of nitrogen (0.3 to 3.8% by mass), sulfur (0.1 to 0.9%), phosphorous (0.01 to 0.3%), potassium (0.5 to 3.4%) and still smaller amounts of chlorine, and bromine.

During complete combustion, the burning of biomass matter produces carbon dioxide (CO_2) and water vapor as the primary products, according to the reaction:

$$CH_2O + O_2 \rightarrow CO_2 + H_2O \qquad (1)$$

where CH_2O represents the average chemical composition of biomass matter. Biomass burning can be considered as an abiotic equivalent of respiration/decay.

In the more realistic case of incomplete combustion in cooler and/or oxygen-deficient fires, i.e., the smoldering phase of burning, carbon is released in the forms of carbon monoxide (CO), methane (CH_4), nonmethane hydrocarbons (NMHCs), and various partially oxidized organic compounds, including aldehydes, alcohols, ketones, and organic acids and particulate black (soot) carbon. Nitrogen is present in biomass mostly as amino groups (R-NH_2) in the amino acids of proteins. During combustion the nitrogen is released by pyrolytic decomposition of the organic matter and partially or completely oxidized to various volatile nitrogen compounds, including molecular nitrogen (N_2), nitric oxide (NO), nitrous oxide (N_2O), ammonia (NH_3), hydrogen cyanide (HCN), cyanogen (NCCN), organic nitriles (acetonitrile (CH_3CN), acrylonitrile (CH_2CHCN)), and propionitrile (CH_3CH_2CN), and nitrates. The sulfur in biomass is organically bound in the form of sulfur-containing amino acids in proteins. During burning the sulfur is released mostly in the form of sulfur dioxide (SO_2) and smaller amounts of carbonyl sulfide (COS) and nonvolatile sulfate (SO_4^{2-}). About one half of the sulfur in the biomass matter is left in the burn ash, whereas, very little of the fuel nitrogen is left in the ash.

Laboratory biomass burning experiments conducted by Lobert *et al.* (1991) have identified the carbon (Table 1) and nitrogen (Table 2) compounds released to the atmosphere by burning. The major gases produced during the biomass burning process include carbon dioxide (CO_2), carbon monoxide (CO), methane (CH_4), oxides of nitrogen (NO_x = nitric oxide (NO) + nitrogen dioxide (NO_2)), and ammonia (NH_3). Carbon dioxide and methane are greenhouse gases, which trap Earth-emitted infrared radiation and lead to global warming. Carbon monoxide, methane, and the oxides of nitrogen lead to the photochemical production of ozone (O_3) in the troposphere. In the troposphere, ozone is an irritant and harmful pollutant, and in some cases, is toxic to living systems. Nitric oxide leads to the chemical production of nitric acid (HNO_3) in the troposphere. Nitric acid is the fastest growing component of acidic precipitation. Ammonia is the only basic gaseous species that neutralizes the acidic nature of the troposphere. Particulates, small (usually about 10 micrometers or smaller) solid particles, such as smoke or soot particles, are also produced during the burning process and released into the atmosphere. These solid particulates absorb and scatter incoming sunlight and hence impact the local, regional, and global climate. In addition, these particulates (specifically particulates 2.5 micrometers or smaller) can lead to various human respiratory and general health problems when inhaled. The gases and particulates produced during biomass burning lead to the formation of "smog." The word "smog" was coined as a combination of smoke and fog and is now used to describe any smoky or hazy pollution in the atmosphere.

Table 1 Carbon and Gases Produced during Biomass Burning (Lobert *et al.*, 1991)

Compound	Mean Emission Factor Relative to the Fuel C (%)
Carbon dioxide (CO_2)	82.58
Carbon monoxide (CO)	5.73
Methane (CH_4)	0.424
Ethane (CH_3CH_3)	0.061
Ethene (CH_2=CH_2)	0.123
Ethine (CH=CH)	0.056
Propane (C_3H_8)	0.019
Propene (C_3H_6)	0.066
n--butane (C_4H_{10})	0.005
2-butene (cis) (C_4H_8)	0.004
2-butene (trans) (C_4H_8)	0.005
i-butene, i-butene ($C_4H_8 + C_4H_8$)	0.033
1,3-butadiene(C_4H_6)	0.021
n-pentane (C_3H_{12})	0.007
Isoprene (C_5H_8)	0.008
Benzene (C_6H_6)	0.064
Toluene (C_7H_8)	0.037
m-, p-xylene (C_8H_{10})	0.011
o-xylene (C_8H_{10})	0.006
Methyl chloride (CH_3Cl)	0.010
NMHC (As C) (C_2 to C_8)	1.18
Ash (As C)	5.00
Total Sum C	94.92 (including ash)

Table 2 Nitrogen Gases Produced during Biomass Burning (Lobert *et al.*, 1991)

Compound	Mean Emission Factor Relative to the Fuel N (%)
Nitrogen oxides (NO_x)	13.55
Ammonia(NH_3)	4.15
Hydrogen cyanide (HCN)	2.64
Acetonitrile (CH_3CN)	1.00
Cyanogen (NCCN) (As N)	0.023
Acrylonitrile (CH_2CHCN)	0.135
Propionitrile (CH_3CH_2CN)	0.071
Nitrous oxide (N_2O)	0.072
Methylamine (CH_3NH_2)	0.047
Dimetylamine ($(CH_3)_2NH$)	0.030
Ethylamine ($CH_3CH_2NH_2$)	0.005
Trimethylamine ($(CH_3)N$)	0.02
2-methyl-1-butylamine ($C_5H_{11}NH_2$)	0.04
n-pentylamine (n-$C_5H_{11}NH_2$)	0.137
Nitrates (70% HNO_3)	1.10
Ash (As N)	9.94
Total sum N (As N)	33.66 (Including ash)
Molecular nitrogen (N_2)	21.60
Higher HCs and particles	20

Areas of biomass burning are varied and include tropical savannas, tropical, temperate and boreal forests and agricultural lands after the harvest. The burning of fuelwood for domestic use is another source of biomass burning. Global estimates of the annual amounts of biomass burning from these sources are estimated in Table 3 (Andreae, 1991). In Table 3, the unit of biomass burned is Tg dm/yr (1 Tg = 10^{12} g = 10^6 metric tons; dm = dry matter (biomass matter)). As already noted, biomass matter is about 45% by weight composed of carbon. Table 3 also gives estimates of the carbon released (TgC/yr) by the burning of this biomass (The total biomass burned is multiplied by 45% to determine the amount of carbon released into the atmosphere during burning). Combining estimates of the total amount of biomass matter burned/yr (Table 3) with measurements of the gaseous and particulate emissions from biomass burning (Tables 1 and 2) permits estimates of the global production and release into the atmosphere of gases and particulates from burning. Estimates of the global contribution of biomass burning are summarized in Table 4 (Andreae, 1991). The data in Tables 3 and 4 clearly indicate that biomass burning is a truly global process of major importance in the global budgets of atmospheric gases and particulates.

Table 3 Global Estimates of Annual Amounts of Biomass Burning and of the Resulting Release of Carbon to the Atmosphere (Andreae, 1991)

Source	Biomass Burned (Tg dm/yr)	Carbon Released (TgC/yr)
Savanna	3690	1660
Agricultural waste	2020	910
Fuel wood	1430	640
Tropical forests	1260	570
Temperate/boreal forests	280	130
World Total	8680	3910

Table 4 Comparison of Global Emissions from Biomass Burning with Emissions from All Sources (Including Biomass Burning) (Andreae, 1991)

Species	Biomass Burning (Tg element[g]/yr)	All Sources (Tg element[g]/yr)	% Due to Biomass Burning
CO_2 (Gross)[a]	3500	8700	40
CO_2 (Net)[b]	1800	7000	26
CO	350	1100	32
CH_4	38	380	10
NMHC[c]	24	100	24
N_2O	0.8	13	6
NO_x	8.5	40	21
NH_3	5.3	44	12
Sulfur	2.8	150	2
COS	0.09	1.4	6
CH_3Cl	0.51	2.3	22
H_2	19	75	25
Tropospheric O_3	420	1100	38
TPM[d]	104	1530	7
POC[e]	69	180	39
EC[f]	19	<22	>86

[a]Biomass burning plus fossil fuel burning.
[b]Deforestation plus fossil fuel burning.
[c]Nonmethane hydrocarbons (excluding isoprene and terpenes).
[d]Total particulate matter (Tg/yr).
[e]Particulate organic matter (including elemental carbon).
[f]Elemental (black-soot) carbon.
[g]Tg element/yr where C, N, S Cl are the elements.

Global Impacts of Biomass Burning

Gaseous and particulate emissions produced during biomass burning and released into the atmosphere impact the local, regional, and global atmosphere and climate in several different ways. Some of these impacts include:

(1) Biomass burning is a significant global source of carbon dioxide and methane. Both are greenhouse gases that can lead to global warming.

(2) Biomass burning is a significant global source of carbon monoxide, methane, hydrocarbons, and oxides of nitrogen. When exposed to solar radiation, these gases lead to the photochemical production of ozone in the troposphere (This chemistry is outlined in a later section in this chapter). Tropospheric ozone is a pollutant and irritant and has a negative impact on plant, animal, and human life.

(3) Methyl chloride and methyl bromide, while only released in trace amounts during biomass burning, may have a negative impact on stratospheric ozone (This chemistry is outlined in a later section in this chapter). Methyl chloride and methyl bromide produced during biomass burning will become even more important as human-produced sources of chlorine and bromine are phased out as a result of the Montreal Protocol banning chlorine and bromine containing gases.

(4) Particulates produced during biomass burning absorb and scatter incoming solar radiation, which impacts climate. In addition, biomass burning-produced particulates will lead to reduced atmospheric visibility. Recent research suggests that atmospheric particulates produced during biomass burning may be able to directly enter the stratosphere via strong vertical currents produced during the fire.

(5) Particulates produced during biomass burning may account as cloud condensation nuclei (CCN) and may impact the formation and distribution of clouds.

(6) Particulates produced during biomass burning, especially particulates of 5-microns or less in diameter, may lead to severe respiratory problems when inhaled.

Calculation of Gaseous and Particulate Emissions from the Vegetation and Peat Fires in Southeast Asia

To assess both the environmental and health impacts of these fires, the gaseous and particulate emissions produced during the fire and released into the atmosphere must be calculated (Levine, 1999). The calculation of gaseous emissions from vegetation and peat fires can be calculated using a form of an expression from Seiler and Crutzen (1980) for each burning ecosystem/terrain:

$$M = A * B * E \quad (2)$$

where M = total mass of vegetation or peat consumed by burning (tons), A = area burned (km^2), B = biomass loading ($tons/km^2$), and E = burning efficiency (dimensionless). The total mass of carbon (M(C)) released to the atmosphere during burning is related to M by the following expression:

$$M(C) = C * M \text{ (Tons of carbon)} \quad (3)$$

C is the mass percentage of carbon in the biomass. For tropical vegetation, C = 0.45 (Andreae, 1991); for peat, C = 0.50 (Yokelson *et al.*, 1996). The mass of CO_2 ($M(CO_2)$) released during the fire is related to M(C) by the following expression:

$$M(CO_2) = CE * M(C) \quad (4)$$

The combustion efficiency (CE) is the fraction of carbon emitted as CO_2 relative to the total carbon compounds released during the fire. For tropical vegetation fires, CE = 0.90 (Andreae, 1991); for peat fires, CE = 0.77 (Yokelson *et al.*, 1997).

Once the mass of CO_2 produced by burning is known, the mass of any other species, X_i ($M(X_i)$), produced by burning and released to the atmosphere can be calculated with knowledge of the CO_2-normalized species emission ratio ($ER(X_i)$). The emission ratio is the ratio of the production of species X_i to the production of CO_2 in the fire. The mass of species, X_i, is related to the mass of CO_2 by the following expression

$$M(X_i) = ER(X_i) * M(CO_2) \text{ (Tons of element } X_i\text{)} \quad (5)$$

For the calculations presented in this paper, X_i = CO, CH_4, NO_x, NH_3, and O_3. It is important to re-emphasize that O_3 is not a direct product of biomass burning. However, O_3 is produced via photochemical reactions of CO, CH_4, and NO_x, all of which are produced directly by biomass burning. Hence, the mass of ozone resulting from biomass burning may be calculated by considering the ozone precursor gases produced by biomass burning.

Values for the CO_2-normalized species emission ratios (ER) for burning tropical vegetation and peat are given in Table 5. The uncertainty in the value of the species emission ratio is typically 30% (Andreae, 1991). These emission measurements were obtained for burning tropical forests in South America, not Southeast Asia. However, studies indicate that the emission ratios from tropical forests in South America should be comparable to those in Southeast Asia (For example, see Andreae, 1991 and Brown and Gaston, 1996). The peat fire emission ratios for gases in Table 5 are based on the measurements of Yokelson *et al.* (1997). These emission measurements were obtained for burning peat from Minnesota and Alaska. Due to the absence in the literature of emission ratio measurements for burning Indonesian peat, values from Yokelson *et al.* (1997) were used. Yokelson *et al.* (1997) did not obtain emission ratios for either NO_x or O_3 from burning peat. To estimate the emission ratio for NO_x from peat fires, the expression for the production ratio of $NH_3/NO_x = 14 \times (1 - MCE)$, was used where MCE is the modified combustion efficiency (Yokelson *et al.,* 1996). For peat fires, MCE = 0.8 and the ratio of $NH_3/NO_x = 2.80$. For NH_3 = 1.28% (Yokelson *et al*., 1997), the value of NO_x is 0.46%. As already noted, CO, CH_4, and NO_x are the chemical precursors in the photochemical production of O_3. For peat fires, the emission ratios for CO, CH_4, and NO_x are 18.15%, 1.04%, and 0.46% (Yokelson *et al.*, 1997). NO_x is the reaction-limiting species in the photochemical production of O_3. To estimate the emission ratio of O_3 for peat fires, we have assumed that the ratio of emission ratio of O_3 to NO_x in forest fires is comparable to that ratio in peat fires. For this assumption, the emission ratio of O_3 from peat fires is found to be 1.04%.

Table 5 Emission Ratios for Tropical Forest Fires and Peat Fires

Species	Tropical Forest Fires	Reference	Peat Fires	Reference
CO_2	90.00%	Andreae (1991)	77.05%	Yokelson *et al.* (1997)
CO	8.5%	Andreae *et al.* (1988)	18.15%	Yokelson *et al.* (1997)
CH_4	0.32%	Blake *et al.* (1996)	1.04%	Yokelson *et al.* (1997)
NO_x	0.21%	Andreae *et al.* (1988)	0.46%	Derived from Yokelson et al. (1997); see text
NH_3	0.09%	Andreae *et al.* (1998)	1.28%	Yokelson *et al.* (1997)
O_3	0.48%	Andreae *et al.* (1988)	1.04%	Derived from Yokelson et al. (1997); see text
TPM[a]	20 tons/kiloton	Ward (1990)	35 tons/kiloton	Ward (1990)

[a]Total particulate matter emission ratios are in units of tons/kiloton (tons of total particulate matter/kiloton of biomass or peat material consumed by fire).

To calculate the total particulate matter (TPM) released from tropical forest fires and peat fires, we use the following expression (Ward, 1991):

$$TPM = M * P \text{ (Tons of carbon)} \tag{6}$$

where P is the conversion of biomass matter or peat matter to particulate matter during burning. For the burning of tropical vegetation, P = 20 tons of TPM per kiloton of biomass consumed by fire; for peat burning, we assume P = 35 tons of TPM per kiloton of organic soil or peat consumed by fire (Ward, 1991).

Total Area Burned and Biomass Consumed During the Fires in Southeast Asia

Perhaps the major uncertainties in the calculation of gaseous and particulate emissions resulting from fires involve poor or incomplete information about four fire parameters: (1) the area burned (A), (2) the ecosystem or terrain that burned (i.e. forests, grasslands, agricultural lands, peat lands, etc.), (3) the biomass loading (B) (i.e. the amount of biomass per unit area of the ecosystem prior to burning), and (4) the fire efficiency (C) (i.e. the amount of biomass in the burned ecosystem that was actually consumed by burning). The area burned can be determined through the use of satellite measurements. Some operational satellite systems to estimate area burned, as well as monitor active fires, are listed in Table 6.

Table 6 Some Operational Satellite Fire Monitoring Systems. The following summary gives the current operational satellite systems used in fire monitoring.

(1) *NOAA(National Oceanic and Atmospheric Administration)/AVHRR (Advanced Very High Resolution Radiometer)*:

Global 1-km imaging systems. Monitors active fires and burned area.

(2) *DMSP (Defense Meteorological Satellite Program)/OLS (Operational Linescan System)*:

Global night-time low light sensor. Monitors active fires.

(3) *GOES (Geostationary Operational Environmental Satellite) Imager*:

Continental high temporal frequency, coarse spatial resolution geostationary imaging. Monitors active fires and smoke.

(4) *ERS (European Remote Sensing Satellite)/ATSR (Along-Track Scanning Radiometer) (European Space Agency)*:

Global 1-km imaging. Monitors active fires and burned areas.

(5) *ERS(European Remote Sensing Satellite /JERS (Japanese Earth Resources Satellite) SAR (Synthetic Aperture Radar) (European Space Agency /NASDA (National Space Development Agency of Japan)*:

Global microwave high resolution system. Monitors burned area.

(6) *LANDSAT (Land Satellite) TM (Thematic Mapper)/MSS (Multispectral Scanner System)*:

Local, high spatial frequency, low temporal frequency. Monitors burned area.

(7) *SPOT (Systeme Pour l'Observation de la Terre (CNES) (Centre National d'Etudes Spatiales)*:

Local, high spatial frequency, low temporal frequency. Monitors burned area.

Indonesia ranks third, after Brazil and the Democratic Republic of the Congo (formerly Zaire), in its area of tropical forest. Of Indonesia's total land area of 1.9 million km^2, current forest cover estimates range from 0.9 to 1.2 million km^2, or 48 to 69% of the total. Forests dominate the landscape of Indonesia (Makarim *et al.*, 1998). Large areas of Indonesian forests burned in 1982 and 1983. In Kalimantan alone, the fires burned from 2.4 to 3.6 million hectare (ha) of forests (Makarim *et al.*, 1998). It is interesting to note that there is an uncertainty of 1.2 million ha or an uncertainty of 50% in our knowledge of the burned area of fires that occurred 16 years ago!

To-date, two different analyses using two different techniques for the determination of the area burned have been reported (Levine *et al.*, 1999;Liew *et al.*, 1998). Levine *et al.* (1999) calculated burned area in Sumatra and Kalimantan for 1997 from 51 fire maps of regions of the highest density of fires prepared by the USDA Forest Service. Levine *et al.* (1999) found a burned area of 12, 847 km^2 in Sumatra and 4221.90 km^2 in Kalimantan, for a total burned area of 17,068.93 km^2 in 1997. The calculated area burned for Sumatra and Kalimanitan for 1997 and 1998 based on USDA Forest Service maps is summarized in Table 7 (Levine *et al.*, 1999).

Table 7 Area Burned in Sumatra, Kalimantan, and Java, Indonesia, October-November, 1997 and February-March 1998. (Based on U. S. D. A. Forest Service Fire Maps; Levine *et al.*, 1999).

Region	Location	Date	Total Area
Sumatra	Jambi	10-15-97	199.39 km^2
	Muaraenim	10-15-97	898.35 km^2
	Sukarami	10-15-97	730.87 km^2
	Jambi	10-19-97	169.86 km^2
	Rengat South	10-19-97	245.71 km^2
	Sukarami	10-19-97	403.86 km^2
	Lahat	10-21-97	319.95 km^2
	Sukarami	10-27-97	4.40 km^2
	Buluranriding River	10-30-97	330.76 km^2
	Talang Gelumbang	10-30-97	947.70 km^2
	Talang Kubuan	10-30-97	2,281.34 km^2
	Kotabumi	11-04-97	693.02 km^2
	North Kotaagung	11-04-97	242.82 km^2
	Kotabumi	11-07-97	73.32 km^2
	Bulau River	11-09-97	258.30 km^2
	East Karangagung	11-09-97	733.23 km^2
	Jambi	11-09-97	35.00 km^2
	Sukadana	11-09-97	169.59 km^2
	Bandar	11-11-97	86.21 km^2

Table7 (continued)

	Kasui	11-11-97	274.53 km^2
	Kotabumi	11-11-97	183.79 km^2
	Sukadana	11-12-97	79.00 km^2
	Sukadana	11-12-97	82.44 km^2
	Bandar	11-13-97	15.29 km^2
	Kotabumi	11-13-97	115.73 km^2
	Surabaja	11-13-97	404.90 km^2
	Bandar	11-14-97	90.93 km^2
	Kasui	11-14-97	165.15 km^2
	Sukadana	11-14-97	37.03 km^2
	Bandar	11-15-97	31.45 km^2
	Sukadana	11-15-97	93.19 km^2
	Kasui	11-16-97	35.80 km^2
	Kotabumi	11-16-97	605.02 km^2
	Surabaja	11-16-97	47.65 km^2
	Bandar	11-17-97	62.63 km^2
	Surabaja	11-17-97	54.81 km^2
	Bujut	11-19-97	124.02 km^2
	Kotabumi	11-19-97	229.04 km^2
	Bujut	11-20-97	47.69 km^2
	Sukadana	11-22-97	381.61 km^2
	Kotabumi	11-23-97	11.34 km^2
	Surabaja	11-23-97	360.47 km^2
	Kasui	11-28-97	113.29 km^2
	Bandar	11-30-97	25.28 km^2
	Surabaja	11-30-97	351.27 km^2
	Kampar	3-06-98	42.60 km^2
	Kampar	3-10-98	44.22 km^2
Kalimantan		*Subtotal*	*12,933.85 km^2*
	Batuwinang	10-27-97	557.28 km^2
	Matua	10-27-97	1,140.87 km^2
	Satui	10-27-97	852.12 km^2

Tabl 7 (continued)

	Bapuju	10-28-97	805.28 km^2
	West Batuwinang	10-29-97	791.05 km^2
	Batuwinang	11-25-97	75.30 km^2
	Nanang	2-07-98	209.27 km^2
	Susang	2-07-98	75.42 km^2
	Santan	2-13-98	328.26 km^2
	Melintang	2-19-98	967.49 km^2
	Susang	2-19-98	19.22 km^2
	Mahakam	2-20-98	422.57 km^2
	Penawai	2-20-98	208.22 km^2
	Sidulang	2-20-98	95.07 km^2
	Saliki	2-21-98	142.88 km^2
	Sambodja	2-21-98	99.97 km^2
	Penawai	2-22-98	152.16 km^2
	Mengangau	2-23-98	209.45 km^2
	Susang	2-23-98	19.30 km^2
	Gitan	2-24-98	306.44 km^2
	Djambu	2-25-98	66.04 km^2
	Bontang	2-27-98	325.93 km^2
	Sedulang	2-27-98	562.44 km^2
	Guntung	2-28-98	973.46 km^2
	Sangatta	3-03-98	670.15 km^2
	Sideman	3-04-98	208.41 km^2
	Rapak	3-13-98	2,050.68 km^2
Java		*Subtotal*	*12,334.73 km^2*
	Malang	10-06-97	76.19 km^2
	Madium	10-22-97	277.77 km^2
	Blitar	10-23-97	9.00 km^2
		Subtotal	*362.96 km^2*
		Total	**25,631.54 km^2**

Liew et al. (1998) analyzed 766 SPOT "quicklook" images with almost complete coverage of Kalimantan and Sumatra from August to December 1997. Liew *et al.* (1998) estimate the burned area in Kalimantan to be 30,600 km^2 and the burned area in Sumatra to be 15,000 km^2, for a total burned area of 45,600 km^2. The Liew *et al.* (1998) area estimate is a factor of 2.7 greater than the area estimate of Levine *et al.* (1998). This is not surprising since the Liew *et al.* (1998) estimate is based on almost complete coverage of Kalimantan and Sumatra, while the Levine *et al.* (1998) estimate is based on the USDA Forest Service maps prepared for only the very highest density fire regions in Kalimantan and Sumatra. For the calculations reported in this paper, we have used the Liew *et al.* (1998) estimate for total burned area in Kalimantan and Sumatra of 45,600 km^2 (This is equivalent to the combined areas of the states of Rhode Island, Delaware, Connecticut, and New Jersey, in the United States). The estimate of Liew *et al.* (1998) represents only a lower limit estimate of the area burned in Southeast Asia in 1997, since the SPOT data only covered Kalimantan and Sumatra and did not include fires on the other Indonesian islands of Irian Jaya, Sulawesi, Java, Sumbawa, Komodo, Flores, Sumba, Timor, and Wetar or the fires in the neighboring countries of Malaysia and Brunei.

What is the nature of the ecosystem/terrain that burned in Kalimantan and Sumatra? In October 1997, NOAA satellite monitoring produced the following distribution of fire hot spots in Indonesia (UNDAC, 1998): agricultural and plantation areas: 45.95%; bush and peat soil areas: 24.27%; productive forests: 15.49%; timber estate areas: 8.51%; protected areas: 4.58%; and transmigration sites: 1.20% (The three forest/timber areas add up to a total of 28.58% of the area burned). While the distribution of fire hot spots is not an actual index for area burned, the NOAA satellite-derived hot spot distribution is quite similar to the ecosystem/terrain distribution of burned area deduced by Liew *et al.* (1998) based on SPOT images of the actual burned areas: agricultural and plantation areas: 50%; forests and bushes: 30%; and peat swamp forests: 20%. Since the estimates of burned ecosystem/terrain of Liew *et al.* (1998) are based on actual SPOT images of the burned area, their estimates were adopted in our calculations.

What is the biomass loading for the three terrain classifications identified by Liew *et al.* (1998)? Values for biomass loading or fuel load for various tropical ecosystems are summarized in Table 8. The biomass loading for tropical forests in Southeast Asia ranges from 5,000 to 55,000 tons/km^2, with a mean value of 23,000 tons/km^2 (Brown and Gaston, 1996). However, in our calculations we have used a value of 10,000 tons/km^2 to be conservative. The biomass loading for agricultural and plantation areas (mainly rubber trees and oil palms) of 5,000 tons/ km^2, is also a conservative value (Liew *et al.*, 1998). Nichol (1997) has investigated the peat deposits of Kalimantan and Sumatra and used a biomass loading value of 97,500 tons/km^2 (Supardi *et al.*, 1993) for the dry peat deposits 1.5 m thick as representative of the Indonesian peat in her study. Brunig (1977) gives a similar value for peat biomass loading. The combustion efficiency for forests is estimated at 0.20 and for peat is estimated at 0.50 (Levine and Cofer, 1999). Based on the discussions presented in this section, the values for burned area, biomass loading, and combustion efficiency, used in the calculations are summarized in Table 9.

Table 8 Biomass Load Range and Burning Efficiency in Tropical Ecosystems (From Scholes, 1996, except as noted).

Vegetation Type	Biomass Load Range (tons/km^2)	Burning Efficiency
Peat[a]	97500	0.50
Tropical rainforests	5000-55000	0.20
Evergreen forests	5000-10000	0.30
Plantations	500-10000	0.40
Dry forests	3000-7000	0.40
Fynbos	2000-4500	0.50
Wetlands	340-1000	0.70
Fertile grasslands	150-550	0.96
Forest/savanna mosaic	150-500	0.45
Infertile savannas	150-500	0.95
Fertile savannas	150-500	0.95
Infertile grasslands	150-350	0.96
Shrublands	50-200	0.95

[a]Brunig (1977), Supardi *et al.* (1993)
[b]Brown and Gaston (1996)

Table 9 Parameters Used in Calculations

(1) Total Area Burned in Kalimantan and Sumatra, Indonesia in 1997: 45,600 km^2	
(2) Distribution of Burned Areas, Biomass Loading and Combustion Efficiency	
A. Agricultural and plantation areas	50%, 5,000 tons/km^2, 0.20
B. Forests and bushes	30%, 10,000 tons/km^2, 0.20
C. Peat swamp forests	20%, 97,500 tons/km^2, 0.50

Results of Calculations: Gaseous and Particulate Emissions from the Fires in Kalimantan and Sumatra, Indonesia, August to December 1997

The calculated gaseous and particulate emissions from the fires in Kalimantan and Sumatra, from August to December 1997, are summarized in Table 10 (It is important to keep in mind that wild fires continued throughout Southeast Asia from January through April 1998 and that the fires covered much more of the region than Kalimantan and

Sumatra). For each of the seven species listed, the emissions due to agricultural/plantation burning (A), forest burning (F), and peat burning (P) are given. The total (T) of all three components (A+F+P) is also given. The "best estimate" total emissions are: CO_2: 191.485 million metric tons of C (MtC); CO: 32.794 MtC; CH_4: 1.845 MtC; NO_x: 5.898 MtN; NH_3: 2.585 MtN; O_3: 7.100 MtO_3; and total particulate matter: 16.154 MtC. The carbon dioxide emissions from these fires is about 2.2% of the global annual production of carbon dioxide from all sources (See Table 4 for global annual production of carbon dioxide, which is 8700 MtC and 191.485 MtC/8700 MtC = 2.2%). The percentage for other gases produced by these fires compared to the global annual production from all sources is: carbon monoxide: 2.98%, methane: 0.48%, oxides of nitrogen: 2.43%, ammonia: 5.87%, and total particulate matter: 1.08%.

Table 10 Gaseous and Particulate Emissions from the Fires in Kalimantan and Sumatra in 1997 (For Total Burned Area = 45,600 km^2). For each species, the best estimate emission value is on first line and the range of emission values in parenthesis under best guess (See text for discussion of emission estimate range and uncertainty calculations). Units of emissions: Million metric tons (Mt) of C for CO_2, CO, and CH_4; Mt of N for NO_x and NH_3; Mt of O_3 for O_3; MtC of particulates; 1 Mt = 10^{12} grams = 1 Teragram, Tg.

	Agricultural/ Plantation Fire Emissions	Forest Fire Emissions	Peat Fire Emissions	Total Fire Emissions
CO_2	9.234	11.080	171.170	191.485
	(4.617 - 13.851)	(5.54 - 16.62)	(85.585 - 256.755)	(95.742 - 287.226)
CO	0.785	0.942	31.067	32.794
	(0.392 - 1.177)	(0.471 - 1.413)	(15.533 - 46.600)	(16.397 - 49.191)
CH_4	0.030	0.035	1.780	1.845
	(0.015 - 0.045)	(0.017 - 0.052)	(0.89 - 2.67)	(0.922 - 2.767)
NO_x	0.023	0.027	0.921	0.971
	(0.011 - 0.034)	(0.013 - 0.040)	(0.460 - 1.381)	(0.485 - 1.456)
NH_3	0.010	0.012	2.563	2.585
	(0.005 - 0.015)	(0.006 - 0.018)	(1.281 - 3.844)	(1.292 - 3.877)
O_3	0.177	0.213	6.710	7.100
	(0.088 - 0.265)	(0.106 -0.319)	(3.35 - 10.06)	(3.55 - 10.65)
TPM	0.460	0.547	15.561	16.568
	(0.23 - 0.69)	(0.273 - 0.820)	(7.780 - 23.341)	(8.284 - 24.852)

Scholes *et al.* (1996) calculated the biomass consumed by burning in 11 different ecosystems in another tropical ecosystem, southern Africa, using equations (2) - (4). Scholes *et al.* (1996) performed a detailed statistical analysis of the errors associated with the calculated values of biomass consumed by fire using a statistical procedure of Nelson (1992), which assumes that all error terms (e) are independent. In the error analysis of Scholes et al. (1996), the total error (E_{total}), which corresponds to the 3-sigma (99%) confidence level, is estimated using the following expression

$$e_{total} = (e^2_{burned\ area} + e^2_{fuel\ load} + e^2_{combustion\ completeness})^{1/2} \qquad (7)$$

Scholes *et al.* (1996) assumed the following uncertainties for each calculation parameter: $e_{burned\ area}$ = 30%, $e_{fuel\ load}$= 30% and $e_{combustion\ completeness}$ = 25%. In the error analysis of the calculations presented in this paper, the error associated with uncertainties in the emission ratio ($e_{emission\ ratio}$) (30%) has also been included. The uncertainties in the calculation parameters in the detailed error of analysis of Scholes *et al.* (1996) were adopted in this study, with the exception of the uncertainty in the burned area of 30%. The burned area determination used in Scholes *et al.* (1996) was based on satellite measurements of active fires that were converted to burned area, which is not a simple one-to-one transformation and introduces errors. The burned area determination used in the present paper was more straight-forward since it was based on direct satellite photography of burned areas using SPOT images (Liew *et al.*, 1998). Hence, there is less uncertainty in this burned area determination and an uncertainty of 10% was assumed (Liew *et al.* (1998) did not give a burned area uncertainty in their paper). Using equation (7), the calculated uncertainty in the emission calculations presented in this paper is 50.2%. The uncertainty range for each species emission is shown in parentheses under the "best estimate" value in Table 10.

However, it is important to re-emphasize that these emission calculations represent lower limit values since the calculations are only based on burning in Kalimantan and Sumatra in 1997. The calculations do not include burning in Java, Sulawesi, Irian Jaya, Sumbawa, Komodo, Flores, Sumba, Timor, and Wetar in Indonesia or in neighboring Malaysia and Brunei.

It is interesting to compare the gaseous and particulate emissions from the 1997 Kalimantan and Sumatra fires with those from the Kuwait oil fires of 1991, described as a major environmental catastrophe. Laursen *et al.* (1992) have calculated the emissions of CO_2, CO, CH_4, NO_x, and particulates from the Kuwait oil fires in units of metric tons per day. The Laursen *et al.* (1993) calculations are summarized in Table 10. To compare these calculations with the calculations presented in this paper for Kalimantan and Sumatra (Table 10), we have normalized our calculations by the total number of days of burning. The SPOT images (Liew *et al.*, 1998) covered a period of 5 months (August-December 1997) or about 150 days. For comparison with the Kuwait fire emissions, we divided our calculated emissions by 150 days. These values are summarized in Table 11. The gaseous and particulate emissions from the fires in Kalimantan and Sumatra significantly exceeded the emissions from the Kuwait oil fires. The 1997 fires in

Kalimantan and Sumatra were a significant source of gaseous and particulate emissions to the local, regional, and global atmosphere.

Table 11 Comparison of Gaseous and Particulate Emissions: The Indonesian Fires and the Kuwait Oil Fires. (Units of emissions: Metric tons per day of C for CO_2, CO, and CH_4; metric tons per day of N for NO_x; metric tons per day for particulates; 1 million metric tons = 10^{12} g = 1 Teragram, Tg).

Species	Indonesian Fires	Kuwait Oil Fires[a]
CO_2	1.28×10^6	5.0×10^5
CO	2.19×10^5	4.4×10^3
CH_4	1.23×10^4	1.5×10^3
NO_x	6.19×10^3	2.0×10^2
Particulates	1.08×10^5	1.2×10^4

[a]Laursen *et al.* (1992)

ATMOSPHERIC CHEMISTRY RESULTING FROM THE GASEOUS AND PARTICULATE EMISSIONS FROM THE FIRES IN SOUTHEAST ASIA

Chemistry of the Hydroxyl Radical (OH) in the Troposphere

The hydroxyl radical (OH) is the major chemical scavenger in the troposphere and it controls the atmospheric lifetime of most gases in the troposphere. The atmospheric lifetime, t, of any gas, x_i, that reacts with the OH radical is given by the following expression:

$$t = 1/k\,[OH] \tag{8}$$

where k is the kinetic reaction rate for the reaction between OH and x_i and [OH] is the concentration of the OH radical (molecules/cm^3).

The concentration of the OH radical is controlled by the balance between its chemical production and destruction. The hydroxyl radical (OH) is formed by the reaction of excited atomic oxygen (O(^{1}D)) with water vapor:

$$O(^1D) + H_2O \xrightarrow{k_1} 2OH \tag{9}$$

Tropospheric excited atomic oxygen (O(^{1}D)) is produced by the photolysis of ozone (O_3). In the troposphere, this photolysis reaction occurs over a very narrow spectral interval, between 290 and 310 nm. The production of excited atomic oxygen decreases as the latitude increases, i.e., less incoming solar radiation for photolysis is available. The bulk of the water vapor in the atmosphere resides in the troposphere. The amount of H_2O in the atmosphere is controlled by the saturation vapor pressure, which decreases with decreasing atmospheric temperature, i.e., as altitude or latitude increases.

The OH radical is destroyed via its reactions with CO and CH_4, both important products of biomass burning:

$$OH + CO \xrightarrow{k_2} H + CO_2 \tag{10}$$

$$OH + CH_4 \xrightarrow{k_3} CH_3 + H_2O \tag{11}$$

The concentration of the OH radical (molecules/cm^3) is determined by dividing the OH production term by its destruction terms:

$$[OH] = 2\, k_1\, [O(^1D)]\, [H_2O]/(k_2\, [CO] + k_3\, [CH_4]) \tag{12}$$

Production of Ozone (O_3) in the Troposphere

In addition to controlling the chemical destruction of OH, the oxidation of CO and CH_4 by OH outlined above initiates the carbon monoxide and methane oxidation schemes, which lead to the photochemical production of O_3 in the troposphere.

Carbon Monoxide Oxidation Chain

$$CO + OH \rightarrow CO_2 + H \tag{13}$$

$$H + O_2 + M \rightarrow HO_2 + M \tag{14}$$

$$HO_2 + NO \rightarrow NO_2 + OH \tag{15}$$

$$NO_2 + h\nu \rightarrow NO + O \tag{16}$$

$$O + O_2 + M \rightarrow O_3 + M \tag{17}$$

Net Reaction:

$$CO + 2O_2 \rightarrow CO_2 + O_3 \tag{18}$$

Note that the key gases in the carbon monoxide oxidation chain leading to the photochemical production of tropospheric ozone, CO, and NO, are both produced by biomass burning.

Methane Oxidation Chain

$$CH_4 + OH \rightarrow CH_3 + H_2O \tag{19}$$

$$CH_3 + O_2 + M \rightarrow CH_3O_2 + M \tag{20}$$

$$CH_3O_2 + NO \rightarrow CH_3O + NO_2 \tag{21}$$

$$CH_3O + O_2 \rightarrow CH_2O + HO_2 \tag{22}$$

$$HO_2 + NO \rightarrow NO_2 + OH \tag{23}$$

$$2 \text{ x } (NO_2 + h\nu \rightarrow NO + O) \tag{24}$$

$$2 \text{ x } (O + O_2 + M \rightarrow O_3 + M) \tag{25}$$

Net Reaction:

$$CH_4 + 4O_2 \rightarrow CH_2O + H_2O + 2O_3 \tag{26}$$

Note that the key gases in the methane oxidation chain leading to the photochemical production of tropospheric ozone, CH_4, and NO, are both produced by biomass burning.

Chemistry of Nitrogen Oxides in the Troposphere

In addition to being a key player in the carbon monoxide and methane oxidation chains leading to the chemical production of ozone in the troposphere, NO also leads to the chemical production of nitric acid (HNO_3), the fastest growing component of acidic precipitation. NO is chemically transformed to nitrogen dioxide (NO_2) and then to nitric oxide via the following reactions:

$$NO + O_3 \rightarrow NO_2 + O_2 \tag{27}$$

$$NO + HO_2 \rightarrow NO_2 + OH \tag{28}$$

$$NO + CH_3O_2 \rightarrow NO_2 + CH_3O \quad (29)$$

$$NO_2 + OH + M \rightarrow HNO_3 + M \quad (30)$$

Chemistry of the Stratosphere

Approximately 90% of the ozone in the atmosphere is found in the stratosphere (15 - 50 km), with about 10% in the troposphere (0 - 15 km). Stratospheric ozone is very important because it absorbs ultraviolet radiation (200 - 300 nm) from the Sun and shields the surface from this biologically lethal radiation. Stratospheric ozone is destroyed via a series of chemical reactions involving chlorine and bromine. Chlorine destroys stratospheric ozone through the following catalytic cycle (Wayne, 1991):

$$Cl + O_3 \rightarrow ClO + O_2 \quad (31)$$

$$ClO + O \rightarrow Cl + O_2 \quad (32)$$

The net reaction is:

$$O + O_3 \rightarrow 2O_2 \quad (33)$$

Bromine destroys stratospheric ozone via the following three reaction cycles (World Meteorological Organization, 1995):

Cycle 1

$$BrO + ClO + h\nu \rightarrow Br + Cl + O_2 \quad (34)$$

$$Br + O_3 \rightarrow BrO + O_2 \quad (35)$$

$$Cl + O_3 \rightarrow ClO + O_2 \quad (36)$$

The net reaction is:

$$2O_3 \rightarrow 3O_2 \quad (37)$$

Cycle 2

$$BrO + HO_2 \rightarrow HOBr + O_2 \quad (38)$$

$$HOBr + h\nu \rightarrow OH + Br \quad (39)$$

$$Br + O_3 \rightarrow BrO + O_2 \tag{40}$$

$$OH + O_3 \rightarrow HO_2 + O_2 \tag{41}$$

Cycle 3

$$BrO + NO_2 + M \rightarrow BrONO_2 + M \tag{42}$$

$$BrONO_2 + M \rightarrow Br + NO_3 \tag{43}$$

$$NO_3 + h\nu \rightarrow NO + O_2 \tag{44}$$

$$Br + O_3 \rightarrow BrO + O_2 \tag{45}$$

$$NO + O_3 \rightarrow NO_2 + O_2 \tag{46}$$

There are two other channels for the reaction between BrO + ClO (Wayne, 1991):

$$BrO + ClO \rightarrow Br + OClO \tag{47}$$

$$BrO + ClO \rightarrow BrCl + O_2 \tag{48}$$

There is also a relevant reaction between BrO and itself (Wayne, 1991):

$$BrO + BrO \rightarrow Br + Br + O_2 \tag{49}$$

Biomass Burning and Atmospheric Nitrogen and Oxygen

Biomass burning is both an instantaneous source (combustion of biomass) and a long-term source (enhanced biogenic soil emissions via increased nitrification and denitrification in soil) of gases to the atmosphere. These gases impact the chemistry of the troposphere and stratosphere, as outlined in this chapter. In addition, biomass burning may also impact atmospheric concentrations and the biogeochemical cycling of nitrogen and oxygen, the two major constituents of the atmosphere. Through the process of nitrogen fixation, molecular nitrogen (N_2) is transformed to the surface in the form of "fixed" nitrogen, i.e., ammonium (NH_4^+) and nitrate (NO_3^-). Nitrogen fixation results from both natural processes (biological fixation in root modules in certain agricultural crops and atmospheric lightning) and human processes (the production of nitrogen fertilizer and high temperature combustion). The world's use of industrially fixed nitrogen fertilizer has increased from about 3 Teragrams (Tg) of N/year in 1940 to about 75 Tg N/yr in 1990 (Levine *et al.*, 1996b). The "fixed" nitrogen in the forms of NH_4^+ and

NO_3^- is returned to the atmosphere mainly in the form of N_2, with smaller amounts of N_2O, and still smaller amounts of NO by denitrification and in the form of NO by nitrification. Burning or "pyro-denitrification" may also be an important source of nitrogen, mostly in the form of N_2, from the biosphere to the atmosphere (Lobert *et al.*, 1991; Levine *et al.*, 1996). The problem is that it is difficult to quantify the amount of N_2 released during burning (Lobert *et al.*, 1991), however, biomass burning or pyrodenitrification may prove to be an important process in the recycling of nitrogen compounds from the biosphere to the atmosphere.

Burning impacts the concentration of atmospheric oxygen in two ways. Carbon released during the burning of biomass combines with atmospheric oxygen to form carbon dioxide. Hence, burning is a sink for atmospheric oxygen. In addition, biomass burning destroys the very source of atmospheric oxygen - its production in the biosphere via the process of photosynthesis in the world's forests.

THE IMPACT OF THE SOUTHEASTERN ASIA FIRES ON THE COMPOSITION AND CHEMISTRY OF THE ATMOSPHERE

A series of recent papers have discussed the impact of the Southeastern Asia fires on the composition and chemistry of the atmosphere (Fujiwara *et al.*, 1999, Hauglustaine *et al.*, 1999, Matsueda and Inoue, 1999, Nakajima *et al.*, 1999, Rinsland *et al.*, 1999, Sawa *et al.*, 1999, and Tsutsumi *et al.*, 1999). The results of these studies are summarised below.

Modeling Ozone and Carbon Monoxide over Indonesia

Hauglustaine *et al.* (1999) used MOZART, a global chemical transport model to investigate the impact of fire emissions calculated by Levine (1999) (presented in Table 10) on the photochemical production of ozone over Indonesia. The calculations indicate that the tropospheric ozone column density increased by 20-25 Dobson Units (DU) and the ozone mixing ratio reached 50 parts per billion by volume (ppbv) in the mid-troposphere in November (These model calculations are consistent with in situ ozone measurements obtained by Fujiwara *et al.*, 1999, described in the next section and satellite measurements of ozone obtained with the Earth Probe/TOMS (Chandra *et al.*, 1998) and the ERS-2/GOME (Burrows *et al.*, 1999)). South of the source region, low ozone mixing ratios of 20-25 ppbv were calculated in the boundary layer due to marine air influence and reduced photochemical activity in the presence of biomass burning aerosols.

Normal, non-fire surface concentrations of carbon monoxide over Indonesia are usually less than about 100 ppbv. The calculations of Hauglustaine *et al.* (1999) indicate increases of carbon monoxide by up 2900 ppbv at the surface over Kalimantan and Suimatra. A perturbation of 50-1000 ppbv was calculated over the entire region of

Indonesia. In the free troposphere, CO increased by 1000 ppbv over Indonesia due to the upward transport prevailing in this region. The calculated high levels of carbon monoxide are consistent with actual measurements of carbon monoxide obtained over Indonesia during the fires by Sawa *et al.* (1999).

Measurements over Indonesia

Tsutsumi *et al.* (1999): Aircraft measurements over Kalimantan, Indonesia, on October 13, 1997, indicated high concentrations of O_3, NO_x, CO, and aerosols. The maximum concentration of O_3 (80.5 ppbv) was found in the middle layer of the smoke haze and very low concentrations (about 20 ppbv) were measured in the lower smoke layer. The authors concluded that the low O_3 concentrations near the surface may be due to the reduction in solar flux due to aerosols high in the haze layer and the loss of O_3 due to large aerosol surface area in the lower haze layer.

Fujiwara *et al.* (1999): Pronounced enhancements of total and tropospheric ozone were observed with a Brewer spectrophotometer and ozonesondes at Watukosek, Indonesia in October 1997. The integrated tropospheric O_3 increased from 20 DU to 55 DU in October 1997. On October 22, 1997, the O_3 concentrations were more than 50 ppbv throughout the troposphere and exceeded 100 ppbv at several altitudes. The authors conclude that the enhanced levels of O_3 measured in October 1997, resulted from O_3 precursors produce during the fires and released into the atmosphere.

Sawa *et al.* (1999) measured carbon monoxide and hydrogen over Kalimantan during the fires. They reported carbon monoxide concentrations in the range of 3000 to 9000 ppbv below an altitude of 2.6 km. Above 4.4 km altitude, the carbon monoxide concentration was 500 to 1200 ppbv.

Measurements between Singapore and Japan

Matsueda *et al.* (1999): Aircraft measurements of carbon dioxide (CO_2), carbon monoxide (CO), and methane (CH_4) were obtained between Japan and Singapore in October 1993, 1996, and 1997. The mixing ratio of all three gases at 9-12 km were enhanced over the South China Sea in 1997 compared with measurements in 1993 and 1996.

Measurements over Hawaii

Rinsland *et al.* (1999): Infrared solar spectral measurements of carbon monoxide (CO), methane (C_2H_6), and hydrogen cyanide (HCN) were obtained for over 250 days between August 1995 and February 1998 above Mauna Loa, Hawaii. Correlated variations of CO, C_2H_6, and HCN and unusual seasonal cycles observed above Mauna Loa, Hawaii, during the second half of 1997 suggest a common origin for these emissions. Back-trajectory model calculations indicate that the source of these anomalous concentrations over Hawaii was the fires in Southeast Asia.

CONCLUSIONS

From August 1997 through early 1998, the entire region of Southeast Asia, but particularly Indonesia and Malaysia, experienced extensive and widespread wildfires. These fires were initiated for land clearing and land-use change, as is the usual custom in this region of the world. However, in 1997, the severe drought conditions resulting from a very severe El Niño Southern Oscillation resulted in very widespread, uncontrolled wildfires that lasted for months, well into 1998. These emissions significantly reduced atmospheric visibility and produced record high levels of atmospheric pollution. These gaseous and particulate emissions also impacted human health. Three different agencies of the United Nations (The United Nations Environmental Program (UNEP), the World Meteorological Organization (WMO), and the World Health Organization (WHO)) conducted studies and workshops and issued reports on the environmental and health impacts of these fires. Many lessons were learned but much work needs to be done to avoid such catastrophes in the future.

Acknowledgments

This research was funded by the Global Change Program of the U. S. Environmental Protection Agency under Interagency Agreement DW 80936540-01-1 and the Director's Discretionary Fund of NASA Langley Research Center.

REFERENCES

1. Andreae, M. O., 1991. Biomass burning: Its history, use, and distribution and its impact on environmental quality and global climate. *Global Biomass Burning: Atmospheric, Climatic, and Biospheric Implications* (J. S. Levine, Ed.). The MIT Press, Cambridge, Massachusetts, pp. 3-21.
2. Andreae, M. O., E. V. Browell, M. Garstang, G. L. Gregory, R. C. Harriss, G. F. Hill, D. J. Jacob, M. C. Pereira, G. W. Sachse, A. W. Setzer, P. L. Silva Dias, R. W. Talbot, A. L. Torres and S.C. Wofsy, 1998. Biomass burning emission and associated haze layers over Amazonia. *Journal of Geophysical Research* **93**: 1509-1527.
3. Blake, N. J., D. R. Blake, B. C. Sive, T. -Y. Chen, F. S. Rowland, J. E. Collins, G. W. Sachse and B. E. Anderson, 1996. Biomass burning emissions and vertical distribution of atmospheric methyl halides and other reduced carbon gases in the South Atlantic Region. *Journal of Geophysical Research* **101**: 24,151-24,164.

4. Brown, S. and G. Gaston, 1996. Estimates of biomass density for tropical forests. *Biomass Burning and Global Change, Volume 1* (J. S. Levine, Ed.). The MIT Press, Cambridge, Massachusetts, pp. 133-139.
5. Brunig, E. F., 1997. The tropical rainforest-A wasted asset or an essential biospheric resource? *Ambio* **6**: 187-191.
6. Burrows, J. P., M. Weber, M. Buchwitz, V. Rozanov, A. Ladstätter-Weißenmayer, A. Richter, R. DeBeek, R. Hoogen, K. Bramstedt, K.-U. Eichmann, M. Eisinger and D. Perner, 1999.The Global Ozone Monitoring Experiment (GOME): Mission concept and first scientific results. *Journal of Atmospheric Sciences* **56**: 151-175.
7. Chandra, S., J. R. Ziemke, W. Min, and W. G. Read, 1998. Effects of 1997-1998 El Niño on tropospheric ozone and water vapor. *Geophysical Research Letters* **25**: 3867-3870.
8. Crutzen, P. J., and M. O. Andreae, 1990. Biomass burning in the tropics: Impact on atmospheric chemistry and biogeochemical cycles. *Science* **250**: 1679-1678.
9. Crutzen, P. J., and J. G. Goldammer (Eds.). 1993. *Fire in the Environment: The Ecological, Atmospheric, and Climatic Importance of Vegetation Fires*. John Wiley and Sons, Chicester, England. 400 pages.
10. Crutzen, P. J., L. E. Heidt, J. P. Krasnec, W. H. Pollock and W. Seiler, 1979. Biomass burning as a source of atmospheric gases CO, H_2, N_2O, NO, CH_3Cl, COS. *Nature* **282**: 253-256.
11. Fujiwara, M., K. Kita, S. Kawakami, T. Ogawa, N. Komala, S. Saraspriya, and A. Suripto, 1999. Tropospheric ozone enhancements during the Indonesian forest fire events in 1994 and in 1997 as revealed by ground-based observations. *Geophysical Research Letters* **26**: 2417-2420.
12. Goldammer, J. G. (Ed.). 1990. *Fire in the Tropical Biota: Ecosystem Processes and Global Challenges*. Springer-Verlag, Berlin, Germany, 497 pages.
13. Goldammer, J. G., and V. V. Furyaev (Eds.). 1996. *Fire in Ecosystems of Boreal Eurasia* Kluwer Academic Publishers, Dordrecht, The Netherlands, 528 pages.
14. Hauglustaine, D. A., G. P. Brasseur, and J. S. Levine. 1999. A sensitivity simulation of tropospheric ozone changes due to the 1997 Indonesian fire emissions. *Geophysical Research Letters*, **26**: 3305-3308.
15. Laursen, K. K., R. J. Ferek, and P. V. Hobbs, 1992. Emission factors for particulates, elemental carbon, and trace gases from the Kuwait oil fires. *Journal of Geophysical Research* **97**: 14,491-14,497.
16. Levine, J. S., W. R. Cofer, D. R. Cahoon, E. L. Winstead, D. I. Sebacher, M. C. Scholes, D. A. B. Parsons, and R. J. Scholes, 1996b. Biomass Burning, Biogenic Soil Emissions, and the Global Nitrogen Budget. *Biomass Burning and Global Change, Volume 1* (J. S. Levine, Ed.). The MIT Press. Cambridge, Massachusetts, pp. 370-380.
17. Levine, J. S., 1999. The 1997 fires in Kalimantan and Sumatra, Indonesia: Gaseous and particulate emissions. *Geophysical Research Letters* **26**: 815-818.
18. Levine, J. S. (Ed.), 1991. *Global Biomass Burning: Atmospheric, Climatic, and Biospheric Implications*. The MIT Press, Cambridge, Massachusetts, 569 pages.

19. Levine, J. S. (Ed.). 1996a. *Biomass Burning and Global Change: Remote Sensing, Modeling and Inventory Development, and Biomass Burning in Africa*. The MIT Press, Cambridge, Massachusetts, 581 pages.
20. Levine, J. S. (Ed.), 1996b. *Biomass Burning and Global Change: Biomass Burning in South America, Southeast Asia, and Temperate and Boreal Ecosystems, and the Oil Fires of Kuwait*. The MIT Press, Cambridge, Massachusetts, 377 pages.
21. Levine, J. S., T. Bobbe, N. Ray, R. G. Witt, and A. Singh, 1999. *Wildland Fires and the Environment: A Global Synthesis*. Environment Information and Assessment Technical Report 99-1, The United Nations Environmental Program, Nairobi, Kenya, 46 pages.
22. Levine, J. S. and W. R. Cofer, 1999. Boreal forest fire emissions and the chemistry of the atmosphere. *Fire, Climate Change and Carbon Cycling in the North American Boreal Forests* (E. S. Kasischke and B. J. Stocks, Eds.), Ecological Studies Series, Springer-Verlag, New York, (In press).
23. Levine, J. S., W. R. Cofer, D. R. Cahoon, and E. L. Winstead, 1995. Biomass burning: A driver for global change. *Environmental Science and Technology* **29**: 120A-125A.
24. Levine, J. S., T. D. Edwards, T. E. McReynolds, and C. W. Dull. 1999. Gaseous and particulate emissions from the fires in Kalimantan and Sumatra, Indonesia. *WMO Workshop on Regional Transboundary Smoke and Haze in South-East Asia* (G. R. Carmichael, Compiler), World Meteorological Organization Report No. 131, Volume 2, Geneva, Switzerland, pp. 261-284.
25. Liew, S. C., O. K. Lim, L. K. Kwoh, and H. Lim, 1998. A study of the 1997 fires in South East Asia using SPOT quicklook mosaics. Paper presented at the 1998 International Geosience and Remote Sensing Symposium, July 6-10, Seattle, Washington, 3 pages.
26. Lobert, J. M., D. H. Scharffe, W.-M. Hao, T. A. Kuhlbusch, R. Seuwen, P. Warneck, and P. J. Crutzen, 1991. Experimental evaluation of biomass burning emissions: Nitrogen and carbon containing compounds. *Global Biomass Burning: Atmospheric, Climatic, and Biospheric Implications* (J. S. Levine, Ed.). The MIT Press, Cambridge, Massachusetts, pp. 289-304.
27. Makarim, N., Y. A. Arbai, A. Deddy, and M. Brady, 1998. Assessment of the 1997 Land and Forest Fires in Indonesia: National Coordination. *International Forest Fire News*, United Nations Economic Commission for Europe and the Food and Agriculture Organization of the United Nations, Geneva, Switzerland, No. 18, January 1998, 4-12.
28. Matsueda, H., and H. Y. Inoue, 1999. Aircraft measurements of trace gases between Japan and Singapore in October of 1993, 1996, and 1997. *Geophysical Research Letters* **26**: 2413-2416,
29. Nakajima, T., A. Higurashi, N. Takeuchi, and J. R. Herman, 1999. Satellite and ground-based study of optical properties of 1997 Indonesian forest fire aerosols. *Geophysical Research Letters* **26**: 2421-2424.
30. Nelson, L.S., 1992. Technical aids: Propagation of error. *Journal of Qualitative Control* **24**: 232-235.

31. Nichol, J., 1997. Bioclimatic Impacts of the 1994 smoke haze event in Southeast Asia. *Atmospheric Environment* **44**: 1209-1219.
32. Rinsland, C. P., A. Goldman, F. J. Murcray, T. M. Stephen, N. S. Pougatchev, J. Fishman, S. J. David, R. D. Blatherwick, P. C. Novelli, N. B. Jones, and B. J. Connor, 1999. Infrared solar spectroscopic measurements of free tropospheric CO, C_2H_6, and HCN above Mauna Loa, Hawaii: Seasonal variations and evidence for enhanced emissions from the Southeast Asian tropical fires of 1997-98. *Journal of Geophysical Research* **104**: 18,667-18,680.
33. Sawa, Y., H. Matsueda, Y. Tsutsumi, J. B. Jensen, H. Y. Inoue, and Y. Makino, 1999. Tropospheric carbon monoxide and hydrogen measurements over Kalimantan in Indonesia and northern Australia during October, 1997. *Geophysical Research Letters* **26**: 1389-1392.
34. Scholes, R. G., J. Kendall, and C. O. Justice, 1996. The quantity of biomass burned in southern Africa. *Journal of Geophysical Research* **101**: 23,667-23,676.
35. Seiler, W. and P.J. Crutzen, 1980. Estimates of gross and net fluxes of carbon between the biosphere and the atmosphere from biomass burning. *Climatic Change* **2**, 207-247.
36. Simons, L. M., 1998. Indonesia's Plague of Fire. *National Geographic* **194**: No. 2, August issue, 100-119.
37. Supardi, A. D. Subekty and S. G. Neuzil, 1993. General geology and peat resources of the Siak Kanan and Bengkalis Island peat deposits, Sumatra, Indonesia. *Modern and Ancient Coal Forming Environments* (J. C. Cobb and C. B. Cecil, Eds.). Geological Society of America Special Paper, Volume 86, pp. 45-61.
38. Tsutsumi, Y., Y. Sawa, Y. Makino, J. B. Jensen, J. L. Gras, B. F. Ryan, S. Diharto, and H. Harjanto, 1999. Aircraft measurements of ozone, NOx, CO, and aerosol concentrations in biomass burning smoke over Indonesia and Australia in October 1997: Depleted ozone layer at low altitude over Indonesia. *Geophysical Research Letters* **26**: 595-598,
39. UNDAC (United Nations Disaster Assessment and Coordination Team), 1998. Mission on Forest Fires, Indonesia, September-November 1997. *International Forest Fire News*, United Nations Economic Commission for Europe and the Food and Agriculture Organization of the United Nations, Geneva, Switzerland, No. 18, January 1998, 13-26.
40. van Wilgen, B. W., M. O. Andreae, J. G. Goldammer, and J. A. Lindesay (Eds.), 1997. *Fire in Southern African Savannas: Ecological and Atmospheric Perspectives*. Witwatersrand University Press, Johannesburg, South Africa, 256 pages.
41. Ward, D. E., 1990. Factors influencing the emissions of gases and particulate matter from biomass burning. *Fire in the Tropical Biota: Ecosystem Processes and Global Challenges* (J. G. Goldammer, Ed.). Springer-Verlag, Berlin, Ecological Studies Volume 84, pp. 418-436.
42. Wayne, R. P., 1991. *Chemistry of Atmospheres* (Second Edition), Oxford University Press, Oxford, England, pp. 160-164.

43. Yokelson, R. J., D. W. T. Griffith, and D. E. Ward, 1996. Open-path Fourier transform infrared studies of large-scale laboratory biomass fires. *Journal of Geophysical Research* **101**: 21,067-21080.

44. Yokelson, R. J., R. Susott, D. E. Ward, J. Reardon and D. W. T. Griffith, 1997. Emissions from smouldering combustion from biomass measured by open-path Fourier transform infrared spectroscopy. *Journal of Geophysical Research* **102**: 18,865-18,877.

Chapter 10

AIR QUALITY DURING HAZE EPISODES AND ITS IMPACT ON HEALTH

***Kathryn Ostermann*[1] *and Michael Brauer*[2]**
[1]Atmospheric Science Programme, The University of British Columbia, Room 217-1984 West Mall, Vancouver, BC V6T 1Z2, Canada
[2]School of Occupational and Environmental Hygiene, The University of British Columbia, 3rd. Floor 2206 East Mall, Vancouver, BC V6T 1Z3, Canada

INTRODUCTION

There are only a limited number of studies directly evaluating the community health impacts of air pollution from vegetation fires, such as those that affected Southeast Asia in 1997 and 1998. In this chapter we describe the available information regarding air pollutant concentrations during "haze" episodes in Southeast Asia and use available data from the literature to discuss the potential population health impacts of the resulting exposures. Several recent review papers have discussed the health impacts and pollutants associated with wood smoke air pollution (Larson and Koenig, 1994; Pierson *et al.*, 1989; Vedal, 1993). Although the emphasis of these reviews was on North American community exposures, many of the conclusions are relevant to the broader understanding of vegetation fire air pollution. This chapter will also discuss the health impacts of exposures to biomass air pollution encountered by forest firefighters and by individuals in developing countries who use biomass for cooking and heating as these exposures demonstrate the plausibility of health impacts associated with exposures to smoke from vegetation fires. In the context of public health impacts associated with haze we also discuss management strategies and efforts to mitigate health effects.

EXPOSURES TO AIR POLLUTION FROM VEGETATION FIRES

Air pollution from vegetation fires, referred to here as biomass smoke, contains a large and diverse number of chemicals, many of which have been associated with adverse health impacts. The major chemicals present in biomass smoke and their sources are listed in Table 1. These chemicals include both particulates and gaseous compounds. Although little is known about the toxicology of biomass smoke as a complex mixture, review of the exposure and health impacts literature, as well as evaluation of the available air monitoring data from the 1997-98 Southeast Asian episodes (Radojevic and Hassan, 1999), indicates that the pollutant variable most consistently elevated in association with biomass smoke is particulate matter.

Table 1 Summary of major biomass pollutants

Compound	Examples	Source	Notes
Inorganic gases	Carbon monoxide (CO)	Incomplete combustion of organic material.	Transported over distances.
	Ozone (O_3)	Secondary product of nitrogen oxides and hydrocarbons.	Only present downwind of fire, transported over distances.
	Nitrogen Dioxide (NO_2)	Oxidation of nitrogen present in biomass amino acids and proteins to NO, which is further oxidized to NO_2.	Reactive – concentrations decrease with distance from fire.
Hydrocarbons	Benzene	Incomplete combustion of organic material.	Some transport – also react to form organic aerosols.
Aldehydes	Acrolein	Incomplete combustion of organic material.	
	Formaldehyde (HCHO)	Incomplete combustion of organic material.	
Particles	Inhalable particles (PM_{10})	Condensation of combustion gases; incomplete combustion of organic material; entrainment of vegetation and ash fragments.	Coarse + fine particles. Coarse particles are not transported and contain mostly soil and ash.
	Respirable Particulates	Condensation of combustion gases; incomplete combustion of organic material.	For biomass smoke, approximately equal to fine particles.
	Fine particles ($PM_{2.5}$)	Condensation of combustion gases; incomplete combustion of organic material.	Transported over long distances; primary and secondary production.
Polycyclic aromatic hydrocarbons (PAHs)	Benzo[a]pyrene (BaP)	Condensation of combustion gases; incomplete combustion of organic material.	Specific species varies with composition of biomass.

Particles in biomass smoke consist of solid and liquid compounds of small diameter composed predominantly of organic and elemental carbon (Pinto *et al.*, 1998). This type of pollution can be classified as particulate matter, which can be divided into several categories. Suspended particulate matter (SPM), total suspended particulate (TSP) and total particulate matter (TPM) represent the sum of all the suspended particles in the atmosphere at a given time. The fraction of particulate matter consisting of particles less than 10 micrometers (μm) in aerodynamic diameter are termed PM_{10}. PM_{10} can be broken down into coarse particulate, comprising particles between 2.5 to 10 μm in aerodynamic diameter, and fine particulate, which is less than 2.5 μm in aerodynamic diameter ($PM_{2.5}$).

The size of particulates produced in vegetation fires is an important factor in understanding potential health risks. Only smaller particles (PM_{10}) are inhaled and have the potential to cause adverse health impacts. Respirable particles ($PM_{2.5}$) penetrate into the lower respiratory tract and may present increased risk. Due to the size distribution of biomass particulates, essentially all will be contained in the $PM_{2.5}$ fraction, while the PM_{10} fraction will include additional particulates from resuspension of soil and ash. Biomass combustion particulate is typically smaller than 1 μm, with a peak in the size distribution between 0.15 and 0.4 μm (Hueglin *et al.*, 1997; Sandberg and Martin, 1975).

MONITORING AIR QUALITY IN SE ASIA

Many countries develop a network of air quality monitoring stations as a means of overseeing pollution levels. A typical air quality monitoring station will routinely measure the concentrations of the major air pollutants: ozone (O_3), carbon monoxide (CO), nitrogen dioxide (NO_2), sulphur dioxide (SO_2) and particulates (usually measured as PM_{10}). Currently there are 22 stations in Malaysia, of which 17 are located in peninsular Malaysia and 5 are located in Sabah and Sarawak while Singapore, (http://www.kjc.gov.my/people/environ/environ.htm), has 15 stations (Radojevic, 1997, 1998). Brunei has one fully equipped air quality station monitoring the five major pollutants, as well as 7 other stations for measuring PM_{10} concentrations, with plans for further expansion (Radojevic, 1997). In Indonesia, different government ministries conduct air pollution measurements independently of one another, in nearly all the capital cities of the 27 provinces (Heil, 1998). Table 2 gives a breakdown of the monitoring that takes place in the countries affected by haze. Partially because of the problem with haze, there has been an increase in the numbers of monitoring stations in the past decade. Specifically, the focus has been on the measurement of particulates, as vegetation fires release these in much higher concentrations than other pollutants. Therefore, better data are available for the more recent haze episodes.

Table 2 Air monitoring networks and reporting in Southeast Asia (WHO, 1998)

	Brunei Darussalam	Indonesia	Malaysia	Philippines	Singapore	Thailand
Monitoring	C	C + M	C	C + M	C + M	C + M
Operation	G	G	P	G	G	G
Index (API / PSI)	Yes	Yes	Yes	No	Yes	No
Concentration reported	No	Yes	No	Yes	No	Yes

C = Continuous; M = Manual; G = Government; P = Private sector

AIR QUALITY STANDARDS AND GUIDELINES

Air quality guidelines and standards have been adopted to protect public health from the negative effects of environmental pollutants (http://www.who.int/peh/air/airindex.htm). These guidelines encompass the routinely measured pollutants, including CO, NO_2, O_3, SO_2 and sometimes PM_{10}. The values given by the United States Environmental Protection Agency (US EPA), as their primary (to protect public health) National Ambient Air Quality Standards (NAAQS), and the World Health Organization (WHO), as well as those of several Southeast Asian countries, are given in Table 3. Averaging times vary depending on the pollutant.

As discussed previously, the particulate matter standards and guidelines are of greatest relevance for air pollution from vegetation fires. Currently, the US EPA has a standard for PM_{10} of 50 $\mu g/m^3$ (annual average) and 150 $\mu g/m^3$ for a 24-hour average. For $PM_{2.5}$ the annual and 24-hour standards are 15 and 65 $\mu g/m^3$, respectively. In its most recent revision of Air Quality Guidelines, the WHO elected not to set a threshold value, but instead has derived a linear relationship between PM_{10} or $PM_{2.5}$ concentrations and various health impacts (http://www.who.int/peh/air/airguides2.htm). This revision is based upon the absence of scientific evidence to support a no-effects threshold concentration for airborne particulate matter. These relationships allow each country to manage particulate air pollution by assessing the health effects associated with different levels of particulate matter.

Table 3 Ambient Air Quality Standards/Guidelines for United States, World Health Organization and some ASEAN countries. All concentrations in μg/m^3 except CO in mg/m^3

Pollutant	Averaging time	Malaysia[a]	Thailand[b]	Indo-nesia[c]	Philip-pines[d]	Singapore[d]	EPA NAAQS[e]	WHO[f]
SO_2	10 min	500	-	-	-	-	-	500
	1 hr	350	-	900	850	-	-	-
	24 hr	105	300	300	370	365	365	125
	annual	-	100	60	-	80	80	50
NO_2	30 min	-	-	-	300	-	-	-
	1 hr	320	320	-	-	-	-	200
	annual	-	-	100	-	100	100	40
CO	15 min	-	-	-	-	-	-	100
	30 min	-	-	-	-	-	-	60
	1 hr	35	50	30	35	40	40	30
	8 hr	10	20	10	10	10	10	10
O_3	30 min	-	-	-	200	-	-	-
	1 hr	200	200	160	-	235	-	-
	8 hr	120	-	-	-	-	157	120
Suspended matter (TSP)	24 hr	260	330	230	180	260	-	-
	annual	90	100	90	-	75	-	-
Particulate <10 μm (PM_{10})	24 hr	150	120	-	-	-	150	No guideline value (impact relationship)
	annual	50	50 -	-	-	-	50	
Particulate <2.5 μm ($PM_{2.5}$)	24 hr	-	-	-	-	-	65	No guideline value (impact relationship)
	annual	-	-	-	-	-	15	
Pb	24 hr	-	10	2	-	-	-	-
	3 months	1.5	-	-	20	1.5	1.5	-
	annual	-	-	-	-	-	-	0.5

[a]Abdullah (1996)
[b]Tiramongkol (1996) except PM_{10} from Tridech (1999)
[c]Lubis (1996) - draft standards
[d]Tabucanon (1996)
[e]http://www.epa.gov/airprogm/oar/oaqps/greenbk/criteria.html
[f]http://www.who.int/peh/air/airguides2.htm

AIR QUALITY INDICES

Air quality indices have been developed as a simple way to inform the general public about pollution conditions. There is currently no standardised air pollutant index for Southeast Asia. In Malaysia, the Air Pollution Index (API) is used. The API is based on the Pollutant Standards Index (PSI) developed by the United States Environmental

Protection Agency (Pinto *et al.*, 1998). The PSI is used in Singapore and Brunei, while Thailand reports the concentration of each pollutant. Indonesia relies on visibility readings as a means to monitor haze (Radojevic, 1997). Table 2 lists a summary of the means of reporting air quality in the countries of Southeast Asia.

The PSI and API are based on the measurement of five pollutants: carbon monoxide, sulphur dioxide, nitrogen dioxide, ozone and PM_{10}. The values for each pollutant are converted to a scale from 0 to 500, with only the highest value being reported. A value of 100 represents the air quality standard for each country. It is important to note that the PSI and API do not take into account the effects of the combination of pollutants. The main difference between the PSI and API comes from the fact that the standards set by the US EPA and the Malaysian Department of Environment differ. The main criteria pollutant during haze events is PM_{10}, as concentrations far exceed those of the other pollutants. Table 4 gives the range of index values for the API and PSI with the corresponding PM_{10} value and cautionary statements.

Table 4 Air quality indices and particulate levels

PSI Value	API Value	24 hr PM_{10} Concentration($\mu g/m^3$)[a]	Health Effect Descriptor	Cautionary Statements[b]
0-50	0-50	0-50	Good	None required
50-100	50-100	51-150	Moderate	None required
100-200	100-286	151-350	Unhealthful	Persons with existing heart or respiratory ailments should reduce physical exertion and outdoor activity. General population should reduce vigorous outdoor activity.
200-300	286-353	351-420	Very Unhealthful	Elderly and persons with existing heart or lung disease should stay indoors and reduce physical activity. General population should avoid vigorous outdoor activity.
300-400	353-428	421-500	Hazardous	Elderly and persons with existing diseases should stay indoors and avoid physical exertion.
>400	>428	>500	Hazardous	General population should avoid outdoor activity. All people should remain indoors, keeping windows and doors closed, and minimize physical exertion.

[a] http://www.epa.gov/oar/aqtrnd97

[b] http://www.epa.gov/airprogm/oar/oaqps/psihold.html

AIR QUALITY DURING HAZE EPISODES

During the 1990s, five separate haze episodes (1990, 1991, 1994, 1997 and 1998) occurred in Southeast Asia. Although not a new phenomenon, as haze in the area dates back to at least the early 1980s and even 1960s (Rindam, 1995), the increasing frequency is of great concern. Figure 1 depicts estimated PM_{10} levels in Sarawak for the period of 1978 to 1997. The haze episodes of 1991, 1994 and 1997 are clearly evidenced by the increases in PM_{10}. This figure is based upon PM_{10} estimated from visibility. A quantitative relationship exists between visibility and the amount of particles in the air, though it is confounded by moisture. An equation for predicting the amount of PM_{10} in the atmosphere can be developed by using this association (Brook, 1998). Data on visibility, humidity and PM_{10} at stations in Sarawak, Malaysia were used to determine a linear equation for PM_{10} (Brook, 1998). Potential sources of error in these estimates are that the correction factors were not derived for particles from vegetation fires, and the subjective nature of visibility measurements. Nonetheless, the equation is useful for identifying peak PM_{10} concentrations.

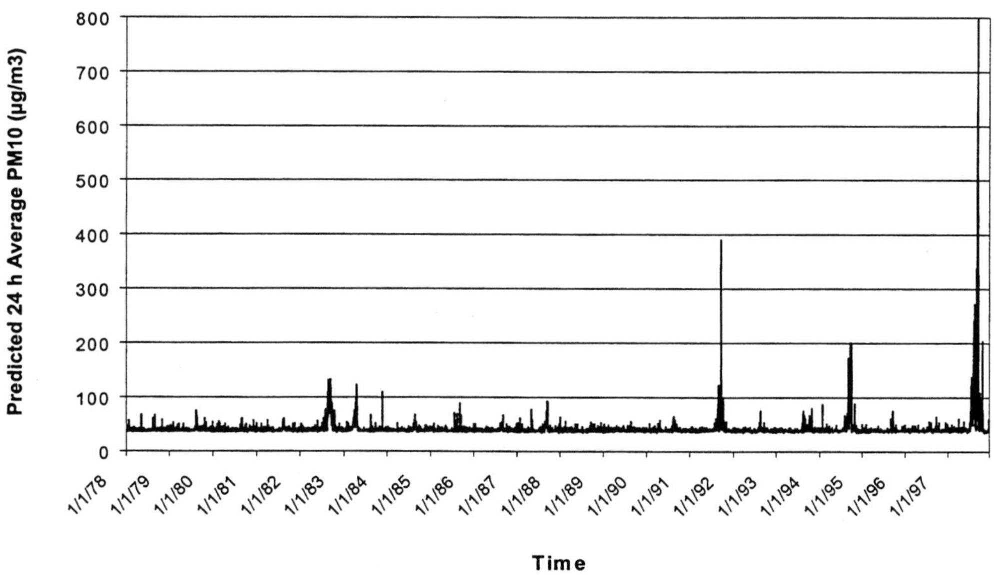

Figure 1 Estimated PM_{10} concentrations in Sarawak from 1978 to 1997 based upon visibility reports at Kuching airport.

Many of the air quality conditions reported in the literature for haze episodes are based on measurements of total particulate. In order to standardise the data, TPM, SPM and TSP are converted to PM_{10}. The ratio of PM_{10} to total particulate varies depending on the stage of combustion, but the United States Environmental Protection Agency estimates that 92.8% of particulates from agricultural burning are less than 10 μm (http://www.epa.gov/nceawww1/ pm_vol1.htm). Also, Ward (1990) suggests that up to 95% of particulate matter is less than 2.5 μm. Therefore, we assume that 90% of total particulate matter is smaller than 10 μm and use this to convert all measured TPM, TSP and SPM values to estimated PM_{10} concentrations. Original measurement values are reported in parentheses.

1990

For a period of two weeks during 1990, haze affected areas of peninsular Malaysia resulting in reduced visibility and an increase in particulate concentrations. The episode began around the middle of August and lasted until the end of that month. A sudden decrease in visibility occurred on 21 August, with values as low as 1 km over the central states of peninsular Malaysia (Tussin, 1995). Levels of particulates increased starting from 15 August. Peak 24-hour values of PM_{10} were equivalent to 426 μg/m^3 (TSP=473.8 μg/m^3) in Petaling Jaya (Tussin, 1995) and 464 μg/m^3 (TSP=516 μg/m^3) in the Klang Valley (Rindam, 1995). These quantities were up to four times the mean levels measured during non-haze periods (Tussin, 1995).

1991

Two separate haze episodes occurred in Malaysia during 1991. The first, which will not be discussed further, took place in June as the result of the injection of ash from the volcanic eruption of Mt Pinatubo (Tussin, 1995; Hassan *et al*, 1995). The second, at the end of September and lasting through most of October, was more severe and was the result of biomass burning from forest fires (Tussin, 1995). The latter episode included two phases separated by a rainy period: from September 27 until October 11, and a less severe event from 22 to 31 October (Tussin, 1995; Hassan *et al*, 1995). The impact of the haze was most evident in Malaysia (Hassan *et al*, 1995). Locations in western Borneo reported visibility of less than 2 km after 2 October (Tussin, 1995). Days of low visibility corresponded to high concentrations of particulate matter (Hassan *et al*, 1995). 24-hour PM_{10} values in Kuala Lampur, Malaysia ranged from 102 μg/m^3 to 254 μg/m^3 (SPM=113 μg/m^3 to 282 μg/m^3) for the first period of 4 to 11 October and 95 μg/m^3 to 153 μg/m^3 (SPM=105 to 170 μg/m^3) for 23 to 30 October (Hassan *et al*, 1995). These values were significantly higher, up to three times the normal mean (Tussin, 1995), than those for the rest of the month (Hassan *et al*, 1995). In Petaling Jaya, peak 24-hour averages were 445 μg/m^3 (SPM=494.5 μg/m^3) on 30 September and 441 μg/m^3 (SPM=489.8 μg/m^3) on 8 October (Tussin, 1995).

1994

After two years of relatively haze-free air quality in 1992 and 1993, fires in 1994 brought haze again to the Singapore-Malaysia-Indonesia region (Nichol, 1997). The area affected by the smoke spanned approximately 3 million square kilometers (Nichol, 1997). The haze lasted from August to October in Indonesia, Malaysia (Hassan *et al*, 1995), Brunei (Radojevic and Hassan, 1999) and Singapore (Chia *et al*, 1995; Nichol, 1997). Most of the available information comes from Malaysia. Visibility was reduced to 1 km in Malaysia with a minimum of 0.3 km in Petaling Jaya, as compared to 30 km under normal conditions (Rindam, 1995). Some measurements also suggest that the haze in 1994 was worse than in previous years in Malaysia (Rindam, 1995). The highest 24-hour average PM_{10} reading reported in Petaling Jaya was 410 $\mu g/m^3$ (TSP=454.5 $\mu g/m^3$) on 30 September (Tussin, 1995), while in Kuala Lampur, PM_{10} reached 409 $\mu g/m^3$ on 5 October (Hassan *et al*, 1995). In Singapore, 24-hour average PM_{10} concentrations peaked above 250 $\mu g/m^3$ (PSI>150) on 13 and 27 September and 29 October (Nichol, 1997). Concentrations of other particulate constituents such as sulphur, potassium, titanium, vanadium, manganese, nickel, arsenic and lead were 3-6 times higher than average during the haze period in late October (Orlic *et al*, 1997). The forest fires, though, did not have a significant influence on the levels of PAHs in Singapore (Chee *et al*, 1997).

1997

The 1997 fires in Indonesia drew worldwide attention as the haze impacted Malaysia, Singapore, southern Thailand and the Philippines and lasted from June through until November. As early as the beginning of May, air quality began to deteriorate in Singapore (Nichol, 1998). 24-hour average concentrations in Singapore, Brunei peaked at approximately 230 $\mu g/m^3$ (PSI=140) on 19 and 29 September (Nichol, 1998). These particulate levels in Singapore were not any more severe than those associated with the 1994 forest fires (Nichol, 1998). In Sarawak, however, 24-hour average PM_{10} peaked as high as 930 $\mu g/m^3$ on 23 September (Nichol, 1998), which represents a value more than 15 times the normal levels (Brauer and Hashim-Hisham, 1998). Due to the location of fires and direction of the wind, Sumatra and Kalimantan in Indonesia were the most severely affected areas (Pinto *et al.*, 1998). Daily averaged concentrations of PM_{10} reached as high as 3546 $\mu g/m^3$ (TPM=3940 $\mu g/m^3$) in Sumatra at the end of September (Heil, 1998). The concentrations were even higher in Kalimantan, on the island of Borneo, with a 24-hour maximum of 3645 $\mu g/m^3$ (TPM=4050 $\mu g/m^3$) (Heil, 1998). Even at the beginning of November in Palembang when the air quality was improving, daily PM_{10} and $PM_{2.5}$ still exceeded the US NAAQS (Pinto *et al.*, 1998). Aircraft measurements performed during this episode indicated that the smoke plume reached an altitude of 4 km and included high concentrations of O_3, NO_x and CO along with aerosols. In the lower layer of the plume, visibility was less than 500m (Tsutsumi *et al*, 1999).

1998

With the persistence of the El Niño that started in 1997 and an abnormally short wet season (Levine *et al*, 1999), forest fires raged again in 1998 and resulted in another, more localized, haze event (Radojevic and Hassan, 1999). In this case, the episode was most acute on Borneo and particularly in Brunei. The haze first emerged on 1 February and remained until 30 April, with especially severe conditions at the beginning of April. Daily average concentrations in the capital, Bandar Seri Begawan, reached nearly 450 $\mu g/m^3$ and the 24 hour guideline for PM_{10} of 70 $\mu g/m^3$ was exceeded 54 times during the period of 1 February to 30 April (Radojevic and Hassan, 1999). The "warning stage," which is associated with a PSI value of 300 (24-hour average PM_{10} concentrations of 420 $\mu g/m^3$), was exceeded on 15 April (Radojevic and Hassan, 1999). During the 1998 episode in Brunei, PM_{10} was the only significant pollutant contributing to the haze (Radojevic and Hassan, 1999). Other gaseous pollutants such as SO_2, O_3 and NO_2 were within acceptable limits, and only the 8 hr guideline for CO was exceeded on several occasions (Radojevic and Hassan, 1999). In Miri, Sarawak, Malaysia, 24-hour average PM_{10} concentrations rose above 600 $\mu g/m^3$ (API=649) on March 30 (WHO, 1998).

Table 5 summarises the available information from the haze events described above.

Table 5 Summary of major Southeast Asian haze events since 1990 reported in the literature. Peak 24-hour average concentrations for each country in each year are given for PM_{10}. No value next to country indicates no measurements available.

Episode Period	Spatial extent	PM_{10} Levels
August 15-31, 1990	Malaysia (Peninsular)	464 $\mu g/m^3$
Sept 27-October 11, 1991 phase I October 22-31, 1991 phase II	Malaysia	445 $\mu g/m^3$
August to October, 1994	Indonesia Malaysia Brunei Singapore	 409 $\mu g/m^3$ 285 $\mu g/m^3$
June - November, 1997	Indonesia Malaysia Singapore Thailand, Philippines, Brunei	3645 $\mu g/m^3$ 930 $\mu g/m^3$ 230 $\mu g/m^3$
1 February - 30 April, 1998	Malaysia (Sarawak) Brunei Indonesia (Kalimantan)	>600 $\mu g/m^3$ ~440 $\mu g/m^3$

ACUTE HEALTH EFFECTS OF HAZE

Indoor Air Pollution in Developing Countries

To understand the potential for health effects resulting from exposure to "haze" we first review the available information regarding exposure in indoor air in developing countries where wood and other biomass is used as a cooking and heating fuel. Respirable particulate levels measured in these settings are typically 1000-2000 μg/m^3 depending upon the specific fuel, ventilation, cooking duration and measurement interval (Smith and Liu, 1993; Wafula *et al.*, 1990; Smith, 1993; Brauer *et al.*, 1996). These levels are 10 – 50 times above those observed in urban areas. In terms of exposure, domestic cooking and heating with biomass clearly presents the highest exposures since individuals are exposed to high levels of smoke on a daily basis for many years. Exposures of this group are typically 70-85 hour-years of exposure.

The health effects of biomass smoke inhalation have been documented in developing countries where women, and in some cases, children spend many hours cooking over unvented indoor stoves. A number of studies have reported associations of health impacts with use of biomass fuels, and a few have directly measured exposure. These studies have been reviewed in detail by Smith (1993) and Chen *et al.* (1990). Exposure to biomass combustion products has been identified as a major risk factor for acute respiratory infections, the leading cause of infant mortality in the developing countries. In addition to the risks of infants, the women who are cooking are also at risk from chronic respiratory diseases as well as adverse pregnancy outcomes (Perez-Padilla *et al.*, 1996). As these exposures (in terms of both concentration and duration) are much higher than would occur as a result of short-term exposure to biomass air pollution associated with forest fires, direct comparisons are difficult to make. For example, individuals who are particularly susceptible may avoid indoor exposures, whereas this is not usually possible for ambient air pollution resulting from vegetation fires. The studies conducted in developing countries indicate the serious consequences of exposure to high levels of biomass air pollution. Increased acute respiratory illness in children associated with biomass smoke exposure is a likely cause of infant mortality while the development of chronic lung disease in adults is associated with premature mortality and substantial morbidity.

Wildland Firefighters

Several studies have evaluated impacts of biomass smoke to another group with high exposures, wildland (forest) firefighters. Exposures of this population are seasonal (4-5 months per year) and highly variable depending upon the number of fires per season, the intensity of the fires and specific job tasks. Exposures of wildland firefighters were recently reviewed by Reinhardt and Ottmar (1997) and appear to be dominated by high particulate (500 – 7000 μg/m^3) and more variable CO exposures (4 – 40 ppmv). Limited monitoring of PAHs has measured low levels (mean exposures <100 ng/m^3 except for phenanthrene at 380 ng/m^3) (Materna *et al.*, 1992). In summary, the exposure measurements of firefighters, while variable, indicate the potential for exposure to carbon monoxide and respirable particulates at levels (above 35 ppmv CO and above 5 mg/m^3 respirable particulate) which have been associated with adverse health impacts.

Epidemiological studies of wildland firefighters clearly indicate an association between exposure and acute effects on respiratory health (lung function, respiratory symptoms) of firefighters (Betchley *et al.*, 1997). Cross-seasonal effects have also been observed in most studies although these effects appear to be relatively small and may be reversible (Letts *et al.*, 1991). It must be noted that firefighters are normally among the most physically fit in the entire population and do not normally suffer from any pre-existing health conditions. The demonstration of health impacts amongst firefighters provides strong evidence that similar effects will be observed within the general population at equivalent or lower levels of exposure.

Community Air Pollution Studies

Particulate Air Pollution

This chapter will not discuss the voluminous particulate epidemiology literature in detail, as a book (Wilson and Spengler, 1996) and several review articles have recently been published (Dockery and Pope, 1994; Pope *et al.*, 1995; Vedal, 1997). Instead, we provide an overview of particulate air pollution epidemiology and discuss available evidence associating biomass particulate air pollution with adverse health outcomes. Many recent studies have indicated that current levels of particulate air pollution are associated with increased daily mortality (Dockery *et al.*, 1992, 1993; Schwartz, 1991, 1993; Pope *et al.*, 1992, 1995; Schwartz and Dockery, 1992a, 1992b; Schwartz *et al.*, 1992; Spix *et al.*, 1993). These studies indicate associations between particle air pollution and increased risk of death, primarily in the elderly and in individuals with pre-existing respiratory and/or cardiac illness (Schwartz, 1994a, 1994b). Recent studies have also suggested an association between particulates and infant mortality (Bobak and Leon, 1992; Woodruff *et al.*, 1997) as well as with low birth weight (Wang et al., 1997). Increased risk of hospital admissions and increased emergency room visits have also been associated with short-term increases in the levels of particle air pollution (Dockery and Pope, 1994; Pope, 1989, 1991; Schwartz, 1994b, 1994c, 1995, 1996; Schwartz *et al.*, 1993). One common feature of the study locales is that the ambient particulates are produced in combustion processes. Studies of naturally-produced particles (such as those generated from windblown soil or volcanic eruptions) show a much smaller impact on health outcomes for an equivalent particle concentration (Hefflin *et al.*, 1994; Buist *et al.*, 1983; Dockery and Pope, 1994). Although inhalable particulates are regulated in many countries, the WHO Air Quality Guidelines declined to recommend specific guidelines for particulate matter as the available studies do not indicate an obvious exposure concentration and duration that could be judged as a threshold (World

Health Organization, 1995). The document argues that the available data suggest a continuum of effects with increasing exposure.

Despite the evidence associating particulate air pollution with acute health impacts, only a limited number of studies have investigated long-term effects. Of these, the most significant are the prospective cohort studies in which the analyses can control for individual differences in risk factors such as smoking. Dockery *et al.* (1993) studied over 8,000 adults in 6 cities with different levels of air pollution over a period of 16 years. The adjusted mortality risk was 26% higher in the most polluted city relative to the least polluted city. Survival decreased with increasing particulate levels (Dockery *et al.*, 1993). In a study of more than 500,000 adults with 8-year follow-up, Pope and colleagues found a significant association between fine particles with cardiopulmonary mortality after controlling for smoking, education and other potential confounding factors (Pope *at al.*, 1995). The adjusted mortality risk was 15 – 25% higher in cities with the highest particulate levels relative to the cities with the lowest levels. Together, these studies indicate that long term exposure to particulate air pollution has a significant impact on survival. The results of the cohort studies of Dockery and Pope (1995) were used to estimate the reduction in life expectancy associated with long-term particulate exposure. When applied to a 1992 life table for Dutch men, the estimated effect of life expectancy is 1.1 years for each 10 $\mu g/m^3$ difference in long-term exposure to $PM_{2.5}$.

In the only cohort study of morbidity associated with long-term particulate exposure, Abbey and colleagues studied a cohort of nearly 4000 non-smoking Seventh Day Adventists in California (Abbey *et al.*, 1991a, 1991b). The relative risks of developing **new** cases of chronic respiratory disease were significantly associated with particulate levels. Significant risks for the development of new cases of chronic bronchitis and obstructive airways disease were found for annual average PM_{10} and $PM_{2.5}$ levels of 20 - 100 $\mu g/m^3$ (Abbey *et al.*, 1995). Increased symptoms severity was associated with annual average concentrations of 20 – 40 $\mu g/m^3$ and 40-50 $\mu g/m^3$ for $PM_{2.5}$ and PM_{10}, respectively. In a more recent analysis, long term differences in PM_{10} concentrations were also associated with lung function decrements (Abbey *et al.*, 1998).

Wood and Other Biomass Smoke

A number of epidemiological studies have evaluated respiratory symptoms and/or lung function in children of North American communities where wood burning is prevalent. In these communities, elevated levels of ambient air pollution are seasonal (3-8 months depending upon the climate) and variable. 24-hour average PM_{10} concentrations as high as 800 $\mu g/m^3$ have been measured in these communities, although peak levels (24-hour averages) of 200 – 400 $\mu g/m^3$ are more common (Heumann *et al.*, 1991; Larson and Koenig, 1994; Vedal, 1993). As wood smoke is generally emitted outdoors and since people spend most of their time indoors, indoor penetration is an important variable for exposure assessment. It is estimated that approximately 70% of outdoor wood smoke particulate penetrates indoors (Larson and Koenig, 1994).

Several early studies focused on the presence of a wood burning stove in the home as a risk factor since wood stoves, especially older varieties, can emit smoke directly into the home (Larson and Koenig, 1994). While these earlier studies strongly suggest that there are adverse impacts associated with wood smoke exposure, their crude exposure assessment precludes more specific conclusions. These studies suggest associations between wood stove use and increased risk of respiratory illness and increased respiratory symptom reporting in children (Dockery *et al.*, 1987; Honicky *et al.*, 1983, 1985; Tuthill, 1984; Butterfield *et al.*, 1989). In a significant improvement from these earlier studies, measured indoor particulate levels revealed an increased risk of hospitalization for acute lower respiratory illness for children living in households that cooked with any wood or had indoor particle concentrations above 65 $\mu g/m^3$ (Robin, 1996). In this study, the indoor particle concentration was positively correlated with cooking and heating with wood but not with other sources of combustion emissions. In the only study to date to evaluate impacts of wood burning on adult asthma, Ostro and colleagues found exposure to indoor combustion sources, including wood stoves, to be associated with increased asthma exacerbation (Ostro *et al.*, 1994).

Several other studies, summarized in Table 6, have evaluated health outcomes in communities where wood smoke is a major source of ambient particulates. Heumann *et al.* (1991) studied lung function of 410 children in high and low exposure areas of Oregon where wood smoke accounts for as much as 80% of the winter period particulate. PM_{10} ranged from approximately 50 – 250 $\mu g/m^3$ in the high exposure area and 20 – 75 $\mu g/m^3$ in the low exposure area. Lung function decreased during the wood burning season for the children in the high exposure area, but not in the low exposure area (Heumann *et al.*, 1991). Two studies in Montana associated acute changes in lung function in a study of 375 children with increased levels of particulates. 24-hour averages ranged from 43-80 $\mu g/m^3$ and 14-38 $\mu g/m^3$ for PM_{10} and $PM_{2.5}$, respectively (Johnson *et al.,* 1990), of which approximately 68% of the particulate was attributed to wood smoke (Larson and Koenig, 1994).

A questionnaire study of respiratory symptoms compared residents of high (mean $PM_{2.5}$ of 55 $\mu g/m^3$) and low (33 $\mu g/m^3$) wood smoke pollution areas of Seattle. Although no significant differences were observed between the high and low exposure areas when all age groups were combined, there were statistically significant higher levels of congestion and wheezing in 1-5 year olds from the high pollution area. This finding supports those of other studies which suggest that young children are particularly susceptible to adverse effects of wood smoke (Browning *et al.*, 1990).

A more comprehensive study in the same high exposure Seattle area, where 80% of the particulates are from wood smoke (Larson and Koenig, 1994) measured significant lung function decrements in the asthmatic subjects in association with increased wood smoke exposure. The highest (nighttime 12-hour average) $PM_{2.5}$ level measured during the study period was approximately 195 $\mu g/m^3$ (Koenig *et al.*, 1993). A companion study found a significant association between PM_{10} levels and asthma emergency room visits throughout Seattle (Schwartz *et al.*, 1993). The mean PM_{10} level during the 1-year study period was 30 $\mu g/m^3$. At this concentration, PM_{10} appeared to be responsible for 12% of the asthma emergency room visits. The authors indicate that, on an annual basis, 60% of the fine particle mass in Seattle residential neighborhoods is from wood burning. A recent study in Santa Clara County, California, an area where wood smoke accounts for

Table 6 Summary of epidemiological studies of ambient biomass air pollution

Population	Study design	Endpoints measured	Results	Reference
All ages > 1 yr	Cross-sectional	Symptoms, respiratory illness	No significant effects. Trend for children 1-5 yr	(Browning *et al.*, 1990)
Children	Longitudinal	Lung function	Decreased lung function during and after wood burning season in exposed community but not in control community.	(Heumann *et al.*, 1991)
Children	Longitudinal	Lung function	Decreased winter lung function in exposed community but not in control community.	(Johnson, *et al.*, 1990)
Children	Longitudinal	Spirometry	Decreased lung function and fine particles in asthmatics.	(Koenig *et al.*, 1993)
All ages	Longitudinal	Emergency room visits	Increased asthma visits with fine particles in areas where wood smoke accounts for 80% of $PM_{2.5}$.	(Schwartz *et al.*, 1993)
All ages	Longitudinal	Emergency room visits	Increased asthma visits with PM_{10} in area where wood smoke accounts for 45% of winter PM_{10}.	(Lipsett *et al.*, 1997)
All ages	Longitudinal	Mortality	Increased daily mortality with PM_{10} in area where wood smoke accounts for 45% of winter PM_{10}.	(Fairley, 1990)
All ages	Longitudinal	Emergency room visits	Increased respiratory visits in community exposed to fire smoke.	(Duclos *et al.*, 1990)
Adult asthmatics	Experimental	Lung function	Decreased lung function following exposure to burning leaves in asthmatics, but not in non-asthmatics.	(From *et al.*, 1992)
Adults with airways obstruction	Prevalence	Symptoms	42% of population reported increased or worsened symptoms during episode of exposure to agricultural burning emissions. 20% reported breathing trouble.	(Long *et al.*, 1998)
Children	Longitudinal	Lung function	Decreased lung function in children during "haze" episode compared to pre-episode measurements.	(Hisham-Hashim *et al.*, 1998)
All ages	Longitudinal	Outpatient visits	Increased visits for asthma, upper respiratory symptoms and rhinitis during "haze" episode periods of elevated PM_{10} in Singapore and Malaysia.	(Brauer and Hisham-Hashim, 1998, World Health Organization, 1998)
All ages	Longitudinal	Emergency room visits	Increased asthma visits with PM_{10} during episode of exposure to biomass burning emissions in Singapore.	(Chew *et al.*, 1995)
All ages	Longitudinal	Emergency room visits	No increase in asthma visits with PM_{10} during episode of exposure to bushfire emissions in Australia.	(Copper *et al.*, 1994)
All ages	Longitudinal	Emergency room visits	No increase in asthma visits with PM_{10} during episode of exposure to bushfire emissions in Australia.	(Smith *et al.*, 1996)

approximately 45% of winter PM_{10}, demonstrated an association between wintertime PM_{10}, increased daily mortality and exacerbations of asthma (Lipsett *et al.*, 1997).

In one of the few studies of air pollution from agricultural burning, 428 subjects with airways obstruction were surveyed for their respiratory symptoms during a 2-week period of exposure to straw and stubble combustion products. During the exposure period, 24-hour average PM_{10} levels increased from 15-40 $\mu g/m^3$ to 80-110 $\mu g/m^3$. The 1-hour levels of carbon monoxide and nitrogen dioxide reached 11 ppmv and 110 ppbv, respectively. Total volatile organic compound levels increased from 30-100 $\mu g/m^3$ before the episode to 100-460 $\mu g/m^3$ during the episode. While 37% of subjects were not bothered by smoke at all, 42% reported that symptoms (cough, wheezing, chest tightness, shortness of breath) developed or became worse due to the air pollution episode and 20% reported that they had breathing trouble. Subjects with asthma and chronic bronchitis were more likely affected (Long *et al.*, 1998). This study indicates that other forms of biomass air pollution, in addition to wood smoke, are associated with some degree of impairment, and suggests that individuals with pre-existing respiratory disease are particularly susceptible.

Several studies have also evaluated the health impacts associated with forest and bush fires. Duclos and colleagues evaluated the impact of a number of large forest fires in California on emergency room visits (Duclos *et al.*, 1990). During the approximately 2½-week period of the fires, asthma and chronic obstructive pulmonary disease visits increased by 40 and 30%, respectively. PM_{10} concentrations as high as 237 $\mu g/m^3$ were measured. During 1994, bush fires near Sydney, Australia, led to elevated PM_{10} levels (maximum hourly values of approximately 250 $\mu g/m^3$) for a 7-day period; ozone levels were not elevated. Two studies of asthma emergency room visits during the bush-fire smoke episode failed to detect any association with air pollution (Copper *et al.*, 1994; Smith *et al.*, 1996). These results appear to conflict with results of studies conducted in North America. Possible reasons are differences in study design and sample size, as well as differences in chemical composition of the particulates and differences in the relative toxicity of the specific particle mixture.

Perhaps the study with the most relevance to the issue of biomass air pollution in Southeast Asia is an analysis of emergency room visits for asthma in Singapore during the 1994 haze episode (Chew *et al.*, 1995). The study, described briefly in a letter to the Lancet, indicates an association between PM_{10} and emergency room visits for childhood asthma. During the haze period, mean PM_{10} levels were 20% higher than the annual average. Although a time series analysis was not conducted, the authors suggest that the association remained significant for all concentrations above 158 $\mu g/m^3$.

Initial reports from surveillance monitoring activities conducted during the 1997 and 1998 episode also indicated effects of health care utilization. In Singapore, for example, there was a 30% increase in hospital attendance for "haze-related" illnesses and a time series analysis indicated that a PM_{10} increase of 100 $\mu g/m^3$ was associated with 12%, 19% and 26% increases in cases of upper respiratory tract illness, asthma and rhinitis, respectively. This analysis did not observe any significant increases in hospital admissions or mortality (World Health Organization, 1998). Similar findings were also observed in Malaysia (Brauer and Hisham-Hashim, 1998; Leech *et al.*, 1998). Preliminary results from an on-going study of the 107 Kuala Lumpur school children conducted found statistically significant decreases in lung function between pre-episode measurements in June-July 1996 and measurements conducted during the episode in September 1997 (Hisham-Hashim *et al.*, 1998). These preliminary results suggest a measurable impact of the 1997 episode on the respiratory function of children. Analyses of the long-term health impacts of the air pollution and the impact with more severe outcomes, such as daily mortality increases, have not yet been conducted.

The epidemiological studies of indoor and community exposure to biomass smoke indicate a consistent relationship between exposure and increased respiratory symptoms, increased risk of respiratory illness and decreased lung function. These studies have mainly been focused on children, although the few studies that evaluated adults also showed similar results. A limited number of studies also indicate associations between biomass smoke exposures and visits to hospital emergency rooms, although the analyses of the Australian bushfires did not. There are also indications from several studies that asthmatics are a particularly sensitive group. No studies have explicitly evaluated the effect of community exposure to biomass air pollution on hospitalizations or mortality, although relationships have been observed in areas where wood smoke is a major contributor to ambient particulates. By analogy with the findings of numerous studies associating increased mortality with urban particulate air pollution mixtures it is reasonable to conclude that similar findings would also be observed in locations exposed to biomass smoke. The particulate mortality studies also do not show evidence for a threshold concentration at which effects are not observed.

Nearly all of the low-level indoor and community biomass smoke studies mentioned above evaluated impacts of concentrations which were much lower than those associated with the 1997 and 1998 Southeast Asian haze episodes. Similarly, the studies of seasonal exposure to wood smoke involved exposure durations that were of comparable length to those experienced in Southeast Asia. Based on these studies, it is reasonable to expect that the Southeast Asian haze episode resulted in the entire spectrum of acute impacts, including increased mortality, as well as sub-chronic (seasonal) effects on lung function, respiratory illness and symptoms. It is not possible at this time to determine the long-term effect, if any, from a single air pollution episode, although repeated yearly occurrences of high biomass smoke exposure should be cause for serious concern. Chronic (several years) exposure to particulate air pollution in urban areas, at much lower levels than experienced in Southeast Asia in 1997, has been associated with decreased life expectancy and with the development of new cases of chronic lung disease.

LONG-TERM HEALTH EFFECTS OF HAZE: CANCER

Cohen and Pope (1995) recently reviewed the evidence associating air pollution with lung cancer. Studies suggest rather consistently that ambient air pollution resulting from fossil fuel combustion is associated with increased rates of lung cancer. Two recent prospective cohort studies observed 30-50% increases in lung cancer rates associated with exposure to respirable particulates, best viewed as a complex mixture originating from diesel exhaust, coal, gasoline and wood burning. The excess lung cancer risk associated with ambient air pollution (relative risks of 1.0 – 1.6) is small compared with that from cigarette smoking (relative risks of 7 – 22) but comparable to the risk associated with long-term environmental tobacco smoke exposure (relative risk of 1.0 – 1.5).

There is little direct information regarding the human cancer risks associated with biomass air pollution. A United States Environmental Protection Agency study found that, despite wood smoke being the major contributor to the mutagenicity of ambient particulate matter, it was 3 times less potent as a mutagen than extractable organics associated with vehicle emissions (Lewis *et al.*, 1988). In an application of this and other work, the estimated lifetime cancer risk associated with 70 years of exposure to air pollution dominated by wood smoke (80%) was calculated to be approximately 1 in 2,000. This calculation assumes lifetime exposure to PM_{10} levels of 25-60 $\mu g/m^3$, of which wood smoke is a major component, for approximately 3 months every year. Benzo(a)pyrene (BaP) is a probable human carcinogen as defined by the International Agency for Research on Cancer (IARC) and it is present in biomass smoke at high levels. Domestic biomass burning in developing countries has been associated with extremely high BaP levels (4000 times the levels in urban air). Smith and Liu (1993) review studies of BaP levels in biomass smoke and discuss studies which have evaluated lung cancer risks. Despite high exposures to a known human carcinogen, there is relatively little evidence for a relationship between lung cancer and biomass smoke exposure. If any effect does exist it is thought to be small, relative to other risk factors such as diet or exposure to air pollution from coal burning. Lung cancer is itself relatively rare in areas of biomass fuel use, even if age-adjusted cancer rates are analyzed (Smith and Liu, 1993). A similar argument is presented for nasopharyngeal cancers, which are also rare in areas of biomass smoke use (Dekoning *et al.*, 1985). Recently, however, a significant relationship between wood stove use and cancer of the mouth, pharynx and larynx has been reported in rural Southern Brazil (Pintos *et al.*, 1998).

Studies conducted in Xuan Wei, China, an area noted for high mortality from respiratory disease and lung cancer, suggested that the high lung cancer rates were strongly associated with the proportion of homes using smoky coal for cooking and heating while no relationship was observed between lung cancer and the percentage of homes using wood (Mumford, 1990). A follow-up study indicated similar associations between lung cancer and coal but not with wood use (Chapman *et al.*, 1988). Indoor PM_{10} concentrations measured during cooking were extremely high (24, 22 and 1.8 mg/m^3 for smoky coal, wood and smokeless coal, respectively). Mutagenicity tests of particulates collected from the various combustion processes indicated that smoky coal was approximately 5 times more mutagenic than wood (Mumford *et al.*, 1987). This study suggests that there was little association between open-fire wood smoke exposure and lung cancer, despite very high exposures with long duration (women generally start cooking at age 12 yr).

The available, although limited, data on biomass smoke and cancer do not indicate an increased risk even at very high levels of exposure. This evidence includes studies of long-term exposure to high levels of biomass smoke from domestic cooking in developing countries. Evidence for a relationship between urban particulate air pollution and lung cancer is also limited, but is suggestive of a small increased risk. There have not been enough studies conducted to evaluate the consistency of any increased risk for different particle sources. However, while biomass smoke is clearly mutagenic, it is much less so than motor vehicle exhaust, on a comparable mass basis.

MANAGEMENT OF HAZE

The management of haze episodes is an important public health measure due to the potential for large numbers of people to be adversely affected. As haze episodes may evolve into potentially complex emergency situations, the development of an effective early warning system involving the collaboration of multidisciplinary groups of scientists, technicians and administrators is critical. For example, monitoring systems involving traditional ground-based air monitoring and remote sensing should be operational and should be used to monitor the development of potentially serious episodes, with information effectively distributed to the public and to public health authorities. Among the critical components of efforts to manage haze episodes is the education of the population regarding the potential health impacts of air pollution produced in vegetation fires. These education efforts must occur prior to the occurrence of an episode and also during episode periods to keep the population informed. The success of any response mechanism will rest on the timely exchange of data, information and experiences, and on the close cooperation between numerous public and private sector institutions. Most haze management plans focus on the following aspects:

- Informing the public and the authorities regarding air quality levels;
- Advising the public on actions to be taken for health protection;
- Ensuring that medical and health supplies are available and establishing facilities to mitigate health impacts;
- Minimizing locally generated air pollution.

Recently, the World Health Organization developed the *Health Guidelines for Episodic Vegetation Fire Events* which are designed to increase awareness of the potentially serious public health impacts of haze and provide a framework for governmental management and response plans (Schwela *et al.*, 1999). These guidelines discuss the development of National Haze Action Plans to ensure full preparedness of the population to the potential health impacts of vegetation fire pollution. These action plans are to be widely publicized before the occurrence of any air pollution episode. Based on this action plan, government departments develop operating procedures and ensure that the population will be aware of any changes made to public services and facilities in an emergency situation. Additionally, the plans include the collection and reporting of surveillance information regarding air pollution related illnesses. The plans also include special educational efforts for susceptible populations such as asthmatics, the elderly, and children to ensure that they are adequately prepared to deal with air pollution episodes and suggest that health authorities, via the media, should proactively address frequently asked questions.

Mitigation Measures

Due to the limited effectiveness of exposure avoidance activities during regional haze episodes, priority emphasis must be given to elimination of the source of the air pollution, which in this case is extinguishing fires or preventing their occurrence. Close interaction between health, environment and meteorological agencies could result in effective forecasting of future air pollution episodes related to vegetation fires. However, despite efforts to prevent and control fires it is acknowledged that other measures may be necessary to help mitigate public health impacts. Following from basic principles of exposure control, if source control is not feasible, then administrative or engineering controls receive priority, followed by personal protective equipment such as dust masks. In this exposure situation, administrative controls might include recommendations to the population to reduce their level of physical activity, while engineering controls include the use and/or enhancement of air conditioning or indoor air cleaning. Reduction of physical activity will certainly reduce the dose of inhaled air pollutants and will likely reduce the risk of health impacts, although no formal studies have been conducted for particulate matter. Other mitigation measures are discussed in more detail in the following sections.

(a) Dust Masks

During the 1997-98 haze episodes, one of the major government and commercial efforts to mitigate public health impacts was the distribution of facial masks. Many different types of masks with variable filtration effectiveness were used. However, even the masks of highest filtration efficiency, approved respirators, must provide an airtight seal around the face in order for these to reduce human exposure to vegetation fire smoke. Further, as all masks are designed for use by adult workers, the effectiveness of even the highest quality masks for use by the general public (including children) has not been evaluated. It is unlikely that any mask will provide more than partial protection; lower quality masks will offer even less protection. In addition, while it is expected that such masks would also filter a high percentage of the smaller (<0.1 μm) particles present in biomass smoke, no performance data are available. Finally, respirators and dust masks are uncomfortable and increase the effort of breathing, especially at moderate and heavy work loads (Jones, 1991).

In the USA, the National Institute for Occupational Safety and Health (NIOSH) certifies three classes of particulate respirators filters (N, R, and P) with different levels of efficiency (95, 99, and 99.97 %). The efficiency indicates the degree to which the filter removes small (0.3 μm) particles. N series particulate respirators are for protection from particles that are free of oil or other severely degrading aerosols and are suitable for vegetation fire smoke. In Europe, the European Committee for Standardization (CEN) classifies 3 types of particulate respirators: P1 80 %, P2 94 % and P3 99.97 %, all of which would be suitable to protect populations from vegetation fire smoke.

In addition to the respirators described above, other types of respiratory protection are available and have been used during haze episodes. Wake and Brown (1988) evaluated one type, nuisance dust masks, for their filtration efficiency. These masks are designed for coarse dusts and not for the fine particles present in biomass smoke. Handkerchiefs and tissues were also tested. Although the smallest particle size used in testing was 1.5 μm, they found no difference between wet and dry handkerchiefs, and in general found the penetration of 1.5 μm particles to be quite high (60-90%) for all dust masks tested. Penetration for handkerchiefs and tissues was 70-97% (Wake and Brown, 1988). Qian and colleagues evaluated dust/mist respirators which met older NIOSH regulations and surgical masks, and compared these to new N95 respirators which, by definition, meet newer NIOSH regulations (Qian *et al.*, 1998). Under flow rate conditions which may be representative of general population use, efficiencies were 98.8, 86 and 80% for the N95 respirators, dust masks and surgical masks, respectively. These efficiencies do not consider face seal leakage. As filter material is loaded, the pressure drop across it increases, encouraging air to bypass the filter material through any leaks that are present. NIOSH has estimated that at least 10-20% leakage occurs in masks not fitted to the wearers' face and measurements have confirmed this problem for dust masks (Tuomi, 1985). Although the N95 respirators have higher collection efficiencies, the dust masks, and even surgical masks will provide a high degree of protection, provided there is an adequate seal around the face and provided that they are changed once loaded.

One dust mask which has been widely used during Southeast Asian haze episodes was tested by Hinds and Kraske (1987) who found this mask type to be 80-90% efficient at normal resting or moderately active respiratory rates and for the particle sizes present in biomass smoke. However, despite this relatively high filtration efficiency, the magnitude of the face seal leakage (up to 100% for sub-micron particles) indicates that fit testing and selection of tight-fitting masks is essential for protection. Surgical masks, in contrast to approved dust masks, are not designed for fit testing or for an adequate face seal.

The Ministry of Environment of Singapore developed recommendations for mask use during biomass air pollution episodes (24 Sep 97, http://www.gov.sg/env/sprd/Rel-Mask.htm). These recommendations suggest that surgical and other similar masks are not useful in preventing the inhalation of fine particles as they are not efficient in the filtration of particles of less than 10 μm, such as those present in biomass smoke. Accordingly, use of these masks may provide a false sense of wellbeing to the users. The recommendations also suggest that respirators (which include some types of dust masks) are able to filter 80% to 99% of particles between 0.2 and 0.4 μm. The recommendations indicate that respirators may be useful, but are uncomfortable and increase the effort of breathing. The recommendations also suggest that during periods of intense air pollution, it would be better for the public to avoid outdoor activity than to put on a mask and stay outdoors for prolonged periods. However, for those who cannot avoid going outdoors, the use of respirators would provide some relief.

(b) Protection by Remaining Indoors

Another major recommendation for populations during haze episodes is to stay indoors. Unfortunately, this recommendation is likely to provide only partial protection, and in some cases, no protection at all. Since the majority of time is spent indoor, exposure indoors is an important variable to consider, even for pollutants generated outdoors. The impact of outdoor particles on indoor levels was discussed in detail by Wallace (1996). Recent research has indicated that the impact of outdoor particles on indoor levels is a function of the particle penetration through the building envelope, the air exchange rate and the particle decay rate. Therefore, the impact of outdoor particles can easily be calculated for any air exchange rate. In typical North American homes (where mean air exchange rates are 0.45-0.55/hr), outdoor air accounts for 75% and 65% of fine and coarse particles, respectively. In general, air conditioned homes typically have lower air exchange rates than homes that use open windows for ventilation. In one study, air conditioned homes had air exchange rates of 0.8/hr, while non air-conditioned homes had rates of 1.2/hr, implying indoor fractions of outdoor $PM_{2.5}$ of 67 and 75%, respectively.

Data from studies conducted in the U.S. for combustion-source particulates indicate that approximately 44% and 88%, respectively, of outdoor particles penetrate indoors during summer for homes with and without air conditioning (Suh *et al.*, 1992). Limited measurements conducted in Singapore in 1994 indicate that during biomass episode periods on average 60% penetrated indoors (Chia, 1995). Simultaneous indoor and outdoor measurements with continuous particle monitors performed in areas impacted by residential wood burning demonstrated a strong correlation between indoor and outdoor levels and an indoor:outdoor ratio of 0.98, presumably due to the high air exchange rates in these homes (Anuszewski, 1998). Little information is available regarding commercial buildings although the infiltration of outdoor particles into commercial buildings is likely to be highly variable as it is dependent upon the air exchange rate and specific characteristics of the ventilation system, including the efficiency of air filters.

(c) Air Cleaners

During the most recent haze episodes, air cleaners were also recommended as a potentially useful mitigation measure. In Singapore, for example, the Ministry of Environment assessed portable air cleaners and found that several models were able to reduce the level of fine particles in a typical living room or bedroom to an acceptable level when there was an intense biomass episode (26 Feb 1998 http://www4.gov.sg/env/sprd/haze-rel-22-98.htm). The Singapore Ministry of Environment also suggests that households can add a special filter to window or split-unit air-conditioners to achieve similar results for particle removal. For central air conditioning systems, electrostatic precipitators, high-efficiency media filters and medium-efficiency media filters can be added so that the particle level in the indoor air can be kept within acceptable levels during a prolonged biomass smoke period.

The effectiveness of air cleaners is also well documented in the scientific literature. Offermann and colleagues found mixing fans, ion generators and small panel-filter devices to be ineffective for respirable particle removal while electrostatic precipitators, extended surface filters and HEPA filter units worked well, with effective cleaning rates (for removal of 98% of respirable particles in a room) of 100 – 300 m^3/hr (Offermann, *et al.*, 1985). Portable air cleaners were also discussed in a US EPA report which indicates that units containing either electrostatic precipitators, negative ion generators, or pleated filters, and hybrid units containing combinations of these mechanisms, are more effective than flat filter units in removing cigarette smoke particles. The report also advises that the use of a single portable unit would not be expected to be effective in large buildings with central heating, ventilating, and air-conditioning (HVAC) systems (USEPA, 1990).

(d) Other Measures

During severe episodes, another protection strategy is the preparation of emergency shelters with effective air conditioning and particle filtration. Susceptible individuals who do not have access to other air-conditioned environments could be allowed free access to emergency shelters located inside large commercial buildings, educational facilities or shopping malls.

The decision to close and curtail business activities will depend upon consideration of traffic, health, environmental, and socioeconomic factors and other local conditions. Depending upon building designs and the presence of air conditioning and filtration, exposure inside schools may be similar to those in homes or businesses. Restrictions on industrial emissions may be warranted depending upon the local air pollution situation and the emission characteristics of particular industries. The emergency evacuation of whole populations to other geographical locations in response to smoke haze is not recommended as a mitigation measure.

(e) Recommendations for Mitigation

As discussed above, the hierarchy for health protection is control or prevention of fires followed by administrative controls such as reduced physical activity and remaining indoors. To enhance the protection offered by remaining indoors, individuals/building managers should take actions to reduce the air exchange rate. Clearly there are comfort and economic costs associated with reduced air exchange, as well as potential health effects due to increased impact of indoor pollution sources. It is not possible at this time to recommend more specific measures which would be feasible to employ on a population-wide basis. There is evidence that air conditioners, especially those with efficient filters, will substantially reduce indoor particle levels. To the extent possible, effective filters should be installed in existing air conditioning systems and individuals should seek environments protected by such systems. There is strong evidence that portable air cleaners are effective at reducing indoor particle levels, provided the specific cleaner is adequately matched to the indoor environment in which it is placed. Fortunately, most air cleaners have been evaluated by manufacturers and their effectiveness in known. Unfortunately, economics will limit the distribution of such devices throughout the population. As with air conditioners, the increased use of such devices by a large segment of the population will have a significant impact on energy consumption, and may in turn have negative impacts on ambient air quality. The least desirable measure is the use of personal protective equipment, such as dust masks. While these are relatively inexpensive and may be distributed to a large segment of the population, at present their effectiveness for general population use must be questioned. Despite this reservation, it is likely that the benefits (even partial) of wearing dust masks will outweigh the (physiological and economic) costs. Accordingly, in the absence of other mitigation techniques, the use of dust masks is warranted. Education of the population regarding specific mask types to purchase, how to wear masks and when to replace them will increase their effectiveness as will the development of new masks designed for general population use.

REFERENCES

1. Abdullah M. 1996. Malaysia: Country Report. In Proceedings of the ASEAN Network on Environmental Monitoring (ASNEM) on the Third ASEAN Workshop on Air Quality Monitoring and Analysis with Emphasis on Polycyclic Aromatic Hydrocarbons, Environmental Research and Training Center (ERTC), Pathumthani, Thailand, 26 February-6 March 1996, pp. 177-188.
2. Abbey D.E., R.J. Burchette, S.F. Knutsen, W.F. McDonnell, M.D. Lebowitz, and P.L Enright, 1998. Long-term particulate and other air pollutants and lung function in nonsmokers. *American Journal of Respiratory and Critical Care Medicine* **158** (1): 289-98.
3. Abbey D.E., P.K. Mills, F.F. Petersen, and W.L. Beeson, 1991a. Long-term ambient concentrations of total suspended particulates and oxidants as related to incidence of chronic disease in California Seventh-Day Adventists. *Environmental Health Perspectives* **94**: 43-50.
4. Abbey D.E., J. Moore, F. Petersen, and L Beeson, 1991b. Estimating cumulative ambient concentrations of air pollutants: description and precision of methods used for an epidemiological study. *Archives of Environmental Health* **46** (5): 281-7.
5. Abbey D.E., B.E. Ostro, F. Petersen, and R.J. Burchette, 1995. Chronic respiratory symptoms associated with estimated long-term ambient concentrations of fine particulates less than 2.5 microns in aerodynamic diameter (PM2.5) and other air pollutants. *Journal of Exposure Analysis and Environmental Epidemiology* **5** (2): 137-59.
6. Anuszewski J., T. Larson, and J. Koenig, 1998. Simultaneous indoor and outdoor particle light-scattering measurements at nine homes using a portable nephelometer. *Journal of Exposure Analysis and Environmental Epidemiology* **8** (4): 483-493.

7. Betchley C., J. Koenig, G. Van Belle, H. Checkoway, and T. Reinhardt, 1997. Pulmonary function and respiratory symptoms in forest firefighters. *Amercian Journal of Industrial Medicine* **31**: 503-509.
8. Bobak M. and D.A. Leon, 1992. Air pollution and infant mortality in the Czech Republic, 1986-88. *The Lancet* **340** (8826): 1010-4.
9. Brauer M., K. Bartlett, J. Regalado-Pineda, and R. Perez-Padilla, 1996. Assessment of particulate concentrations from domestic biomass combustion in rural Mexico. *Environmental Science and Technology* **30** (1): 104-109.
10. Brauer M. and J. Hisham-Hashim, 1998. Indonesian fires: Crisis and reaction *Environmental Science and Technology* **32**: 404A-407A.
11. Brook, J. 1998. Prediction of past PM_{10}. In: The Study of Haze. Draft final report prepared for the State Government of Sarawak. W.F. Baird & Associates Coastal Engineers Ltd. Ottawa, Ontario, Canada.
12. Browning K., J. Koenig, H. Checkoway, T. Larson, and W. Pierson, 1990. A questionnaire study of respiratory health in areas of high and low ambient wood smoke pollution. *Pediatric Asthma, Allergy and Immunology* **4** (3): 183-191.
13. Buist A., L. Johnson, W. Vollmer, G. Sexton, and P. Kanarek, 1983. Acute effects of volcanic ash from Mount Saint Helens on lung function in children. *American Reviews of Respiratory Disease* **127** (6): 714-719.
14. Butterfield P., G. LaCava, E. Edmundson, and J. Penner, 1989. Woodstoves and indoor air: The effects on preschoolers' upper respiratory systems. *Journal of Environmental Health* **52** (3): 172-173.
15. Cassano P., G. Hu, J. Chen, and R. Peto, 1994. Cooking fuels and pulmonary function in the Peoples Republic of China. *American Journal of Respiratory and Critical Care Medicine* **149** (4): A663.
16. Chapman R., J. Mumford, X. He, W. Jiang, and R. Yang, 1988. The epidemiology of lung cancer in Xuan Wei, China: current Progress, issues, and research strategies. *Archives of Environmental Health* **43** (2): 180-185.
17. Chee K.K., M.K. Wong, and H.K. Lee, 1997. Microwave-assisted solvent extraction of air particulates for the determination of PAHs. *Environmental Monitoring and Assessment* **44** (1-3): 391-403.
18. Chen B.H., C.J. Hong, M.R. Pandey, and K.R. Smith, 1990. Indoor air pollution in developing countries. *World Health Statistics Quarterly* **43** (3): 127-38.
19. Chew F., B. Ooi, J. Hui, R. Saharom, D. Goh,and B. Lee, 1995. Singapore's haze and acute asthma in children. *The Lancet* **346** (Nov. 25): p. 1427.
20. Chia H., K. Chia, P. Ooi, T. Ng, K. Goh, and H. Lee, 1995. Effect of the Recent Haze in Singapore on the Frequency of Attacks Among a Group of Known Asthmatics. Institue of Environmental Epidemiology, Ministry of Environment, Singapore, Singapore, pp 87-93.
21. Cohen A.J. and C.A. Pope 3rd, 1995. Lung cancer and air pollution. *Environmental Health Perspectives* **103** (Suppl 8): 219-24.
22. Copper C, M. Mira, M. Danforth, K. Abraham, B. Fasher, and P. Bolton, 1994. Acute exacerbations of asthma and bushfires. *The Lancet* **343** (June 11): p.1509.

23. Dekoning H., K. Smith, and J. Last, 1985. Biomass fuel combustion and health. *Bulletin of the World Health Organization* **63** (1): 11-26.
24. Dockery D, J. Spengler, and F. Speizer, 1987. Association of health status with indicators of indoor air pollution from and epidemiologic study in six U.S. cities. International Conference on Indoor Air Quality and Climate, Berlin, Institue for Water, Soil and Air Hygiene.
25. Dockery D.W., C.A. Pope , X. Xu, J.D. Spengler, J.H. Ware, M.E. Fay, B.G. Ferris Jr., and F.E. Speizer, 1993. An association between air pollution and mortality in six U.S. cities. *New England Journal of Medicine* **329** (24): 1753-9.
26. Dockery D.W. and C.A. Pope, 3rd., 1994. Acute respiratory effects of particulate air pollution. *Annual Reviews of Public Health* **15**: 107-32.
27. Dockery D.W., J. Schwartz, and J.D. Spengler, 1992. Air pollution and daily mortality: associations with particulates and acid aerosols. *Environmental Research* **59** (2): 362-73.
28. Duclos P., L. Sanderson, and M. Lipsett, 1990. The 1987 forest fire disaster in California: Assessment of emergency room visits. *Archives of Environmental Health* **45** (1): 53-58.
29. Ellegard A. 1996. Cooking Fuel smoke and respiratory symptoms among women in low-income areas in Maputo. *Environmental Health Perspectives* **104** (9): 980-985.
30. Fairley D. 1990. The relationship of daily mortality to suspended particulates in Santa Clara County, 1980-1986. *Environmental Health Perspectives* **89**: 159-168.
31. From L., L. Bergen, and C. Humlie, 1992. The effects of open leaf burning on spirometric measurements in asthma. *Chest* **101** (5): 1236-1239.
32. Hassan M.N., A.Z. Abidin, M.K. Yusoff, A.W. Ghazali, A. Muda, and M.P. Zakaria, 1995. Damage costs of the 1991 and 1994 haze episodes in Malaysia. In: Proceedings of the International Symposium on Climate and Life in the Asia Pacific (K.U. Sirinanda, Ed.), University of Brunei Darussalam, pp. 105-124.
33. Heil A. 1998. Air pollution caused by large scale forest fires in Indonesia 1997. Paper presented at South East Asian Land/Forest Fires: Science and Policy Workshop, Centre for Remote Imaging, Sensing and Processing (CRISP), 30 April -2 May 1998, Singapore.
34. Hefflin B., B. Jalaludin, E. McClure, N. Cobb, C. Johnson, L. Jecha, and R. Etzel, 1994. Surveillance for dust storms and respiratory diseases in Washington State, 1991. *Archives of Environmental Health* **49**: 170-174.
35. Heumann M., L. Foster, L. Johnson, and L. Kelly, 1991. Wood smoke air pollution and changes in pulmonary function among elementary school children. Paper presented at the 84th Annual Meeting of the Air and Waste Management Association, Vancouver, BC, Air and Wate Managment Association.
36. Hinds W. and G. Kraske, 1987. Performance of dust respirators with facial seal leaks: I. *Experimental American Industrial Hygiene Association Journal* **48** (10): 836-841.
37. Hisham-Hashim .J, Z. Hashim, J. Jalaludin, S. Lubis, and R. Hashim, 1998. Respiratory function of elementary school children exposed to the 1997 Kuala Lumpur haze. *Epidemiology* **9** (4) Supplement: S103.

38. Honicky R., C. Akpom, and J. Osborne, 1983. Infant respiratory illness and indoor air pollution from a woodburning stove. *Pediatrics* **71** (1): 126-128.
39. Honicky R., J. Osborne, and C. Akpom, 1985. Symptom of respiratory illness in young children and the use of wood-burning stoves for indoor heating. *Pediatrics* **75** (3): 587-593.
40. Hueglin C., C. Gaegauf, S. Kunzel, and H. Burscher, 1997. Characterization of wood combustion particles: Morphology, mobility, and photoelectric activity. *Environmental Science and Technology* **31**: 3439-3447.
41. Johnson K., R. Gideon, and D. Loftsgaarden, 1990. Montana air pollution study: Children's health effects. *Journal of Official Statistics* **5** (4): 391-408.
42. Jones J. 1991. The physiological cost of wearing a disposable respirator. *American Industrial Hygiene Association Journal* **52** (6): 219-225.
43. Koenig J., T. Larson, Q. Hanley, V. Rebolledo, K. Dumler, H. Checkoway, S. Wang, D. Lin, and W. Pierson, 1993. Pulmonary function changes in children associated with fine particulate matter. *Environmental Research* **63** (1): 26-38.
44. Larson T. and J. Koenig, 1994. Wood smoke: Emissions and noncancer respiratory effects. *Annual Review of Public Health* **15**: 133-156.
45. Leech J., R. Burnett, S. Cakmak, M. Arif, and G. Chang, 1998. The Sarawak September haze episode. *American Journal of Respiratory and Critical Care Medicine* **157** (3): A260.
46. Letts D., A. Fidler, S. Deitchman, and C. Reh, 1991. Health Hazard Evaluation Report No. HETA-91-152-2140, U.S. Department of the Interior, National Park Service, Southern California, Hazard Evaluations and Technical Assistance Branch, NIOSH, U.S. Department of Health and Human Services, Cincinnati, Ohio.
47. Levine J.S., W.R. Cofer III, D.R. Cahoon Jr, and E.L. Winstead, 1995. Biomass burning: A driver for global change. *Environmental Science and Technology* **29** (3): 120-125.
48. Levine J., T. Bobbe, N. Ray, R.G. Witt, and A. Singh, 1999. Wildland Fires and the Environment: A Global Synthesis. UNEP/DEIAEW/TR.99-1. United Nations Environment Program, Nairobi, Kenya.
49. Lewis C., R. Baumgardner, and R. Steven, 1988. Contribution of woodsmoke and motor vehicle emissions to ambient aerosol mutagenicity. *Environmental Science and Technology* **22** (8): 968-971.
50. Lipsett M., S. Hurley and B. Ostro, 1997. Air pollution and emergency room visits for asthma in Santa Clara County, California. *Environmental Health Perspectives* **105** (2): 216-222.
51. Liu D., I. Tager, J. Balmes, and R. Harrison, 1992. The effect of smoke inhalation on lung function and airway responsiveness in wildland fire fighters. *American Review of Respiratory Disease* **146** (6): 1469-1473.
52. Long W., R.B. Tate, M. Neuman, J. Manfreda, A.B. Becker, and N.R. Anthonisen, 1998. Respiratory symptoms in a susceptible population due to burning of agricultural residue. *Chest* **113** (2): 351-7.
53. Lubis S.M. 1996. Indonesia: Country Report. In Proceedings of the ASEAN Network on Environmental Monitoring (ASNEM) on the Third ASEAN Workshop

on Air Quality Monitoring and Analysis with Emphasis on Polycyclic Aromatic Hydrocarbons, Environmental Research and Training Center (ERTC), Pathumthani, Thailand, 26 February-6 March 1996, pp. 153-176.

54. Materna B., R.J. Jeffery, P. M. Sutton, N. Rothman and R.J. Harrison, 1992. Occupational exposures in California wildland fire fighting. *American Industrial Hygiene Association Journal* **53** (1): 69-76.
55. Mumford J. 1990. Human lung cancer risks due to complex organic mixtures of combustion emissions. *Recent Results in Cancer Research* **120**,106-121.
56. Mumford J.L., X.Z. He, RS Chapman, S. R. Cao, D.B. Harris, X.M. Li, Y.L. Xian, W.Z. Jiang, C.W. Xu, and J.C. Chuang, 1987. Lung cancer and indoor air pollution in Xuan Wei, China. *Science* **235** (4785): 217-220.
57. Nichol J. 1997. Bioclimatic impacts of the 1994 smoke haze event in Southeast Asia. *Atmospheric Environment* **31** : 1209-1219.
58. Nichol J. 1998. Smoke haze in SE Asia: A predictable recurrence. *Atmospheric Environment* **32**: 2715-2716.
59. Offermann F., R. Sextro, W. Fisk, D. Grimsrud, W. Nazaroff, A. Nero, K. Revzan, and J. Yater, 1985. Control of respirable particles in indoor air with portable air cleaners. *Atmospheric Environment* **19** (11): 1761-1771.
60. Orlic I., B. Welan, F. Watt, S.M. Tang, 1997. Air pollution in Singapore: Its multielemental aspect as measured by nuclear analytical techniques *Environmental Monitoring and Assessment* **44**: 455-470.
61. Ostro B., M. Lipsett, J. Mann, M. Wiener, and J. Selner, 1994. Indoor air pollution and asthma: Results from a panel study. *American Journal of Respiratory and Critical Care Medicine* **149**: 1400-6.
62. Perez-Padilla R, J. Regalado, S. Vedal, P. Pare, R. Chapela, R. Sansores, and M. Selman, 1996. Exposure to biomass smoke and chronic airway disease in Mexican women. *American Journal of Respiratory and Critical Care Medicine* **154**(3) Part 1: 701-706.
63. Pierson W., J. Koenig, and E.J. Bardana, 1989. Potential adverse health effects of wood smoke. *Western Journal of Medicine* **151** (3): 339-342.
64. Pinto J., L.D. Grant, and T.A. Hartlage, 1998. Report on US EPA Air Monitoring of Haze from S.E. Asia Biomass Fires. EPA/600/R-98/071. USEPA, National Center for Environmental Assessment, Research Triangle Park, NC.
65. Pintos J., E.L. Franco, L.P. Kowalski, B.V. Oliveira, and M.P. Curado, 1998. Use of wood stoves and risk of cancers of the upper aero-digestive tract: A case-control study. *International Journal of Epidemiology* **27** (6): 936-40.
66. Pope C.A., 3rd. 1991. Respiratory hospital admissions associated with PM10 pollution in Utah, Salt Lake, and Cache Valleys. *Archives of Environmental Health* **46** (2): 90-7.
67. Pope C.A., 3rd., D.V. Bates, and M.E. Raizenne, 1995. Health effects of particulate air pollution: Time for reassessment? *Environmental Health Perspectives* **103** (5): 472-80.
68. Pope C.A., 3rd., M.J. Thun, M.M. Namboodiri, D.W. Dockery, J.S. Evans, F.E. Speizer, and C.W. Heath Jr., 1995. Particulate air pollution as a predictor of

mortality in a prospective study of U.S. adults. *American Journal of Respiratory and Critical Care Medicine* **151** (3 Pt 1): 669-74.

69. Pope C.A., 3rd. 1989. Respiratory disease associated with community air pollution and a steel mill, Utah Valley. *American Journal of Public Health* **79** (5): 623-8.
70. Pope C.A., J. Schwartz, and M.R. Ransom, 1992. Daily mortality and PM10 pollution in Utah Valley. *Archives of Environmental Health* **47** (3): 211-7.
71. Qian Y., K. Willeke, S. Grinshpun, J. Donnelly, and C. Coffey, 1998. Performance of N95 respirators: Filtration efficiency for airborne microbial and inert particles. *American Industrial Hygiene Association Journal* **59** (2): 128-132.
72. Radojevic M. 1997. The haze in Southeast Asia. *Environmental Scientist* **6** (6): 1-3
73. Radojevic M. 1998. Burning issues. *Chemistry in Britain* **34** (12): 38-42
74. Radojevic M. and H. Hassan, 1999. Air quality in Brunei Darussalam during the 1998 haze episode. *Atmospheric Environment* **33** (22): 3651-3658.
75. Rashid M. and R.F. Griffiths, 1995. Trends of atmospheric fine and coarse particulates in Kuala Lampur, Malaysia. *Environmental Technology* **16**: 25-34.
76. Reh C. and S. Deitchman, 1992. Health Hazard Evaluation Report No. HETA-88-320-2176, U.S. Department of the Interior, National Park Service, Yellowstone National Park, Wyoming, Hazard Evaluations and Technical Assistance Branch, NIOSH, U.S. Department of Health and Human Services, Cincinnati, Ohio.
77. Reinhardt T. and R. Ottmar, 1997. Smoke Exposure Among Wildland Firefighters: A Review and Discussion of Current Literature. PNW-GTR-373. U.S. Department of Agriculture, Forest Service. Pacific Northwest Research Station.
78. Rindam, M. 1995. The haze phenomena in Malaysia: Problems and challenges. In: Proceedings of the International Symposium on Climate and Life in the Asia Pacific (K.U. Sirinanda, Ed.), University of Brunei Darussalam, pp. 74-84.
79. Robin L. 1996. Wood-burning stoves and lower respiratory illness in Navajo children. *Pediatric Infectious Disease Journal* **15** (10): 859-865.
80. Sandberg D. and R. Martin, 1975. Particle sizes in slash fire smoke. Portland, OR, U.S. Department of Agriculture, Forest Service, Pacific Northwest Research Station. Research Paper PNW-199.
81. Schwartz J. 1991. Particulate air pollution and daily mortality in Detroit. *Environmental Research* **56** (2): 204-13.
82. Schwartz J. 1993. Air pollution and daily mortality in Birmingham, Alabama. *American Journal of Epidemiology* **137** (10): 1136-47.
83. Schwartz J. 1994a. Air pollution and daily mortality: a review and meta analysis. *Environmental Research* **64** (1): 36-52.
84. Schwartz J. 1994b. Air pollution and hospital admissions for the elderly in Birmingham, Alabama. *American Journal of Epidemiology* **139** (6): 589-98.
85. Schwartz J. 1994c. PM10, ozone, and hospital admissions for the elderly in Minneapolis-St. Paul, Minnesota. *Archives of Environmental Health* **49** (5): 366-74.
86. Schwartz J. 1994d. What are people dying of on high air pollution days? *Environmental Research* **64** (1): 26-35.
87. Schwartz J. 1995. Short term fluctuations in air pollution and hospital admissions of the elderly for respiratory disease. *Thorax* **50** (5): 531-8.

88. Schwartz J. 1996. Air pollution and hospital admissions for respiratory disease. *Epidemiology* **7** (1): 20-8.
89. Schwartz J. and D.W. Dockery, 1992a. Increased mortality in Philadelphia associated with daily air pollution concentrations. *American Reviews of Respiratory Disease* **145** (3): 600-4.
90. Schwartz J. and D.W. Dockery, 1992b. Particulate air pollution and daily mortality in Steubenville, Ohio [published erratum appears in American Journal of Epidemiology 1995, January 1, **141**(1):87]. *American Journal of Epidemiology* **135** (1): 12-9; discussion: 20-5.
91. Schwartz J., D. Slater, T.V. Larson, W.E. Pierson, and J.Q. Koenig, 1993. Particulate air pollution and hospital emergency room visits for asthma in Seattle. *American Reviews of Respiratory Disease* **147** (4): 826-31.
92. Schwela D., J. Goldammer, L. Morawska, and O. Simpson, 1999. Health Guidelines for Episodic Vegetation Fire Events. Institute of Environmental Epidemiology, Ministry of the Environment, Singapore, World Health Organization.
93. Smith K. 1993. Fuel combustion, air pollution exposure and health: The situation in developing countries. *Annual Review of Energy and Environment* **18**: 529-566.
94. Smith K. and Y. Liu, 1993. *Epidemiology of lung cancer. In: Indoor Air Pollution in Developing Countries*. (J. Samet Ed.) Marcel Dekker, New York, pp. 151-184.
95. Smith M., B. Jalaludin, J. Byles, L. Lim, and S. Leeder, 1996. Asthma presentations to emergency departments in western Sydney during the January 1994 Bushfires. International *Journal of Epidemiology* **25** (6): 1227-1236.
96. Spix C., J. Heinrich, D. Dockery, J. Schwartz, G. Volksch, K. Schwinkowski, C. Collen, and H.E. Wichmann, 1993. Air pollution and daily mortality in Erfurt, east Germany, 1980-1989. *Environmental Health Perspectives* **101** (6): 518-26.
97. Suh H., J. Spengler, and P. Koutrakis, 1992. Personal exposures to acid aerosols and ammonia. *Environmental Science and Technology* **26**: 2507-2517.
98. Tabucanon M.S. 1996. Cooperation and Technological Development under ASEAN Network on Environmental Monitoring (ASNEM). In Proceedings of the ASEAN Network on Environmental Monitoring (ASNEM) on the Third ASEAN Workshop on Air Quality Monitoring and Analysis with Emphasis on Polycyclic Aromatic Hydrocarbons, Environmental Research and Training Center (ERTC), Pathumthani, Thailand, 26 February-6 March 1996, pp. 20-35.
99. Tiramongkol P. 1996. Thailand: Country Report. In Proceedings of the ASEAN Network on Environmental Monitoring (ASNEM) on the Third ASEAN Workshop on Air Quality Monitoring and Analysis with Emphasis on Polycyclic Aromatic Hydrocarbons, Environmental Research and Training Center (ERTC), Pathumthani, Thailand, 26 February-6 March 1996, pp. 224-244.
100. Tridech S. 1999. Air pollution control strategies in Thailand. Poster presented at the International Conference cum Workshop on Air Quality Management, University of Brunei Darussalam, November 15-19, 1999.
101. Tsutsumi Y., Y. Sawa, Y. Makino, J.B. Jensen, J.L. Gras, B.F. Ryan, S. Diharto, and H. Harjanto, 1999. Aircraft measurements of ozone, NOx, CO, and aerosol

concentrations in biomass burning smoke over Indonesia and Australia in October 1997: Depleted ozone layer at low altitude over Indonesia. *Geophysical Research Letters* **26** (5): 595-598.

102. Tuomi T. 1985. Face seal leakage of half masks and surgical masks. *American Industrial Hygiene Association Journal* **46** (6): 308-312.
103. Tussin M. 1995. Thick haze in Malaysia in the 90's. In: Proceedings of the International Symposium on Climate and Life in the Asia Pacific (K.U. Sirinanda, Ed.), University of Brunei Darussalam, pp. 56-73.
104. Tuthill R. 1984. Woodstoves, formaldehyde, and respiratory disease. *American Journal of Epidemiology* **120** (6): 952-955.
105. USEPA. 1990. Residential Air Cleaning Devices: A Summary of Available Information. EPA 400/1-90/002. United States Environmental Protection Agency, Indoor Air Division. Washington, DC,
106. Vedal S. 1993. Health Effects of Wood Smoke. Report to the Provincial Health Officer of British Columbia. The University of British Columbia, Vancouver, BC.
107. Vedal S. 1997. Ambient particles and health: Lines that divide. *Journal of the Air and Waste Management Association* **47** (5): 551-81.
108. Wafula E.M., F.E. Onyango, H. Thairu, J.S. Boleij, F. Hoek, P. Ruigewaard, S. Kagwanja, H. De Koning, A. Pio and E. Kimani, 1990. Indoor air pollution in a Kenyan village. *East African Medical Journal* **67** (1): 24-32.
109. Wake D. and R. Brown, 1988. Measurements of the filtration efficiency of nuisance dust respirators against respirable and non-respirable aerosols. *Annals of Occupational Hygiene* **32** (3): 295-315.
110. Wallace L. 1996. Indoor particles: a review. *Journal of the Air and Waste Management Association* **46** (2): 98-126.
111. Wang X, H. Ding, L. Ryan, and X. Xu, 1997. Association between air pollution and low birth weight: a community- based study. *Environmental Health Perspectives* **105** (5): 514-20.
112. Ward DE. 1990. Factors influencing the emissions of gases and particulate matter from biomass burning. In *Fire in the Tropical Biota: Ecosystem Processes and Global Challenges* (J.G. Goldhammer Ed.) Springer, Berlin, pp. 418-436.
113. Wilson R. and J. Spengler, 1996. *Particles in Our Air: Concenrations and Health Effects*. Harvard University Press, Cambridge, MA.
114. Woodruff T.J., J. Grillo, and K.C. Schoendorf, 1997.The relationship between selected causes of postneonatal infant mortality and particulate air pollution in the United States. *Environmental Health Perspectives* **105** (6): 608-12.
115. World Health Organization, Regional Office for Europe. 1995. Update and revision of the air quality guidelines for Europe. Copenhagen, World Health Organization, Regional Office for Europe: 1-29 EUR/ICP/EHAZ 94 05/PB01.
116. World Health Organization, Regional Offices for Southeast Asia and the Western Pacific. 1998. Biregional Workshop on Health Impacts of Haze-Related Air Pollution. (WP)EHE(O)/ICP/EHH/004-E. Report Series No. RS/98/GE/17(MAA). Kuala Lampur, Malaysia.

Websites:

Malaysian Meteorological Service. Environment Meteorological Services http://www.kjc.gov.my/people/environ/environ.htm

United States Environmental Protection Agency. Air Quality Criteria for Particulate Matter Chapter 5. http://www.epa.gov/nceawww1/pmvol1.htm

United States Environmental Protection Agency. Measuring Air Quality The Pollutant Standards Index. http://www.epa.gov/airprogm/oar/oaqps/psihold.html

United States Environmental Protection Agency. National Air Quality and Emissions Trends Report, 1997. http://www.epa.gov/oar/aqtrnd97

United States Environmental Protection Agency. National Ambient Air Quality Standards (NAAQS) http://www.epa.gov/airprogm/oar/oaqps/greenbk/criteria.html

World Health Organization. Sustainable Development and Healthy Environments Protection of the Human Environment. Air Quality management. http://www.who.int/peh/air/airguides2.htm

World Health Organization. Sustainable Development and Healthy Environments Protection of the Human Environment. WHO Guidelines for air quality. http://www.who.int/peh/air/airindex.htm

Chapter 11

THE INDONESIAN FIRES AND HAZE OF 1997: THE ECONOMIC TOLL*

David Glover
Economy and Environment Program for Southeast Asia,
Tanglin P.O. Box 101, Singapore 912404

INTRODUCTION

In the middle of 1997, forest fires burning in Indonesia began to affect neighbouring countries, spreading thick clouds of smoke and haze to Malaysia and Singapore. Seasonal rains in early December brought a brief respite but soon after, the dry conditions and fires returned. By 1998, Brunei and to a lesser extent, Thailand, Vietnam and Philippines had also felt the haze. By the time the 1997-98 fire episode was finally over, some 8 million hectares of land would have burned (*Straits Times*, Nov. 23, 1998) while countless millions of people suffered the effects of air pollution. The fires - deliberately set for the most part, and exacerbated by the drought conditions of El Niño - were one of the century's worst environmental disasters.

This chapter tries to assess the extent of the damage caused by the fires and haze and to express these in terms that are readily understandable: dollars and cents. Drawing on research carried out while the fires of 1998 were still raging, it looks at the damages suffered by Indonesia, Malaysia and Singapore during the first outbreak of fires, in 1997.

---

* This chapter presents findings from a study by the Economy and Environment Program for Southeast Asia (EEPSEA) and the Worldwide Fund for Nature-Indonesia (WWF). Principal researchers were Priscilla Hon, Mohd Shawadid, Jamal Othman, Jack Ruitenbeek, David James, David Glover, Timothy Jessup and Martin Hardiono. The study covers the period August-December,1997. We are grateful to the Centre for Remote Imaging and Sensing (CRISP) at the National University of Singapore for providing the maps used in estimating the extent of fire damage in Indonesia. This chapter is a summary of D. Glover and T. Jessup, eds., *Indonesia's Fires and Haze: the Cost of Catastrophe.* Singapore: Institute for SE Asian Studies, 1999.

From October 1997 to May 1998, the Economy and Environment Program for SE Asia (EEPSEA) and the World Wide Fund for Nature, Indonesia Program (WWF) engaged in a rapid assessment and valuation of the damages and publicized the findings widely in the world media and in discussions with government and non-governmental organizations. The purpose of the project was to show - in an objective but dramatic fashion - the extent of the damages and in so doing to encourage prompt action to prevent further fire outbreaks. The study's findings were widely quoted and played an important role in policy discussion in the region.

For the project's voice to be heard before further fire outbreaks occurred, the study had to be truncated. It assessed only the damages for the 1997 episode - not those that occurred in 1998. The total damages for the entire El Niño-related event were considerably higher than those reported here. To provide a rough idea, the Centre for Remote Imaging and Sensing estimates that 5-6 million ha. of land burned in 1997, and an additional 3 million in 1998 (*Straits Times* 23 November 1998); if damages varied in the same proportion, then the total estimable damages for the entire El Niño period may have been some 50-60% higher than those reported here.

Furthermore, the study did not attempt to value every possible damage. In some cases, data were unavailable; in others, there is not widespread agreement even among economists on what estimation methods could be devised. This is particularly true for damages like loss of life or biodiversity. Many of the costs of the fires and haze were therefore omitted in this project; the reader is reminded that the estimates presented are lower bounds; in reality, the damages must have been higher.

In spite of this, the damage estimates are significant, amounting to some US $4.5 billion. It is not only the total size of the damages that is impressive but their variety. Those assessed in this study included (from fire): damage to timber, agriculture, a wide range of direct and indirect forest benefits, fire-fighting costs, and release of carbon, affecting climate change; and (from haze): short-term health costs, tourism losses and some losses of production. The damages assessed, and their values, are defined and described below. First, however, we explain the rationale for the study.

WHY VALUATION?

The valuation of environmental damages can be useful in several ways. First, it can allow more complete and more accurate cost-benefit analysis (CBA) of projects or policy measures. At one time, projects were typically subject to analysis of only those costs and benefits to which the market attaches monetary value. Environmental damages were neglected because environmental goods and services are generally not bought or sold in markets and thus have no obvious price. Such neglect led - and often still leads - to bad decisions.

More recently, it has become common to add an environmental impact assessment (EIA); the EIA can assess the benefits or damages left out of the CBA. But the decision maker is then left with two non-comparable assessments: a CBA in monetary terms and an EIA in physical terms (e.g. tons of pollution emitted or hectares of trees lost). If the CBA and the EIA come to opposite conclusions, the decision maker is left without guidance as to how to proceed. By estimating monetary values for the environmental impacts, the results of the EIA can be integrated into an "extended CBA" that provides an unambiguous conclusion.

Second, valuation can tell us the relative importance of an environmental improvement or degradation and how the impacts are distributed across the affected population. This is impractical if environmental impacts are measured only in physical terms. In the case of Indonesia's forest fires, for example, knowing that 5 million hectares of vegetation were burned by fire and 70 million people affected by haze cannot tell us which effect was more important. That judgement is possible only when the two effects have been put into the same unit of measurement (e.g. dollars).

Third, valuation can draw attention to environmental problems and make their importance more palpable. Even if monetary values are not used in formal cost-benefit analysis, they can make it easier for policy makers and the general public to appreciate the scale of a problem, since they can more readily be compared to other damages or alternative uses of resources.

In this study, the second and third uses of valuation predominate. Our aims were mainly to draw attention to the magnitude of the disaster; put it in perspective relative to other disasters and other uses of resources; identify which components of the damage were largest; and see which countries suffered most.

RESEARCH METHODS

A) General Considerations

All values were:

(a) calculated first in local currency;

(b) in present value terms (i.e. many losses occur one time only; others recur. Income streams or environmental services that could provide recurring benefits are converted to a one-time only [present value] equivalent);

(c) in net terms (i.e. damages are net benefits foregone; net benefit = gross value of the foregone good or service minus the cost of producing or extracting it; this is equivalent to value added; or profit minus normal rate of return to capital; or economic rent);

(d) an attempt to approximate consumer surplus foregone, rather than actual expenditures (e.g. some people were able to obtain medical treatment or evacuate an affected area. Other people were similarly affected but were unable. Actual expenditures for treatment are therefore extrapolated to the entire affected population).

In cases where it was not feasible to conduct new surveys, the benefit transfer (BT) approach was used. This involves the transfer of values from existing studies to the new study site, with appropriate adjustments for the size of the affected area, income levels and other factors.

Various BT values and other adjustment factors are mentioned below. These are derived from various sources, including World Bank and the ADB Workbook on *Economic Evaluation of Environmental Impacts* (1996). Wherever possible, BT values were "reality-checked" against local conditions.

Valuation is not appropriate or adequate for depicting the significance of some damages. Some of these issues are:

• Magnitude of damage relative to ability to bear its cost: wealthy people can sustain larger losses than very poor people, so dollar figures are not necessarily a good measure of suffering.

• Valuing loss of life is difficult and controversial. In this study, we assume that such losses are significant but incalculable.

B) Estimation Methods for Haze Damages

This section outlines a common methodology prescribed for the three country studies. Methods were adapted to local conditions and data availability in each country during application; the adjustments are described in detail in the country reports.

The period covered was August 1 - October 31, 1997. In principle, the study should compare the situation with and without haze. In practice this involved a comparison of August-October'97 to a "normal" August-October. The "normal" values used here were either: (a) Aug-Oct'96; (b) average of Aug-Oct over the past 5 years; or (c) projected trend of Aug-Oct over past 5 years, depending on what was most appropriate in a given case.

Care was taken to separate the effects of the haze from those of the drought and the ASEAN financial crisis.

1. Short-term Health Costs: "Adjusted Cost of Illness" Approach

Three steps to obtain an adjusted COI are outlined below:

a) Estimate treatment cost

(i) estimate hospital and clinic admissions for haze-related ailments per 10,000 population for Aug-Oct' 97. Use "haze-related ailments" as defined by each country's health service. If there is no such definition, use the Malaysian definition: upper respiratory ailments, asthma, bronchitis and conjunctivitis;

(ii) estimate same for Aug-Oct'96 or average of Aug-Oct. over previous 5 years;

(iii) subtract (ii) from (i) to get "excess" admissions;

(iv) adjust for affected but untreated population. The ratio of untreated to treated case varies from country to country but is in the range of 3 or 4 to 1. The ratio for each country can be found in standard health sector studies by WB or ADB.

(v) adjust for treatment costs beyond hospital visits (mainly medicines). As per (iv), there is a standard adjustment factor that varies by country.

(vi) "shadow price" i.e. add the value of any government subsidies for treatment. Alternatively, use the price of a visit to a private clinic.

(vii) if necessary, extrapolate to area outside that where hospital data were collected (use visits per 10,000 ratio in (i)).

(viii) if possible, get cross-section data on affected and unaffected areas as a check on time series in (i).

(ix) get adult/child breakdown on hospital data. This was not used in valuing treatment costs, but in estimating lost workdays below.

These steps were modified for individual countries, depending on data constraints. For Singapore, they were followed largely as outlined. In Malaysia, data on hospital/clinic admissions were matched with pollution levels to produce a dose-response function. This was extrapolated to areas of Malaysia where data on admissions were unreliable. The dose-response function was also transferred to Indonesia, where a map of cumulative haze intensity was overlaid on a population map to estimate the number of people exposed to haze pollution of various levels.

b) Estimate workdays lost

Use hospital/clinic visits by adults (men and women) as a proxy for workdays lost. Adjust visits to workdays lost by a factor suggested by local doctors.

If feasible, adjust for any double counting, if people frequently go first to clinic and then to a hospital on the same day for the same sickness episode. This would not necessarily affect treatment cost, but would affect workdays lost.

Multiply each workday lost by the average or minimum daily wage, depending on which is most suitable in a given country. Say which one you used. Do this for all adults, male and female. If employees continue to receive wages while home sick, workdays lost are considered a loss to employer.

c) Adjust COI for discomfort (to approximate WTP)

Add (a) treatment cost + (b) workdays lost to get (c) cost of illness (COI).

Cost of illness has been found to seriously underestimate "total" damage from an illness, as measured by an individual's willingness to pay (WTP) to avoid it (this is because in spite of treatment and sick leave, the individual still suffers discomfort). The ratio of WTP to COI varies with the ailment. Some ranges of values can be found in ADB Workbook (1996:188). For asthma, it is about 2:1.

This "Adjusted C0I", for lack of a better term, is the value to be used for short- term health damages. These are the short-term health damages only. Long-term, cumulative damages are not valued.

2. Haze-related Production Losses:

These could include rural and urban activities such as reduced crop yields resulting from reduced sunlight. In practice, the only losses measurable were: a) foregone profits in Malaysia from fishing due to reduced visibility (fishing days foregone multiplied by expected profit per day); b) reduced industrial & commercial activity due to the 10 day state of emergency in Kuching (% of GNP foregone).

3. Tourism Losses:

Estimate reduced tourist arrivals from non-ASEAN sources (to control for effect of '97 ASEAN economic crisis): compare Aug-Oct'97 to "normal Aug-Oct".

4. Airline and Airport Losses:

To obtain the losses incurred from airport closures due to poor visibility, one would need data on cancelled flights, expressed in mileage lost, multiplied by the airline's average profit per mile. To this should be added any profits foregone from operation of the airports themselves.

C) Estimation Methods for Fire Damages

The estimation methodology consists essentially of multiplying the areas burned in August-December, 1997 and multiplying those by per hectare values for various vegetation types and land uses. The per hectare values are taken from existing data on Indonesia and, failing that, from comparable ecosystems elsewhere with appropriate adjustments. Economic damages are in net terms (i.e. profit foregone, not total revenue foregone). Discounting of future costs was done at a rate of 10%.

1. Area Burned: Estimates are based on a total area burned of 5 million ha.,distributed as follows: 20% forest, 50% agriculture/plantation, 30% other (unproductive). These figures are derived primarily from satellite mapping studies of Sumatra and Kalimantan by the National University of Singapore's Centre for Remote Imaging, Sensing and Processing (CRISP), with adjustments by EEPSEA and WWF for areas burned outside those provinces.

2. Timber: Timber values take into account estimates of timber stock by the government of Indonesia, as well as growth estimates of forests and net international prices. A net price of $\$50/m^3$ was used. This was cross-checked with an alternative estimation method based on land values and found to yield consistent results.

3. Agriculture: Two key variables that enter into this calculation are the agricultural land value and the production lost in terms of years of output. Differences in productivity between plantations and smallholdings were factored in. We have assumed that, after burning, full agricultural productivity would be re-established in 3 years, with partial productivity being re-established in years 1 and 2 after the burns. This is consistent with average productive cycles of mixed crops (combination of annuals and perennials and tree crops).

4. Direct Forest Services: A Benefit Transfer approach was used, drawing on average world values of tropical rainforest ecosystems, applying them only to the forest area in the sample (i.e. 1 million ha.). The principal source was Costanza *et al* (1997). Values for culture, timber and climate control/regulation and genetic resources were removed to avoid double counting with independent estimates described elsewhere. This yielded a net value lost of $530/ha./yr. It was assumed that non-timber forest products would be re-established over a period of 5 years.

5. Indirect Forest Services: A similar procedure to that described for Direct Forest Services was applied and yielded a net value lost of $1481/ha./yr. It was further assumed that the losses applied only to the area “effectively burnt” of forest which, consistent with the 'combustion factor' in CRISP estimates, was 50% of actual forested area. It was assumed that indirect forest services would be re-established over 2 years.

6. Biodiversity Losses: The approach used here is to value "capturable biodioversity" from Indonesia's perspective. It is not the full value of international value of biodiversity. The figure takes a value of $300/km^2/yr. as an average of values found from various studies of willingness to pay to preserve tropical rainforest of various qualities.

7. Fire-fighting Costs: This includes all documented costs for fire-fighting beyond "normal year" expenses. It includes contributions of personnel and cash from within and outside Indonesia.

8. Carbon Release: Carbon dioxide and methane emission estimates in the CRISP study were increased by the ratio of total area burned (5 million ha.) to area assessed by CRISP (4.56 million ha.). Such emissions increase global warming, which in turn is assumed to cause economic damage. Previous studies for the Intergovernmental Panel on Climate Change have put a value of up to US$30 on the damage caused by a ton of carbon emitted; figures up to this amount are commonly used in international negotiations. In this study, a conservative figure of $10/ton was used.

FINDINGS

Estimates for both fire and haze damage are shown in Tables 1 and 2. Damages excluded: long-term health damages, reduced crop productivity, aesthetic value of reduced visibility, avertive expenditures, accidents, loss of life, evacuations, loss of confidence by foreign investors.

Table 1 Fire and haze-related damages from the 1997 Indonesian forest fires (in USD millions)

Fire-related Damages			
Type of Loss	Lost to Indonesia	Lost to Other Countries	**Total**
Timber	493.7	-	493.7
Agriculture	470.4	-	470.4
Direct Forest Benefits	705.0	-	705.0
Indirect Forest Benefits	1077.1	-	1077.1
Capturable Biodiversity	30.0	-	30.0
Fire Fighting Costs	11.7	13.4	25.1
Carbon Release	-	272.1	272.1
Total Fire	**2787.9**	**285.5**	**3073.4**

Haze-related Damages (summary)			
Type of Loss	Lost to Indonesia	Lost to Other Countries	Total
Short term health	924.0	16.8	940.8
Tourism	70.4	185.8	256.2
Other	17.6	181.5	199.1
Total Haze	**1012.0**	**384.1**	**1396.1**

Total Fire & Haze	**3799.9 (85%)**	**669.6 (15%)**	**4469.5**

Notes to Table 1

1. Where Asian currency values were used, they were converted to dollars at the July 1997 exchange rates of USD 1 = Rp 2500 / RM 2.5 / SGD 1.4.
2. Fires and haze affected 5 million hectares in Indonesia and 70 million people throughout the region.
3. Agriculture losses include those to plantations and smallholdings. They do not include possible haze damage via reduced photosynthesis, pollination, and so on.
4. Direct forest benefits include non-timber forest products such as food, local medicines, raw materials and recreation.
5. Indirect forest benefits include storm protection, water supply and regulation, erosion control, soil formation, nutrient cycling and waste treatment.
6. "Capturable biodiversity" refers to the potential income lost to Indonesia from international conservation expenditures i.e. the amount that international agencies and NGOs have shown they are willing to pay to conserve tropical forests. It does not reflect the intrinsic value of species whose extinction has been hastened; the potential value of ecotourism or internationally marketed pharmaceuticals; the human cultural diversity of indigenous forest-based cultures; or other benefits too difficult to value. These losses are shared by Indonesia and the rest of the world.
7. Carbon release: Release of carbon from fires will contribute to climate change, which will in turn result in economic damage. This figure reflects the amount of damage that the 1997 release is expected to cause.
8. Haze-related damages not estimated here include long-term health damages, reduced crop productivity, aesthetic value of reduced visibility, avertive expenditures, accidents, loss of life, evacuations, loss of confidence by foreign investors.

Table 2 Haze-related damages from the 1997 forest fires (detailed) in millions.

Damage	Indonesia		Malaysia		Singapore		**Total**
	Rp	USD	RM	USD	SGD	USD	USD
Short-Term Health Damages	2,310,000	924	20.1	8.0	12.5	8.8	940.8
Industrial Production Losses	U	U	393.5	157.4	N	N	157.4
Tourism	176,000	70.4	318.5	127.4	81.8	58.4	256.2
Airline & Airport Losses	44,000	17.6	0.5	0.2	9.7	6.9	24.7
Fishing Decline	U	U	40.6	16.2	N	N	16.2
Cloud Seeding	U	U	2.1	0.8	N	N	0.8
TOTAL	2,530,000	1,012	794.3	310.0	104.0	74.1	**1,396.1**

Notes:
N = negligible or not applicable
U = unknown: data unavailable
small discrepancies in totals reflect rounding
** July 1997 Exchange Rates : USD 1 = Rp 2500 / RM 2.5 / SGD 1.4*

DISCUSSION OF FINDINGS

1. *The total damages from fires and haze (nearly $4.5 billion):*

• exceed the damages assessed for purposes of legal liability in the Exxon Valdez and Bhopal disasters combined;
• exceed the amount of funds needed to provide all of Indonesia's 120 million rural poor with basic sanitation, water and sewerage services;
• are more than double the total foreign aid received by Indonesia annually;
• are equivalent to about 2.5% of Indonesia's GNP;
• were suffered largely by Indonesia itself.

2. *For haze damages alone ($1.4 billion):*

• The resources lost to Malaysia as a result of the haze could have financed all of the federal government's social programs for the last three years.
• Malaysia's expenditures on cloud seeding alone would have been enough to establish 320 ha. nature park and maintain it for 15 years.

• Singapore's tourism losses alone could have fully funded the country's Community Chest, comprising 50 charities and benefiting 180,000 people, for three years.

• Indonesia could have used its lost resources to provide basic sanitation, water and sewage services for 40 million people (about 1/3 of the country's rural poor).

• The total losses to the 3 countries from haze could have financed the provision of such services to 56 million Indonesians.

3. Tourism accounts for a large share of losses. These losses are in foreign currency and are acutely felt during the current financial crisis.

4. Losses beyond these may be incurred in the future. For example:

(a) People may suffer long-term health damage, including increased risk of cancer.

(b) Tourists and foreign investors may begin to associate the region with pollution, resulting in future losses of hard currency.

REFERENCES

1. Asian Development Bank, 1996, *Economic Evaluation of Environmental Impacts: Workbook,* Manila: Asian Development Bank.
2. Constanza, R., R. d'Arge, R. de Groot, S. Farber, M. Grasso, B. Hannon, K. Limburg, S. Naeem, R.V. O'Neill, J. Paruelo, R.G. Raskin, P. Sutton and M.van den Belt, 1997,"The Value of the World's Ecosystem Services and Natural Capital", *Nature* **387**: 253.
3. *Straits Times,* 23 November 1998, "Fires raze 3 million hectares of forests in Borneo".

Chapter 12

POLICY IMPLICATIONS AND GOVERNMENT RESPONSES TO THE FIRES AND HAZE OF 1997 AND 1998

Peter Eaton

Department of Public Policy and Administration, University of Brunei Darussalam, Tungku Link BE 1410, Brunei Darussalam

INTRODUCTION

Fire and haze has been a matter of public concern in Southeast Asia for many years. In particular, the great forest fires of 1982 and 1983 caused widespread damage in East Kalimantan and Sabah, and further fires badly affected Indonesia in 1986, 1991 and 1994, resulting in haze which spread to Malaysia and Singapore. In each case governments responded by promising greater regulation of land clearing, more effective fire control measures, and greater international cooperation through the Association of South East Asian Nations (ASEAN). However, these policies proved to be inadequate to deal with the fires which broke out when conditions of drought associated with the El Niño Southern Oscillation (ENSO) occurred again in 1997 and 1998. The most widespread fires during this period occurred in the Indonesian provinces of South Sumatra, Riau, Jambi and Lampung and Central, South and West Kalimantan in 1997, and in East Kalimantan, Sabah, Brunei and northern Sarawak in 1998, but there were also serious outbreaks in Irian Jaya, Sulawesi, Thailand and the Phillipines island of Palawan. Smoke from the fires in Sumatra and Borneo caused widespread air pollution over much of the southern part of Southeast Asia on a scale that prompted not only immediate emergency relief and remedial action but also required long-term programmes providing more permanent solutions.

In spite of the gravity of the situation, governments in the region often seemed slow to respond and to release information on the extent of the air pollution. In Malaysia, the government forbade university lecturers from making public statements on the haze for fear of deterring tourism and later ministers were to criticise the press for continuing to raise the issue (*The Sun*, 3 June 1998). It was in fact the media and non-government organisations (NGOs) that played a significant role in publicising the extent of the problem, creating public awareness and in stimulating government action. A series of tragic accidents due to poor visibility caused by the haze, such as the Garuda plane crash near Medan, the collision of ships in the Straits of Malacca, and the sinking of river boats in Sumatra and Kalimantan emphasised the gravity and urgency of the situation, as did reports of the increase of haze-related illnesses and deaths.

In the formulation of policy, air quality monitoring and the pollution index played a significant role in stimulating different stages of government action and recommendations to the public on necessary precautions. Countries which already had comprehensive pollution monitoring systems, for example Singapore, were better informed at an earlier stage than those such as Brunei where the haze led to the acquisition and establishment of such a system. In addition, in 1997 and 1998, high resolution remote sensing made possible much more exact location of fires enabling the identification of estates and plantations responsible for starting them. However, in spite of the availability of monitoring systems, experience of earlier El Niño effects on the region, and forecasts of drought, very few precautionary measures were taken in advance to prevent or mitigate the fires and the consequent pollution. When governments were forced to take action they responded with a mixture of emergency regulations and programmes to be formalised and presented in national haze action plans.

Programmes to prevent fire involved widespread mobilisation of resources, and the adoption of a variety of technologies. Different strategies to induce rainfall ranged from calls to prayer to cloud seeding. Existing legislation and sanctions were strengthened and additional regulations were introduced to prevent open burning. New government agencies were created to coordinate fire prevention and haze mitigation efforts. States of emergency were declared; schools, airports and places of employment were closed. Masks were issued for protection, with priority being given to schoolchildren. There were proposals for mass evacuation, especially from the most affected urban centres; these were of doubtful practicability and ultimately did not prove necessary, although individuals moved temporarily from areas of dense haze.

The haze of 1997 and 1998 extended to southern parts of Thailand, Vietnam and the Philippines, and forest fires also occurred in six provinces of the Philippines and parts of Thailand, but the countries most affected by the haze itself were Indonesia, Malaysia, Brunei and Singapore. This chapter is especially concerned with examining the policy responses, plans, programmes and their implementation in these four countries. Subsequent sections are concerned with the National Haze Action Plans, fire regulation, fire-fighting, air quality, administration, international dimensions, and land policies.

NATIONAL HAZE ACTION PLANS

National Haze Action Plans were prepared by all ASEAN governments following meetings of their Environment Ministers in 1997 and the formulation of a Regional Haze Action Plan. These plans provide statements and guidelines on policy, outline the aims and objectives, describe programmes and strategies, define the necessary institutional changes and strengthening, and assess the resources needed for implementation.

The individual plans all have common stated objectives in that they aim to prevent and mitigate land and forest fires and to minimise haze pollution, but vary in their emphasis on different strategies and measures needed. Not surprisingly, the plans for states such as Indonesia and Singapore provide something of a contrast. The Indonesian plan shows particular concern with land clearing and conversion, and an awareness of the need for stricter enforcement, innovation and change; the Singapore plan reflects that of an urban society where forest fire risk is minimal, and air pollution is closely monitored and regulated. However, certain concerns and strategies are common to most of the plans. These are the control and suppression of land and forest fires; closer control and regulation of open burning, vehicle and industrial emissions; ambient air monitoring and development of early warning systems; inter-agency cooperation and support; education and public awareness. In addition, plans provide for intensified enforcement and surveillance, new legislation and stricter sanctions. Some plans also outline operational procedures and provide indications of resource needs.

REGULATION OF BURNING

Because many of the fires had been started by their deliberate use to clear land and burn waste, prohibitions on such clearing practices in dry periods were applied to prevent further fires. Similar bans had been imposed by the Indonesian government after the 1994 fires, for example Decree no. 38 of 1995 by the Director General of Plantations had prohibited land clearing by burning, but these restrictions had not proved very effective (*Straits Times*, 3 June 1995). The new legislation, the Basic Environmental Law, no. 23 of 1997, provided more precise definitions of responsibilities and stipulated stricter sanctions with maximum periods of imprisonment of ten years and fines of up to 500 million rupiahs (US$684,930). Agencies responsible for enforcement included the Environmental Impact Management Agency (BAPEDAL), the police, forestry officials, and national and provincial prosecutors' offices.

In 1997 high resolution remote sensing made it easier to identify the guilty. Shifting cultivators had often been blamed as convenient scapegoats in the past, but a series of dramatic images produced by the National University of Singapore Centre for Remote Imaging, Sensing and Processing (CRISP) which were published in the media (*Straits Times*, 1 October 1997) and made available on the Internet showed major fires associated with plantation and estates in Sumatra and Kalimantan. The Singapore Environment

Ministry was already sending satellite images to BAPEDAL to assist in the location of fires.

The Indonesian Minister for the Environment, Sarwono, admitted that large firms accounted for 80 per cent of forest fires (*Straits Times*, 10 September 1997) and President Suharto proclaimed an indefinite ban on land clearing by fire. This was followed by the publication of a list of fifty companies operating in Riau, West and East Kalimantan provinces that were suspected of using fire to clear land after the ban. Later this list was increased to 176 companies. These were ordered to disprove the allegations or face revocation of their permits. In the case of timber firms, 144 licences were revoked, although later most of these were restored after the companies agreed not to clear land and waste wood through burning.

The 1997 environmental legislation also contained provisions for class actions, enabling NGOs to file class actions and bring civil suits against forest concession holders responsible for burning. In 1998 the Indonesian Forum for Environment (WALHI), a coalition of environmental groups, took action against eleven firms in South Sumatra using evidence from satellite images; two were found guilty and ordered to pay court expenses, replant trees and set up fire-fighting squads but they were not ordered to pay the large fines WALHI had demanded. Another successful civil suit was brought by thirteen youth groups in North Sumatra against the Association of Indonesian Forest Entrepeneurs (APHI), an organisation headed by Bob Hasan, a close associate of President Suharto and for a short period his Minister of Trade and Industy, for irresponsible burning. By mid-1998 the Indonesian government had cancelled the licences of 154 logging, forestry and plantation companies for open burning. Otherwise, government enforcement was limited and other legal sanctions rarely applied, especially when the issue lost some of its urgency with the end of the drought and dispersal of the haze, and the Indonesian government found itself facing financial and political crises.

In addition to the Indonesian prohibitions on the use of fire for clearing land, the four countries most affected by the haze found it necessary to regulate other emissions in order to prevent further deterioration of air quality. In particular, laws on the burning of waste were enforced more strictly and penalties made more severe. In Malaysia, regulations under the Environmental Quality Act were used to prevent burning, and in 1998 these were amended to make landowners and tenants responsible for fires on their land, not the employees or contractors working on it; maximum fines were also increased to RM 500,000 (US$131,579). Although no open burning was to be allowed during haze periods when the Air Pollutant Index (API) was above 100, at other times exceptions were permitted when burning was used to control diseases and pests, for research and training in combating fires, for burning paddy stalks and straw, removing sugar cane leaves prior to harvesting, and shifting cultivation subject to a maximum area of two hectares per household.

Enforcement of the Malaysian regulations became stricter and a large number of cases of open burning and factories emitting excessive smoke were reported during 1997 and 1998, with oil palm factories, industrial estates and plantations being major culprits. Aerial surveillance by police helicopters was also used and this use of “the eye in the sky" at times proved an effective means of detecting illegal burning. Sanctions varied from warnings and fines to short periods of imprisonment.

Brunei does not yet have any specific environmental quality or air pollution legislation, but the Penal Code was amended to include an Emergency Order making open burning during prescribed periods punishable by "unlimited" fines and imprisonment for up five years. The Minor Offences Act was also invoked; this made any burning "to the annoyance, inconvenience, or danger to the public" an offence.

In Singapore land and forest fires had not occurred during 1997 and 1998 and were not perceived as a major risk. However, National Parks Board officers were responsible for surveillance in reserves and parks, and the Civil Defence Force has responsibility for dealing with fires in these areas. No new legislation or sanctions were necessary to prevent air pollution as Singapore already had strictly enforced regulations under the Clean Air Act prohibiting open burning and controlling industrial emissions.

FIRE-FIGHTING

When fires had broken out and been located the authorities were faced by the challenge of mobilising resources to contain and extinguish the fires and determining the appropriate methods and technologies for performing these tasks. Traditional methods of combating forest fires tend to be labour intensive, and in all the countries affected it was necessary to reinforce the local fire services with personnel from other branches of the public service, in particular the army and forest departments. In Indonesia, 50,000 members of the armed forces were mobilised in 1997 to meet manpower shortages in fighting the fires in Sumatra and Kalimantan and 1,420 firemen were also sent from Malaysia. Estate and plantation workers were also involved, but participation by local rural communities was often limited to the protection of their own gardens and homes. More devolution of responsibility to these communities and traditional village leaders, and providing them with basic training and equipment could have been very effective in detecting and preventing the spread of fires at an early stage. In Malaysia, volunteer fire-fighting units tended to be restricted to urban areas, but there were attempts to increase the numbers in rural areas. The role of NGOs had been mainly research, providing information and creating environmental awareness, but they became increasingly involved; for example, WWF Malaysia was active in fire-fighting in the Kinabatangan area of Sabah.

A variety of methodologies were applied to combat the fires. Conventional methods of using pumps and hosing fires with water were used where possible, but were often hindered by the inaccessibility of the sites and the lack of availability of water sources. To combat the latter problem temporary water reservoirs were sometimes created and water diverted into them. In Malaysia large containers of water were transported to remote fire-affected areas by Royal Malaysian Airforce helicopters. Where water was available, flooding was also attempted to extinguish ground fires, especially those affecting peat. Spraying with foam was also used to suppress ground fires. Fire-breaks were also dug, bulldozed or cut hurriedly in an attempt to contain the spread of forest burning. Where possible, water-filled ditches also acted as breaks and sources of water, and these were effective in preventing the spread of peat fires.

A major problem of the fire-fighting services was that equipment and techniques used in urban areas were difficult to adapt to the forest and ground fires. For example, protective clothing used in towns often proved too hot and bulky in the forests; and equipment that was dependent on road access was of little value. There was a need for more training, both for the fire services and auxiliaries, and training centres were proposed for Malaysia and Central Kalimantan. Inventories of fire-fighting equipment were drawn up by the different countries, and requests for assistance were made at national and international levels.

Helicopters were used to drop water to extinguish fires in inaccessible areas, although the capacity of the buckets used to carry the water was limited. Helitankers had a much greater capacity but were expensive; those rented by the Brunei government from the American company Erikson cost a minimum of US$ 32,000 per day. Helicopters belonging to the armed forces were generally used for carrying water and personnel, although in 1998 Malaysia purchased two Russian M1-17 helicopters for use by the Fire and Rescue Services. Fixed winged planes were also used for "water bombing" forest fires, and both Australia and the United States provided assistance to Indonesia. In addition to remote sensing by satellite, aerial surveillance by both military and civil planes was also important in giving warning of new fires and Global Positioning Systems (GPS) were used to fix their exact location.

Various attempts at rainmaking were made to suppress fires, disperse the haze and end the drought. Cloud seeding was used to encourage rain with silver iodide and salt being dropped in cumulus clouds. In the period of intense haze, between 8 September and 6 November 1997, Malaysia carried out 252 cloud seeding operations at an estimated cost of US$ 831,600 (Glover and Jessup, 1999). Cloud seeding efforts had a success rate of 41 per cent in Peninsular Malaysia, but were less effective in Borneo and Sumatra. Other hi-tech rainmaking methods proposed but not attempted in Malaysia included a plan for moving clouds with a plastic pipe suspended from a helicopter and another to create artificial cyclones with Russian satellite technology. Traditional methods sometimes seemed more effective. In Brunei, the religious authorities conducted special mass prayers to ask for rain (*Borneo Bulletin*, 25 March 1998) and were later rewarded by scattered showers.

Governments also recognised the need for new operational procedures for the detection and suppression of fires. In Indonesia, separate sub-regional fire-fighting arrangements were developed for Sumatra and Kalimantan. An emergency central command post (POSKO) was established to collect and disseminate information on fires and hot spots to provincial and district governments throughout the country. Actual fire control procedures and management were made the responsibility of fire control teams in the provinces (PUSDALKARHUTLA) and districts (SATLAKS), and the units in the field (SATGAS). Malaysia drew up a contingency action plan indicating how resources should be mobilised following the detection of fires and defining the responsibilities of different agencies: the Departments of the Environment and Forestry, local authorities, the army, police and fire service. In Brunei, coordination was provided by a fire-fighting operations centre at the fire services departmental headquarters, but decision-making was decentralised to eight field operation centres or command posts.

AIR QUALITY

In spite of the vast destruction caused by the fires and the loss of forest and wildlife resources, it was the air pollution that was caused which had the most widespread effects on the people of Southeast Asia, their welfare, health and safety. It was the haze that most affected urban populations and prompted calls on governments for action. The transboundary air pollution also created awareness of the international nature of the problem and the need for cooperation. Unlike some environmental problems, the haze could not be hidden; everybody was aware of it. What the general public were unsure of was the effect it was having on their health and what measures could be taken to mitigate these effects. Here, the governments had to respond, if in some cases rather slowly, with information, education and warning programmes. The pollution index acquired particular significance and readings were published daily by the media, and efforts were made to explain the significance of the readings and precautions that had to be taken (see Chapter 10). In Brunei, for example, Ministry of Health guidelines described Pollution Standard Index (PSI) readings of 101-200 as “unhealthy”, those of 201-300 as “very unhealthy”, and readings of 300-400 as "hazardous", and recommended reduced outdoor activities for the young, elderly and those with heart and lung diseases. If readings reached over 400 all people were advised to stay indoors with doors and windows closed.

The different pollution rankings also provided guidelines in determining policy responses to reduce pollution and protect health. In Malaysia, increasing curbs on private transport and support for public transport were proposed to cut down on vehicle emissions, while industries and power stations were to switch to low-sulphur content fuel. When the API levels reached over 300 school and working hours were to be reduced, if readings continued over 400 for 24 hours a state of emergency was suggested, and if levels over 500 persisted towns were to be evacuated. In fact, API figures reached much higher values than this in Kuching, and although there were proposals for its evacuation, as there had been in Sumatran towns, the scale and logistics of moving the population deterred further action, although a state of emergency was declared in Sarawak from 19th to 28th September 1997. Generally, governments’ responses to extreme haze conditions were to close schools and, as in the Sarawak state of emergency, work-places, to prohibit any open burning, issue health warnings and distribute protective masks. Most of the masks distributed by the health authorities tended to be the surgical and industrial types which provided protection against larger particles, but were not so effective against other pollutants (see Chapter 10).

Concern for the effects of haze on health led to a close monitoring of the incidence of respiratory diseases in hospitals and clinics and to studies of the relationship between haze occurrence and illness. It also created more public awareness of the dangers of air pollution. In Brunei, the crisis rapidly resulted in the establishment of an air quality monitoring network, and data on air quality was made available more regularly to the public, as it was in other countries of the region. In Malaysia and Brunei haze action "hotlines" were established to answer requests for information on pollution levels.

Various attempts were also made to mitigate the effects of air pollution by excluding or dispersing it. Domestically there was a demand for air purifiers and air conditioners designed for that purpose (see Chapter 10). At the municipal level, Kuala Lumpur was particularly concerned with air quality as it was to host the Commonwealth Games in September 1998, and it experimented with giant air purifiers and with outdoor sprinklers from high buildings, the Petronas twin towers and the Kuala Lumpur tower being the buildings selected for this purpose. This application of a water curtain had some success in reducing suspended particulates, and sulphur and carbon dioxide, but was not considered cost-effective, especially at a time of water shortage.

ADMINISTRATION

The implementation of the policies to prevent and extinguish fires and to mitigate air pollution provided an administrative challenge which the bureaucratic structures of the ASEAN countries were not well adapted to meet. The hierarchical structures and closely defined procedures of most government departments made coordination of efforts difficult.

In Indonesia, the environmental ministry has been described as "constitutionally not a policy-implementation agency of government" and one which has to leave the implementation of programmes to other departments and ministries which "often have their own agendas and which may have fundamental reservations about the desirability of some environmental programmes" (Boardman and Shaw, 1995: 102). During the 1997 fires, Sarwono, the Indonesian Environment Minister, complained that his department was "fighting a lone battle" in combating the fires and haze; he spoke of a "late and passive" response from government agencies and attributed the poor coordination between ministries and departments to "the very minimal support" from government and military officials" (*Straits Times*, 24 September 1997).

In Malaysia, the Department of Environment was organised to implement the regulations of the Environmental Quality Act but not to control forest and peat fires. There was uncertainty about responsibility between the Ministries of Primary Industries and of Science, Technology and Environment as to which was supposed to be the lead agency in implementing the national forest fire-fighting plan which had been formulated earlier in 1995.

The different national governments have all attempted to improve cooperation between agencies by creating new coordinating bodies. In Indonesia, the National Coordinating Body for Land and Forest Fire Control (TKNPKHL) was established in 1995 following the fires the previous year. This body was headed by the Minister for Environment and the Director General of Forest Protection and Nature Conservation

(PHPA) and consisted of representatives from concerned ministries and agencies. According to the Indonesian National Haze Action Plan, it was responsible for organising local capability and coordination to anticipate and prepare for fires, formulating regulations and guidelines for land clearing without burning, developing fire control techniques and management, operating a fire detection and early warning system; and communicating and cooperating with international agencies and foreign countries affected by smoke and haze. BAPEDAL, the Environmental Impact Management Agency, acts as the secretariat for this team and plays an important role in acting as an emergency crisis centre, and collecting and disseminating information on the fires. Other crisis centres (POSKO) at national level operated within the Forestry an Internal Affairs ministries, both with responsibilities for fire control and coordination, and at the Metereological and Geophysical Agency and the Space and Aeronautics Agency which received, analysed and distributed information from NOAA satellite images, and the Singapore and Malaysian meteorological services.

In Malaysia, a National Committee on Haze was established to formulate strategies to control and mitigate the impacts of haze and promote scientific studies into haze phenomena. It was chaired by the Minister of Science, Technology and Environment and again contained representatives from relevant government agencies. The National Committee was also supposed to coordinate the work of other committees, and to work with the National Committee on Disaster Relief and Management.

In Brunei, a National Committee on Haze was formed in 1997; it was jointly chaired by the Ministers of Development and Health (the latter was later to be dismissed by the Sultan). Brunei has no environment minister or department, only a relatively small environment unit operating within the Ministry of Development. The National Committee had a coordinating role and contained representatives of relevant government agencies. Most other ASEAN countries had similar committees although in Singapore, Thailand and the Philippines they were known as task forces. The lead agencies tended to be environmental ministries or departments, although in Thailand it was the health ministry.

INTERNATIONAL DIMENSIONS

The spread of transboundary air pollution placed a strain on ASEAN diplomatic relationships and procedures which had traditionally precluded criticism of fellow member states and interference in their internal affairs. There was considerable pressure from the Singapore and Malaysian governments on Indonesia to take more effective action to prevent the fires (*Sunday Times*, 12 October 1997), suggestions of compensation payments from Indonesia to its neighbours, and a public apology from President Suharto to other ASEAN states for the suffering caused by the haze (*Straits Times*, 17 September 1997). In spite of these tensions ASEAN continued to provide a structure for international cooperation and coordination at regional level. There are periodic meetings of its Environment Ministers and also Ministerial Meetings on Haze. At a different level, a

Haze Technical Task Force, mainly consisting of senior officials was also established. Provisions for bilateral cooperation in fighting the fires were contained in a Memorandum of Understanding between Malaysia and Indonesia, and Malaysia sent firemen to assist in Sumatra and Kalimantan.

In June 1995, the Environment Ministers had agreed on an ASEAN Cooperation Plan on Transboundary Pollution which contained broad policies and strategies for cooperation and exchange of information, which largely remained unimplemented. In 1997 it was decided a new Regional Haze Action Plan was necessary, and this was drawn up by the Environment Ministers at their meeting in December 1997. The plan consisted of prevention, monitoring and mitigation components which were assigned to Malaysia, Singapore and Indonesia respectively. It aimed to strengthen national policies and capacities for fire prevention and the haze mitigation, and also to develop the region's early warning and monitoring system, especially with regard to information on the outbreak of fires and predictions on the spread of haze. In this dissemination of information the ASEAN Specialised Meteorological Centre (ASMC) was to play a key role and provide data to individual countries on satellite imagery, winds, rainfall, visibility and air quality. In addition, the Asian Development Bank (ADB) was to provide technical assistance and consultancy services in support of the plan.

During 1997 and 1998 many countries from outside the region provided financial assistance, technical help and equipment. International agencies such as the United Nations Environment Programme (UNEP) also offered aid. Also, there were already several other foreign assisted projects in operation; which had been started in response to Indonesian requests for help following earlier fire and haze episodes. These included the bilateral Indonesian-German 'Integrated Forest Fire Management Project' in East Kalimantan, the European Union 'Forest Fire Prevention and Control Project' in Sumatra, and Japanese International Cooperation Agency (JICA) fire management projects in West Kalimantan and Sumatra. These provided comprehensive data on the causes of the fires and also proposals for their management and control, and in some cases also had a training and awareness content.

Another possibility for more effective international cooperation in dealing with transboundary pollution caused by the haze would be the application of the international Convention on Long-Range Transboundary Air Pollution. This convention was originally designed to address the problem of acid rain; it was signed in 1979 under the auspices of the United Nations Economic Commission for Europe (UN/ECE), and its protocols have only involved European countries, where they have been successful in achieving significant reductions in the emissions of sulphur, nitrogen oxides, volatile organic compounds (VOCs), heavy metals, and persistent organic pollutants (POPs). Under these agreements the parties committed themselves to reducing emissions to certain levels by target dates. The success of the convention has led to suggestions that it should be extended to Asia (Bashkin and Park, 1998), but its effectiveness in preventing transboundary haze in Southeast Asia would ultimately depend on Indonesia's willingness and ability to control the use of fire in clearing land. Compliance with any international agreements is very much dependent on the local capacity to implement them, and this seems significantly lacking. It is a situation where sanctions are difficult to apply. The creation of an equal access legal regime in the region has been suggested (Lebel and Murdiyaso, 1998); this would allow transboundary litigation against polluters for loss of revenue caused by haze, but there are obvious difficulties in apportioning individual corporate guilt, especially at the international level. It is interesting that when fires caused renewed fear of haze early in 2000, the International Association of Crude Palm Oil Buyers threatened to boycott Indonesian palm oil, and it is this type of deterrent that could eventually prove effective in influencing land management practices.

The international significance of the fires and haze must also be considered in relation to the two international agreements signed at the United Nations Conference on Environment and Development at Rio de Janeiro in 1992: the Framework Convention on Climate Change and the Convention on Biological Diversity. Carbon dioxide and methane emissions from the fires have contributed to the problem of global warming on a scale that threatens to negate many of the reductions promised during the subsequent Kyoto Protocol negotiations. In the case of biodiversity, the impacts on wildlife species and the destruction and fragmentation of their habitats, threatens to undermine many of the conservation programmes being established, often with international assistance from NGOs, the World Bank and other aid organisations. In this respect it is ironic that the World Bank, especially through the Global Environmental Facility, is supporting projects aimed at preserving biodiversity, while the International Monetary Fund (IMF) is advocating policies that will lead to greater deforestation. There is clearly also a need for coordination at the international level.

LAND AND RESOURCE POLICIES

We have seen that much of the policy-making associated with the fire and haze was essentially reactive rather than proactive. Its effectiveness is difficult to evaluate because ultimately both issues were dependent on the external climatic factors of wind direction and rainfall. All the states involved failed to prevent the outbreak and spread of fires and in most cases were slow to react to the resulting air pollution. When they did respond it was often with new or stricter legislation, organisational change, greater mobilisation of resources, and better information and monitoring. However, these strategies seem to have had only little success in controlling the Indonesian fires, and during 1999 and 2000, satellite images showed a reoccurrence of hot-spots in Sumatra and Kalimantan.

The failure of policies to prevent the fires in Indonesia appears to be largely one of non-implementation. The policy of zero burning and the legislation to enforce it have never really been put into effect at the local level. Decentralisation has been difficult and slow because many of the programmes have had to be channelled through the offices of the provincial government. The successful execution of policies was dependent upon the coordinated efforts of a number of agencies which often proved uncooperative or inefficient. The issue also lost much of its urgency when the fires died out and the country became preoccupied with other problems such as the financial crisis, unemployment, poverty, and ethnic, religious and separatist conflicts. After the fall from power of Suharto, the enforcement of policies to prevent fires became even more difficult amidst the cries for *reformasi*.

The policies themselves were inadequate in that they did not address some of the wider issues involved, especially those related to property rights and land use. Many government programmes had been concerned with the conversion of forest to other forms of land use, in particular oil palm estates, and timber and pulp plantations. These programmes had been encouraged by external bodies such as the IMF and perceived as means of stimulating foreign investment and increasing exports. It was also significant that relatives and close associates of President Suharto were involved in both the oil palm and timber industries. The fact that Malaysian and Singapore firms were also involved in joint ventures with Indonesian palm oil companies (EPI, 1998: 32) may also have blunted their governments' criticisms of the Indonesian fires.

The clearing of land for estates and plantations generally involved the use of fire which was regarded as the cheapest and speediest method, and in the past had been the traditional tool used by smallholders and shifting cultivators. However it was the scale of the burning and the lack of controls, together with the excessively dry conditions, that resulted in the fires that burnt out of control during 1997 and 1998. A policy of zero burning in land clearance has been advocated by NGOs and then later taken up by the ASEAN committee of environment ministers, but is generally regarded by land developers as an expensive alternative, and its implementation would require either strict supervision or economic incentives to utilise the waste.

The transmigration programme also contributed to the problem. The resettlement of large numbers of people from the densely populated islands of Java, Madura and Bali to the less developed outer islands had resulted in extensive land clearing in both Sumatra and Kalimantan, often in areas not suitable for permanent cultivation (Cleary and Eaton, 1996: 94-100). In 1997 some of the worst fires were associated with the "One Million Hectares" rice growing scheme in Central Kalimantan, a transmigration project involving the drainage of extensive areas of peat swamp. As in the case of other attempts to convert peat swamp to agriculture elsewhere in the region, decreasing water levels and the use of fire in land preparation caused serious ground fires that were difficult to extinguish (FAO, 1999: 275).

Forestry policy and practice also contributed to fire risks. Under the Suharto government large areas had been allocated to state and private companies as logging concessions, and although companies were supposed to follow the Indonesian Selective Cutting System, in practice many smaller trees were damaged and destroyed. The relatively short periods of the concessions also did not favour careful forest management and protection. A European Union Fire Response Group Report (1998: 22) found that "the density of fires within areas designated as logging concessions was significantly higher than outside such areas". A policy of taxing and then banning log exports had been adopted in order to favour forest based industries, resulting in the development of excessive mill capacity and an industry that was inefficient in having the low log conversion rate in Asia (Osgood, 1994). The result of destructive logging and wasteful conversion was an increase in waste wood, often disposed of by burning or providing fuel for wild fires. The low prices of wood going to the processing mill provided little incentive to use or sell scrap wood or protect standing timber. Another cause of fires was clearing secondary forest and scrub for the plantations established to meet the needs of the new forest-based industries.

The policies of forest exploitation, plantations, transmigration and promotion of large estates often resulted in the marginalisation of indigenous communities, especially in Kalimantan. Their rights to land and forest products under customary or *adat* law were frequently ignored, as were thir traditional conservation practices. They were left with few incentives to preserve the forest and often a resentment of the outsiders destroying its resources. Poffenberger (1990: 17) has described the situation as one in which "conflicts over forest land ... result from economic development policies in which indigenous people have little or no role, and from government agencies' inability to understand indigenous land management and tenure practices or to respond to their diversity. Until forest communities and their indigenous use are incorporated into national forestry plans and use systems, conflicts and mismanagement will continue."

Land tenure disputes and uncertainty over land boundaries may be causes of fires and also impediments to their control. An FAO meeting reported: "One of the major factors that increase fire risk ... is the lack of transparent and effective mechanism for assigning land rights that can be accepted by all segments of society. This can lead to open access situations in which no one feels responsible about fire control, or to conflicts that are sometimes settled by intentional burning of crops or forest. Even if tenure is clear, accurate land ownership maps are needed to enforce burning regulations." (FAO, 1999: 24)

CONCLUSION

If fires and haze on the scale of those in 1997 and 1998 are to be prevented in the future there would need to be changes in Indonesian land use and forestry policies with a reduction in the conversion of large areas of forest to estates and plantations. Future planning and decision making need to take into account the long-term environmental values of the tropical rain forest (Eaton, 1996). Indigenous rights should be identified, formalised and protected. It is only when local communities feel responsible for their land and forests that effective fire prevention at village level will occur. With this responsibility there is the added need to allocate more resources for fire-fighting at local level.

The fires and haze provided a lesson in the difficulties of implementing environmental policies in Indonesia and to a lesser extent in some of its neighbouring countries. The administrative systems often proved unsuitable for dealing with the environmental problems they faced. Their rigid hierarchical sectorial structures made cooperation and rapid responses difficult. Government environmental departments or agencies often lacked the power and resources of the more established ministries whose development policies had been responsible for much of the land clearing. In Indonesia, in spite of presidential directives and stricter legislation prohibiting burning there was a lack of enforcement and compliance at local level, and effective decentralisation proved difficult.

The haze itself had not only physically reduced transparency, it had also revealed a lack of transparency in government processes. Eventually one positive result of the disaster was the necessity to make more information available to the public, especially with regard to pollution levels and their effect on health. International cooperation and developments in remote sensing and pollution monitoring also resulted in more data being accessible both to the governments of the region and the community at large. Early warning systems have been improved and there is a recognition of the need for modelling of both fire behaviour and transboundary air pollution. As more relevant scientific knowledge is becoming available, it is important that it should be communicated to the policy makers so that future fire and haze episodes can be countered by prompter, more effectively coordinated and better informed government responses at all levels.

References

1. Bashkin V, and Park S.U., 1998. *Acid Deposition and Ecosystem Sensitivity in East Asia,* Nova Science Publishers, New York.
2. Boardman, R. and Shaw, T.M., 1995. 'Protecting the Environment in *Environmental Policies in the Third World,* Greenwood Press, Westport. pp. 85-107.
3. *The Borneo Bulletin*:
   'Contractor jailed for burning waste during haze', 19 November 1997.
   'United Action to battle the haze', 25 March 1998.
4. Cleary, M. and P. Eaton, 1996. *Tradition and Reform: Land Tenure and Rural Development in South-East Asia*, Oxford University Press, Kuala Lumpur.
5. Eaton, P., 1996. 'Socio-economic values and tropical rain forest conservation' in D.S. Edwards, W.E. Booth and S.C. Choy (eds.), *Tropical Rainforest Research – Current Issues*, Kluwer Academic Publishers, Dordrecht. pp. 417-423.
6. Economy and Environment Program for Southeast Asia (EEPSEA) and Worldwide Fund for Nature (WWF) Indonesia, 1998. "Study of the Economic Value of 1997 Haze Damages in Southeast Asia", EEPSEA, Singapore.
7. Environmental Investigation Agency (EIA), 1998. *The Politics of Extinction,* EIA, London.
8. European Union Fire Response Group, 1998. *Assessment of the 1997 Fires in Indonesia*, Ministry of Forest and Estate Crops, Jakarta.
9. Food and Agricultural Organisation (FAO), 1999. *FAO Meeting on Public Policies Affecting Forest Fires,* FAO, Rome.
10. Glover, D. and T. Jessup (eds.), 1999. *Indonesia's Fires and Haze: the Cost of Catastrophe*, Institute of Southeast Asian Studies, Singapore.
11. Lebel, L. and D. Murdiyarso, 1998. 'Fires in Southeast Asia', International Human Dimensions Program Newsletter, No.3.

12. Osgood, D., 1994. 'Government failure and deforestation in Indonesia' in M. Poffenberger, (ed.) *Keepers of the Forest,* Kumarian Press, West Hartford. pp.217-225.
13. Poffenberger, M., 1990. 'Evolution of Forest Management Systems in Southeast Asia' in M. Poffenberger (ed.), *Keepers of the Forest,* Kumarian Press, West Hartford. pp.7-26.
14. *The Straits Times*:
    'Indonesia to crack down on firms burning forest to clear land', 3 June 1995.
    'Suharto bans land clearing by burning', 10 September 1997.
    'Suharto apologises for the worsening haze conditions', 17 September 1997.
    'Jakarta has done "too little, too late" ', 24 September 1997.
    'What satellite pictures show', 1 October 1997.
15. *The Sun*:
    'Enough about the haze, Sabaruddin tells press', 3 June 1998.
16. *The Sunday Times*:
    'Diary of a disaster: People kept in the haze far too long', 12 October 1997.

Chapter 13

FROM DENIAL TO DEBATE – AND BACK AGAIN! MALAYSIAN PRESS COVERAGE OF THE AIR POLLUTION AND 'HAZE' EPISODES, JULY 1997-JULY 1999

James McLellan

Department of English Language and Applied Linguistics, University of Brunei Darussalam, Tungku Link BE 1410, Brunei Darussalam

INTRODUCTION

This chapter discusses how the topic of the air pollution caused by drought and forest fires in Indonesia, Malaysia and Brunei was treated in the Malaysian print media over a period of two years from 1997 to 1999. It aims to complement the analysis of government reaction and policy in the chapter by Peter Eaton. Like informal language use in multilingual societies, the 'haze' is a phenomenon over which governments have shown themselves unable to exercise any serious measure of control. Yet, also like language, its effects are immediately observable to citizens and residents of the countries in question, and a cause for extreme public concern.

This chapter studies a corpus of texts from the mainstream English- and Malay-language press in Malaysia, which show an initial concern with downplaying the devastating effects of the haze and with maintenance of the country's image as a tourist destination, then an increased desire to address and debate issues of public health and environmental degradation caused by the air pollution. Later, in mid-1999, there is a marked return to the 'denial mode', signalled by the closure of the government-controlled website that had given daily updates of the Air Pollution Index figures for main urban centres in Malaysia.

TERMINOLOGY: THE LEXIS OF THE HAZE EPISODE

The choice of the word 'haze' to describe the noxious atmospheric pollution that periodically affects the Southeast Asian region is euphemistic, since the term normally denotes a naturally occurring climatic condition in which visibility is affected, as in 'heat haze', 'hazy sunshine'.

The same applies to the Malay term '*jerebu*', which has a similar denotational meaning. The Malay term appears to be a recent coining, as it is not found in either the first (1970) or the second (1984) editions of the authoritative "*Kamus Dewan*" dictionary, and only makes its appearance in the third edition (1996).

Other lexical choices such as 'smog' or 'air pollution' are not used with any frequency in the Malaysian press. This reflects a tendency, often found in Malaysian public discourse, of not 'calling a spade a spade'. Negative connotations and implications can be avoided, it appears, if one avoids the use of words and phrases that are deemed 'sensitive' or offensive; (see Buruma, 1992: 108-138 for further discussion of sensitive issues in the context of Malaysia).

A comparable example is that of the Malaysian Culture, Arts and Tourism Minister telling tour agencies to avoid use of the word 'monsoon':

Text 1

"He said that the word "monsoon" should not be used as it gave the impression that there was continuous rain during the season and it would drive away tourists.

That period should instead be described as the rainy season, he told reporters..."
(*Borneo Post*, 24 May 1993)

This tendency reached a high point in April 1998, when the following report appeared in the *New Straits Times*:

Text 2

Stop describing it as haze, TV stations warned

KUALA LUMPUR, Sun. - Information Minister Datuk Seri Mohamed Rahmat today warned private television stations, particularly newly launched ntv7, not to blow the haze problem out of proportion. If they persisted in doing so, he said, he would revoke their operating licences or not renew them.

"The operators of the television stations are not thinking of the country's interests," he said. "They are just thinking of returns on their investments." ntv7, which began transmission on April 7, had been reporting on the poor visibility which it described as haze.

Mohamed denied that there was any haze in Malaysia presently and requested that the local media "focus on the positive, and not the negative". "They can report on the fires and action taken to put them out, but they should not label it as haze."

Mohamed said it was acceptable if television stations showed footages of poor visibility with the explanation that it was caused by fires. When asked about the high Air Pollutant Index readings over the past month, Mohamed attributed them to smoke caused by forest and peat fires, and other pollutants. "As long as all the fires are not put out, it will be hazy," he told a Press conference after launching "The Tough Gets Going Fund" [*sic*] at Palace of the Golden Horses this morning.
On April 10, the Meteorological Services Department had been reported as saying that the recent haze was caused by bush, plantation, fores and peat fires.
(*New Straits Times*, 20 April 1998)

On the day that this statement was made, the Air Pollution Index in Miri, Sarawak, was over 200. Datuk Seri Mohamed Rahmat was replaced as Information Minister in May 1999.

THE DISCOURSE OF THE HAZE

This section attempts to show the preoccupations underlying press statements by Malaysian ministers and others, as exemplified in Text 2 above. The corpus of Malaysian news texts includes frequent reference to:

- maintaining the country's good image,
- not discouraging tourists from visiting,
- not wishing to upset relations with Indonesia in particular and within ASEAN as a whole.

Further examples of these preoccupations are found in texts 3 and 4:

Text 3

All-out effort to prevent haze during Games

Gen. (Rtd.) Tan Sri Hashim Mohamed Ali said today that fires, irrespective of whether they were in Malaysia or Indonesia, would be put out the moment they were detected to prevent haze.
(*New Straits Times*, 24 June 1998)

"Games" in text 3 is a reference to the Commonwealth Games, held in (haze-free) Kuala Lumpur in September 1998.

Text 4

Jerebu tidak jejaskan pelancongan di Sabah

Masalah jerebu yang melanda beberapa kawasan di Sabah tidak menjejaskan industri pelancongan di negeri ini.

[Translation: *Haze is not a threat to tourism in Sabah*

The haze problem which has affected several districts in Sabah does not threaten the tourist industry in this state.]
(*Utusan Malaysia*, 31 March 1998)

Text 4 is extracted from a report of proceedings in the Sabah State Assembly, quoting a reply from the (then) Chief Minister to a question from the Leader of the Opposition.

On the other hand, the English and Malay print media also ran a series of feature articles outlining scientific, climatological and public health aspects of the haze crisis. Examples include "*Bagaimana jerebu terjadi*" (translation: 'How haze occurs', *Utusan Malaysia*, 20 September 1997), "*Manusia perlu alam indah*" (tr.: 'People need a beautiful environment', *Utusan Malaysia*, 8 October 1997), and "Devastating effect of fires on wildlife and ecosystem" (*New Straits Times*, 28 April 1998).

Making specific reference to the haze, the Malaysian Minister of Education was reported publicising a Cabinet directive barring academic researchers "from talking to the media on sensitive issues that could distort the country's good image".

Text 5

" 'If we make statements that the haze here is bad and it has long-term effects on health, citizens and foreign visitors will have doubts about Malaysia.....
The Government is very concerned of [*sic*] the long-term effects on health, but we cannot have statement after statement which can be picked up by the foreign media that could create a very distorted picture that can affect the country' Najib explained.
'We do not want tourists to cancel bookings and visitors to spread stories about Malaysia.' "
(*New Straits Times*, 4 November 1997)

A story appearing on another page of the same issue of the *New Straits Times* reports a speech made by the Group Chief Editor, A. Kadir Jasin, in which he points out the irony of the Minister of Education's position, as noted above, in the light of the efforts to create a knowledge-based society and make the country into a centre of academic excellence:

Text 6

Attempts to gag scientists and researchers from talking to the mass media on important issues like the haze and coxsackie are regrettable as it negates efforts to create a knowledge-based society.
The irony is that those supposed to transform the country into a regional centre for academic excellence are the ones stifling academic freedom with their directive, New Straits Times group chief editor Datuk A. Kadir Jasin said today....
(*New Straits Times*, 4 November 1997)

This serves to show that even in Malaysian news organs that are considered to reflect and promote the government's point of view, there is no lack of open debate, and no reluctance to take issue with government ministers. In mitigation it can be noted that this particular political commentator at this point in time had more latitude for expressing views contrary to those of the government than any other Malaysian journalist.

An analytical report, located in the features rather than the news section of the *New Straits Times* (Life and Times, *New Straits Times*, 10 August 1998), addresses the issue of different treatment of the forest fires in Southeast Asia and those occurring in Florida, USA, late in 1998 by the US and Malaysian media. Under the headline "Western press bias smoked out" the writer, Pang Hin Yue, concludes that western reporting of the Southeast Asian fires and haze was "negative", since it ran stories of foreign expatriates fleeing and tourists being advised against visiting the region. However, US reporting of the Florida fires did not contain such messages. The feature writer's aim in this text is to offer a critical discourse analysis, in the manner of Fairclough (1995), exposing the underlying bias in media reporting. The present writer has a similar aim in contributing this chapter.

Another seminal text for the analysis of media coverage of the haze crisis appeared in the Singapore *Sunday Times* (12 October 1997, Review section). Strictly speaking this falls outside the scope of this chapter on Malaysian media coverage, but it is included here since it deals with that most sensitive of sensitive issues, inter-ASEAN relations. This feature article gives prominence to Singapore's role in haze and forest fire monitoring through the use of satellite photography and remote-sensing equipment, and dissemination of information through websites (such as that of the Meteorological Department, http://www.gov.sg/metsin/, a major source of up-to-date information for those resident in the region, especially since the closure of the Malaysian Government website -. The *Sunday Times* feature details the initial reluctance to name and blame Indonesia as the major source of the fires and haze, in line with long-established ASEAN principles of polite diplomacy and non-interference:

Text 7

"The 'Asean way' of holding discreet behind-the-scenes discussions, while keeping the Asean public in the dark, may work in sorting out political, trade or diplomatic kinks.
But it does not work when it comes to a disaster like the haze, a public health hazard everyone will see and smell the minute he [*sic*] wakes up."
(*The Sunday Times*, 12 October 1997, Review section)

Indonesia is in many ways the linchpin of ASEAN, and recent political upheavals there, at a time of regional economic crisis, have called into serious question the harmony and cohesion which the organisation likes to portray as its *modus operandi*. Cotton (1999) discusses the challenges posed by the haze disaster to the ASEAN mode of regional engagement: citing some of the same reports referred to in this chapter, he comments, "These examples illustrate that the Malaysian Government is unwilling to permit free discourse in an untrammeled civic space on a natural problem of this magnitude." (Cotton 1999, p.348).

In Malaysia especially, there is a '*takut kepada Abang*' (tr.: fear of big brother) syndrome in relation to Indonesia, caused by the huge difference in the population of the two countries, by geopolitical considerations such as shipping lanes, and rendered all the more poignant by the perceived shared language and cultural and religious values and by issues such as the mass migration of workers, legally and otherwise, from Indonesia to Malaysia. A letter published in the *New Straits Times* under the headline "Act on the haze now before it's too late", asks if Malaysians are "too timid to face our neighbour" (*New Straits Times*, 2 March 1998).

What is perhaps most surprising to the outside observer here is the proportional prominence given to questions of the country's image and the fear of discouraging tourist arrivals, as against public health and safety issues in the Malaysian news media reports.

MARGINALISING THE EAST

Kuching, capital city of the East Malaysian State of Sarawak, experienced the most severe pollution and recorded API figures in excess of 800 between 19 and 27 September 1997, with a state of emergency being declared for Sarawak State on Friday 19 September. Even the impractical notion of evacuation of Kuching City was raised in politicians' speeches at this time. This emergency was given due prominence in both West and East Malaysian-based newspapers during this period.

During the second haze episode, however, from February to May 1998, the East Malaysian states of Sabah and Sarawak, on the island of Borneo, were marginalised in the reports emanating from the seat of the Malaysian Federal Government in Kuala Lumpur. Although this period also saw serious fires in parts of Kalimantan (Indonesian Borneo), especially East Kalimantan, a significant difference was that this time the haze in East Malaysia was caused by local burning, so the blame could not be laid solely at the door of the Indonesian authorities (*New Straits Times*, 10 April 1998).

Throughout this period the website giving daily API readings (http://www.jas.sains.my/doe/api.html) ran a hypertext footer proclaiming "Clear air and blue skies in Malaysia" which showed haze-free photographs of the Kuala Lumpur skyline. Whilst these images may have given a partially accurate account of the conditions in West Malaysia, the Air Pollution Index (API) figures for Miri in northeastern Sarawak reached a height of 633, well into the range defined as 'hazardous', at 4.00 p.m. on 30th March 1998 (*Borneo Post*, 1 April 1998), when all schools in the Miri Division were ordered to be closed (*New Straits Times*, 31 March 1998). The API reached 615 on 8 April 1998 (*Borneo Bulletin*, 9 April 1998). This critical situation was seriously under-reported in the West-Malaysian-based mainstream print media.

Tables 1 and 2 summarise analysis of news coverage in the English-language *Borneo Post*, printed in Kuching and Miri, Sarawak and in Kota Kinabalu, Sabah, compared with the Malay-language *Berita Harian*, published in Kuala Lumpur, during the week of 1-8

April 1998. In the *Borneo Post* the page 1 lead is usually from Kuching or Kuala Lumpur, whilst page 28, the back page, is devoted to reports emanating from Miri. The *Berita Harian*'s page 1 lead is usually the main item of national news; a separate section of one or two pages is devoted to reports from Sabah and Sarawak.. These two tables show clearly the under-reporting of the haze situation, especially in the West-Malaysia-based *Berita Harian* at this time in the north-western part of Sarawak State.

Table 1 *Borneo Post*

Date	Highest API [Miri]	p.1 Lead Story	p. 28 Lead Story
1/4/98	563	--- (= *not haze-related*)	Open burning crackdown
2/4/98	234	Fire-fighting funds for Miri	Over 20 water tanks for Ba'kelalan
3/4/98	330	------	Schools may reopen
4/4/98	482	------	Gov't doing its best
5/4/98	529	------	Pujut fire: 5 missing
6/4/98	500+	Cash for fire victims	---------
7/4/98	500+	------	Pumps/tanks for Limbang/Lawas
8/4/98	615	------	---------

Table 2 *Berita Harian* (headlines translated from Malay)

Date	Highest API [Miri]	p.1 lead story	Sabah/Sarawak page lead story
1/4/98	563	--- (= *not haze-related*)	---------
2/4/98	234	------	S'wak tells Miri factories to reduce operations
3/4/98	330	------	---------
4/4/98	482	------	---------
5/4/98	529	------	---------
6/4/98	500+	------	Haze: Marine Dept. tells fishermen not to put to sea
7/4/98	500+	------	School classes in Miri postponed again
8/4/98	615	------	---------

BACK TO DENIAL?

In May 1999 the Malaysian Meteorological Services Department, acting on a Cabinet directive, stopped disseminating daily updates of the API figures through their website and the print media, continuing to give out only visibility figures for selected Malaysian towns. This decision was not reported in the mainstream media at all, until the following story appeared in mid-June 1999:

Text 8

Ministry to issue statements only if air quality unhealthy

PETALING JAYA, Tues. - The Science, Technology and Environment Ministry will only issue statements to the media if the air pollutant index is classified as unhealthy. Deputy Science, Technology and Environment Minister Datuk Abu Bakar Daud said this was to prevent certain parties from sensationalising the haze issue.

He said the Ministry felt that adverse publicity about the 1997 haze incident in the country had dented the nation's image, especially in the West. "Due to this, we decided not to publish the API as we do not want irresponsible parties to take advantage of the situation and exaggerate the condition of the haze in the country."

He denied allegations reported by a foreign wire service that there had been a gag order authorised by the Ministry on the publication of the air quality readings.

Abu Bakar said this after launching a Made-in-Malaysia silencer for controlling the noise from air-conditioning and ventilation systems in office buildings, hotels, shopping complexes and hospitals.

Abu Bakar said the public could gain access to air quality readings through the Ministry's website (http://www.jas.sains.my/doe/api.html) which used to issue updated API readings once a day.

However, he said the website had not been updated since May 22 because of a computer problem.

(*New Straits Times*, 16 June 1999)

A week later the Science, Technology and Environment Minister finally admitted the closure of the website:

Text 9

Law: Raise API issue at environs council meeting
By Esther Tan

KUALA LUMPUR, Tues. - The Government's decision to stop publicising the daily air pollutant index readings may be raised at the Environmental Quality Council meeting, Science, Technology and Environment Minister Datuk Law Hieng Ding said today the World Wide Fund for Nature Malaysia (WWF) should raise the issue if it was dissatisfied with the Government's explanation.

"WWF is represented in the EQC. They can talk about it there," he said when commenting on statements by WWF about the move to stop issuing daily API readings to the media and the people.

The EQC is an advisory body to the Science, Technology and Environment Minister and is represented [*sic*] by academicians, Government agencies, ministries, scientists, non-governmental organisations and industry.
WWF had said that lack of transparency would also arouse suspicion that the Government had something to hide. Law had earlier explained that the Government was compelled to do this to prevent certain parties from "misinterpreting" the API and sensationalising the haze issue. Instead of providing the API readings, he said the Department of Environment would only provide the qualitative measurement to describe the air quality as "good," "moderate" or "unhealthy."
On the air quality, Law said there was no need to worry about haze this year as the weathermen have forecast a prolonged La Nina impact. "The impact of La Nina is longer than expected. It will last till the end of the year. Therefore, there is no need to worry about haze."
"We can expect slightly more rain than normal in Peninsular Malaysia and Sarawak in the next few months. Sabah will have more than normal rain," he said.
On the slight hazy conditions, he said the air quality was classified as "good" in most areas. The number of hotspots in Indonesia has been reduced drastically…
(*New Straits Times*, 23 June 1999)

In March 1998 Malaysians had been assured that such a step would not be taken: the Malay-language *Utusan Malaysia* carried this report:

Text 10

"JAS nafikan enggan mendedahkan bacaan IPU

Beliau (Pengarah Jabatan Alam Sekitar, Abdul Aziz Rasol) berkata JAS tidak pernah menyembunyikan bacaan IPU seperti didakwa oleh orang ramai…."
(*Utusan Malaysia*, 31 March 1998)

(tr.: *Environment Department denies wanting to conceal API readings.*

He (The Director of The Sarawak Environment Department Abdul Aziz Rasol) said the Department had never suppressed the API readings, as charged by the public…)

Datuk Abu Bakar Daud (see text 8) lost his seat in the Malaysian Parliament in the November 1999 general election. Datuk Law Hing Ding (text 9) retained his.

CONCLUSION

Subsequent to the suppression of the air pollution data on the Malaysian *Jabatan Alam Sekitar*'s website, interested parties are now more than ever dependent for air pollution readings on websites maintained by the Singapore Meteorological Service, reprinted daily in the Singapore *Straits Times* newspaper, and on other websites such as those of BAPEDAL, the Indonesian Environmental Impact Managing Agency (http://www.bapedal.go.id) and the World-wide Fund for Nature Indonesia (http://www.wwf.or.id).

The haze episodes thus take their place in a series of 'natural' calamities which also includes the 'coxsackie'/ enterovirus outbreak, responsible for the deaths of some 19 children in Sarawak in 1997-8, and the '*nipah*' virus, spread from pigs to humans throughout Malaysia in 1998-9, which caused over 100 deaths. The treatment of all these issues in the official print media reflect the mainstream Malaysian media's concern to play down the threats to public health and safety, and their anger at the foreign media (especially though not exclusively British, American and Australian) which seek to play them up.

This analysis of a limited corpus of news reports from the Malaysian print media, both Malay and English, has attempted to show evidence for a pattern of initial denial of the haze problem and extreme concern for the country's image, especially as a tourist destination, followed by a tendency towards more open debate and address of public health and environmental issues surrounding the air pollution. Subsequently though, there has been a reversion to the denial mode, signalled in particular by the closure of the Malaysian Environment Department's haze monitoring website. This attitude persists: at the time of writing and revising this chapter (March 2000) fires in Sumatra and Kalimantan have started once again, causing more haze and increased concern on the part of Malaysia, Brunei and especially Singapore. In the Singapore *Sunday Times* (12 March 2000), in a box alongside an article datelined Kuala Lumpur and entitled "Haze danger in Straits of Malacca", the following text appears:

Text 11

Please....

Malaysia's Environment Minister yesterday told the local media not to play up the air-quality index for fear that it would keep tourists away. Datuk Law Hing Dieng urged Malaysian not to worry too much as the haze situation was under control.
(*Sunday Times*, 12 March 2000)

Acknowledgement

The writer wishes to acknowledge the kind assistance of Professor Peter Eaton in drawing his attention to several of the press reports quoted in this chapter, and to thank him for his considerable editorial input.

References

1. *Berita Harian* (Kuala Lumpur) 1-8 April 1998.
2. *Borneo Bulletin* (Bandar Seri Begawan, Brunei Darussalam)
   'Air Pollution critical in Miri', 9 April 1998.
3. *Borneo Post* (Kuching, Sarawak, East Malaysia):
   'Use the appropriate words in tourism', 24 May 1993.
4. *Borneo Post* 1-8 April 1998.
5. Buruma, I. (1992) *God's Dust*. London: Virago. Ch. 3: "Sensitive Issues."
6. Cotton, J. (1999) The "haze" over Southeast Asia: challenging the ASEAN mode of regional engagement. *Pacific Affairs Vol. 72 No. 3*, Fall 1999, pp. 331-351.
7. Fairclough, N. (1995) *Critical Discourse Analysis*. Harlow: Longman.
8. *Kamus Dewan* (1st Edition 1970) Kuala Lumpur: Dewan Bahasa dan Pustaka.
9. *Kamus Dewan* (2nd Edition 1984) Kuala Lumpur: Dewan Bahasa dan Pustaka.
10. *Kamus Dewan* (3nd Edition 1996) Kuala Lumpur: Dewan Bahasa dan Pustaka.
11. *New Straits Times* (Kuala Lumpur, West Malaysia)
    'Researchers banned from talking on findings that can distort image', 4 November 1997.
    'Attempts to gag scientists against nurturing knowledge', 4 November 1997.
    'Act on the haze now before it's too late'. Letter to the Editor, 2 March 1998.
    'Schools in Miri closed due to worsening haze', 31 March 1998.
    'Haze back in various parts of country', 10 April 1998.
    'Stop describing it as haze, TV stations warned', 20 April 1998.
    'Devastating effect of fires on wildlife and ecosystem', 28 April 1998.
    'All-out effort to prevent haze during Games', 24 June 1998.
    'Western press bias smoked out', by Pang Hin Yue, 10 August 1998.
    'Ministry to issue statements only if air quality unhealthy', 16 June 1999.
    'Law: raise API issue at environs quality council', 23 June 1999.
12. *The Sunday Times* (Singapore):
    'Dairy of a disaster: people kept in the dark for too long', 12 October 1997.
    'Haze danger in Straits of Malacca', 12 March 2000.
13. *Utusan Malaysia* (Kuala Lumpur, Malaysia):
    'Bagaimana jerebu terjadi', 20 September 1997.
    'Manusia perlu alam indah', 8 October 1997.
    'JAS nafikan enggan mendedahkan bacaan IPU', 31 March 1998.
    'Jerebu tidak jejaskan pelancongan di Sabah', 31 March 1998.

Chapter 14

CONCLUSIONS

Miroslav Radojević[1] and Peter Eaton[2]

[1]Department of Chemistry, University of Brunei Darussalam, Tungku Link BE1410, Brunei Darussalam

[2]Department of Public Policy and Administration, University of Brunei Darussalam, Tungku Link BE 1410, Brunei Darussalam

THE PROBLEM

The fires that broke out in Sumatra and Borneo in 1997 and 1998, and the resulting haze which covered much of Southeast Asia were a major environmental disaster. The effects on health, tropical forest ecosystems and global climate change were all far reaching. In his chapter on the economic costs of the damage, Glover (Chapter 11) estimates the total costs to have been US$4.5 billion, but it should be remembered that these figures refer only to 1997, and would have been much higher if they had included 1998 when northern and eastern Borneo were badly affected.

The fires and haze were not new phenomena and have been reported periodically in Borneo since the nineteenth century (Chapter 3). The fires usually occur during periods of drought related to the El Niño Southern Oscillation (ENSO), as discussed by Harger (Chapter 2), but although these provide the climatic conditions, the cause of nearly all the fires has been human activities. In particular, the widespread use of fire in clearing land for forest plantations, oil palm estates and other agricultural projects was often a major factor. The different causes of the fires are comprehensively analysed in the chapter by members of the European Union Forest Fire Prevention and Control Project (Chapter 4) who discuss the role of fire in land management and forest conversion, and emphasise that the worst fires were not in areas of virgin forest, but in logged-over degraded areas and also where attempts were made to drain and clear wetlands. They conclude that the solution to Indonesia's fires lies in improved local land use planning and management. Similar conclusions are reached in the chapter on government policies (Chapter 12) which also examines the different national action plans and their implementation.

During 1997 and 1998, remote sensing was important in fire detection and haze monitoring (Chapter 5); by precisely locating the source of fires it also enabled the guilty to be identified. Satellite imagery was also used to provide data for the ASEAN Specialised Meteorological Centre. Another interesting development was the use of the Internet to disseminate information; this enabled NGOs to play an active educational and advocatory role. The haze also caused the public to demand more information than was being released by governments and the media (Chapter 13), ultimately resulting in a greater transparency and accountability.

The present book summarises the current state of knowledge regarding the nature, causes and effects of forest fires and associated haze in Southeast Asia. The contributing authors also point out gaps in our knowledge and areas where more research is needed in order to improve our understanding of this important air pollution phenomenon and develop more effective management practices and policies. The following sections indicate areas where more research is needed.

CHEMISTRY OF FOREST FIRES AND HAZE

Until recently, there has been little concerted effort to understand the nature and extent of haze pollution from forest fires in Southeast Asia or its impacts. There is almost no chemical information on the earlier episodes, and more comprehensive studies have started to emerge only during the 1997 and 1998 haze episodes. The most detailed studies of the chemical composition of haze have been carried out in Singapore and Brunei due to the availability of sophisticated monitoring equipment and specialists with expertise in air pollution. Some chemical studies have also been conducted in Malaysia, however, in Indonesia, where most of the forest fire hot-spots are located and where the haze is most severe, there have been virtually no significant studies into the chemistry of fires or haze. This is mainly due to the lack of expertise and the absence of rather expensive instrumentation required for monitoring the haze.

The research carried out so far into the chemistry of forest fires and haze has been reviewed in Chapter 7. Although these preliminary studies have yielded some very useful data and greatly improved our understanding of the haze phenomenon, more investigations need to be carried out in the future. In particular, there is a need to obtain more precise estimates of emission factors for various pollutants. These emission factors could serve as input parameters in computer models used to predict the various environmental impacts of forest fires. Both laboratory and field studies are required.

Future studies should ideally involve researchers from the region (Malaysia, Indonesia, Singapore and Brunei) as well as scientists from developed nations (Europe, US, Australia, Japan). Such a joint effort would contribute to regional capacity building in the field of air pollution and atmospheric chemistry, something that is sadly lacking at the moment. Currently, few institutions in the region have the capability, both in terms of experienced personnel and instrumentation, to undertake such studies.

HEALTH EFFECTS

There is a need to investigate mortality statistics in Southeast Asia in order to examine whether there is any increase in mortality during haze episodes, and if so, what is the relationship between mortality and pollutant concentrations. Several studies in the United States and Canada have concluded that up to 60,000 people die each year due to breathing particles at or below legally allowed limits. Also, the incidence of diseases should be investigated over the entire area affected by haze, and the health of exposed populations should be monitored for many years to come. Previous epidemiological studies of exposure to biomass smoke include: indoor exposure in developing countries using biomass for cooking and heating, wildland firefighters, and indoor and outdoor exposure in North American communities where wood burning is prevalent (Chapter 10). These indicate a relationship between exposure to biomass smoke and respiratory symptoms, increased risk of respiratory illness and decreased lung function and they suggest that asthmatics and children are particularly sensitive to biomass pollution.

In addition to criteria pollutants, many organic micropollutants (e.g. benzene, aldehydes, polynuclear aromatic hydrocarbons) are also present in the haze. Some of these pollutants are known carcinogens, mutagens and teratogens, and these could have long-term effects on the health of exposed populations. Although the concentrations of organic micropollutants in haze have been determined in some studies, there is a lack of ambient exposure standards for many of these substances. There can be considerable uncertainty in current air quality recommendations for organic micropollutants because of the following factors:

(1) In the absence of any epidemiological evidence on the general population, these recommended values have been derived from occupational exposure studies. Occupational exposure studies are typically single pollutant studies and the effects of a number of micropollutants acting together is not considered. Furthermore, occupational studies involve medically fit adults who make up the working population. This group is significantly different from the general population which also includes non-working individuals: pregnant women, children, the elderly, and the ill. These groups are especially susceptible to air pollution and they may not be adequately accounted for in extrapolations from occupational studies to the general population. Epidemiological studies of indoor exposure to biomass smoke in developing countries do not suggest an increased incidence of cancer, however, the number of such studies is limited.

(2) In the case of haze, not only do many pollutants act together, they also occur in association with very high levels of particulates which are respirable.

Impacts on Atmospheric Chemistry and Rainwater Acidity

Forest fires are important sources of several atmospheric pollutants, some of which could affect global climate either through warming (e.g. greenhouse gases) or cooling (e.g. suspended particles) effects. Emissions of suspended particles and several atmospheric gases, including CO, CO_2, CH_4, and NO_x, from the 1997 forest fires in Indonesia exceeded the emissions of these pollutants from the Kuwaiti oil fires of 1991 (Chapter 9). Forest fires could impact on atmospheric chemistry in several ways (Chapters 6, 7, 8, and 9):

- by emitting primary gaseous pollutants (CO, NO, hydrocarbons, etc.) which could react in the atmosphere to produce O_3 and other secondary pollutants;
- by emitting suspended particles which could act as a medium for surface-catalysed reactions, and
- by affecting the intensity of solar radiation, which in turn could affect the photochemistry of the atmosphere.

The relative importance of these potential effects is still unclear and further research is needed. While some studies suggest enhanced tropospheric O_3 as a result of forest fires, other studies have failed to reveal significant increases in ground-level O_3 concentrations, and some studies even indicate a decrease in surface O_3 concentrations during haze episodes. Greater O_3 concentrations at higher altitudes may result from photochemical reactions of primary gaseous pollutants (CO, NO, etc.) diffusing upward out of the top of the smoke plumes. Photochemical activity may be significantly reduced within the smoke plumes due to the reduction in solar flux because of high particle concentrations, and therefore enhanced levels of ground-level O_3 are generally not observed during haze.

Several authors have suggested that forest fires may lead to rainwater acidification due to the emission of acidifying gases such as SO_2, NO_x and organic acids although most of the results obtained so far suggest that acidification of rainwater is not an important consequence of forest fires in Southeast Asia (Chapter 7). This may be due to the neutralisation of acidity by basic components (e.g. Ca) in smoke particles. However, there is a need for more extensive rainwater monitoring in the region affected by haze.

Haze Monitoring

Air quality is determined at a limited number of air quality monitoring stations (AQMSs) throughout South East Asia. So far there are 23 such stations in Malaysia, 15 in Singapore, one in Brunei and new stations are being set up in response to recent haze episodes. These stations monitor continuously the concentrations of criteria pollutants: CO, NO_2, SO_2, O_3 and PM_{10} (particles <10 µm in diameter) using reference or equivalent

US Environmental Protection Agency (EPA) methods. AQMSs tend to be fully automated and they employ instrumental methods. Indonesia does not yet have an adequate air quality monitoring system and the haze is monitored generally by visibility readings. Past air quality data reported for Indonesia consists of SO_2 and NO_2 measurements based on antiquated methods which are no longer used elsewhere. In addition to the fully equipped AQMSs there are a number of PM_{10} only monitors scattered throughout the region; some of these can provide continuous PM_{10} measurements. Indonesia has acquired several continuous PM_{10} monitors in response to the haze episode of 1997.

Unfortunately, most of the AQMSs are located in highly populated urban centres as they were originally set up to monitor air quality degradation due to motor vehicle and industrial emissions. It is quite difficult to assess the impacts of forest fires and regional haze on the basis of air quality data from cities in the region (e.g. Jakarta, Kuala Lumpur, Singapore) because of the considerable interference from industrial and vehicular sources, which can give rise to very high pollution levels even in the absence of haze from forest fires. There are very few monitoring stations in Sumatra and Kalimantan where most of the forest fire hot spots responsible for the haze are located.

Although the number of AQMSs in the region is growing, the monitoring network is still far from adequate. Several improvements can be made to improve the haze monitoring capability in the region. There is a need to increase the number of AQMSs in the region, especially in Indonesia. Some of the future AQMS should be set up at rural locations and in areas close to the forest fires, where the health effects are most severe. AQMSs should also monitor $PM_{2.5}$, in addition to PM_{10} which is currently being determined. Particles <2.5 μm in diameter can penetrate deep into the pulmonary region consisting of bronchioles and alveoli, and they are therefore called "respirable". Since these particles can remain deeper and longer in the lungs than the "inhalable" PM_{10} particles, they could have more detrimental effects on health. Currently, $PM_{2.5}$ is not measured at AQMSs in Southeast Asia. Also, there is a need for more extensive measurements of organic micropollutants (VOCs, PAHs) present in haze. Compounds in the gas and particle phases should be determined both during haze episodes, and in the absence of haze so as to establish background levels.

Since there can be considerable spatial and temporal variation in haze intensity, and since people tend to spend more times indoors and inside motor vehicles than outdoors, especially during haze episodes when health advisories recommend staying indoors, there is a need for more personal exposure studies.

Haze Management

Recent studies of the haze have suggested that the Pollutant Standards Index (PSI) and air quality management practices based on the PSI (or API in Malaysia) are not effective for dealing with haze from forest fires. While these indicators are adequate for dealing with urban air pollution caused by industrial and vehicular emissions, new approaches are required for air quality management during haze episodes. The PSI (or API) may not be a suitable system for reporting air quality during haze episodes due to physicochemical differences between urban air pollution, for which the PSI was developed, and haze from forest fires. Urban air quality is generally degraded by emissions from motor vehicles and industrial sources. Furthermore, the PSI does not extend beyond a value of 500 due to the absence of such highly polluted conditions in the US where PSI standards were originally developed, nor does it account for short-term exposure to extremely high PM_{10} concentrations of 1 mg/m^3 or more, as in the case of haze in Southeast Asia.

In view of the differences between haze from biomass fires and urban air pollution, action plans adopted by countries in SE Asia are largely ineffective in dealing with regional haze. These plans are based on the United States EPA's recommendations and generally refer to urban air pollution and not to haze. While it is reasonable to adopt such plans with regard to urban air quality, this approach is ineffective during periods of forest fires and haze. For example, prescribed responses such as the reduction of industrial and vehicular emissions will do little to improve air quality during haze episodes.

Currently there are no air quality standards for $PM_{2.5}$ (particles < 2.5 μm in diameter) in the region. The United States EPA is proposing the introduction of $PM_{2.5}$ standards of 65 $\mu g/m^3$ and 15 $\mu g/m^3$ for 24 hour and one year averaging periods, respectively. Since haze particles produced by forest fires are almost entirely <2.5 μm in diameter, consideration should be given to adopting a $PM_{2.5}$ air quality standard by Southeast Asian governments.

TROPICAL ECOSYSTEM DIVERSITY

The chapter by Davies (Chapter 8) indicates the need for research on the effects of the fires and increasing pollution on tropical ecosystem diversity and functioning. Continued investigations into the regeneration of forests after fires are needed in order to determine rehabilitation programmes and the avoidance of further environmental degradation. It is still difficult to assess the effect of the fires on wildlife, but surveys indicate a decrease in endangered species such as the orangutan (*pongo pygmaeus*) as well as other forms of life such as small mammals, amphibians and insects with limited tolerance ranges and dispersal abilities. Further monitoring is needed to determine what has been the effect on the rich biodiversity of the region. Studies in the Kutai Park after the 1982-3 fires indicate that the numbers of most affected wildlife species recovered, although in many cases slowly, but this was in a protected environment where the vegetation was allowed to recover; if the fires lead to changes in land use and further forest fragmentation the impact may be far greater.

Another cause for concern is to determine the effect of repeated fires on the stability of forest ecosystems, and to identify the critical threshold at which the changes become irreversible, the forest no longer recovers, and species disappear. Because there are so many gaps in our knowledge of both the impacts of fire and haze, and the tropical rainforest processes and range of species affected, it is still difficult to predict the significance and drastic nature of the changes that may result if further episodes of fire and haze recur in Southeast Asia.

PUBLIC POLICY IMPLICATIONS

Fires and haze have been reoccurring problems for Indonesia and its neighbours for the past twenty years and previous government policies to prevent and mitigate them have generally proved unsuccessful. In particular, implementation of policies and coordination between government agencies has been poor. Greater decentralisation is needed and there should be more allocation of responsibilities for fire control and prevention to local bodies. Chapters 4 and 12 of this book both emphasise the need for action at the local level, and this has also been a feature of the recommendations of aid agencies acting in cooperation with the Indonesian government.

Most of the affected ASEAN countries reacted to the haze problem by introducing new legislation or amending existing statutes and regulations to provide for stricter penalties for open burning. Although these had some effect during the worst haze phases, especially in Malaysia, legal sanctions do not seem to have prevented clearing land by burning in Indonesia. Policies of zero burning have been advocated at ministerial level and by NGOs, but have been regarded by land developers as uneconomic and unrealistic. Enforcement of legislation has been difficult in the past because of difficulties in identifying those who are guilty, although the use of remote sensing has made it possible to define much more accurately where fires have been started. A major challenge is to make legal instruments more effective in the Indonesian judicial and political environment. Possibilities of transboundary litigation and compensation payments have been discussed but at present are unlikely within the ASEAN framework.

The transboundary nature of haze provided a stimulus for attempts at international cooperation which resulted in the Regional Haze Action Plan, the development of an early warning and monitoring system, and technical and manpower assistance in fire-fighting operations. The UN/ECE Convention on Long-range Transboundary Air Pollution with its different protocols, which has been successful in reducing acid rain and other forms of air pollution in Europe, has been suggested as a model for a Southeast Asian regional treaty, but the application of such a treaty would need to take into account the nature of the problem which is not caused by industrial emissions which can be controlled and reduced by abatement technologies. The contribution of the fires to the greenhouse effect and global climatic change is another cause for international concern and action, but again it is difficult to apply the 1992 United Nations Framework Convention on Climate Change and its subsequent protocols to the situation in Indonesia, especially in view of its developing country status under the Convention. Existing international legal regimes seem to be limited in their capacity to provide mechanisms for promoting or enforcing measures to reduce the Southeast Asian fire and haze, and there remains a need to develop flexible and innovative policies that will make any international agreements realistic and effective.

One area for policy change that has been identified in this book is in relation to land use and development. Programmes of converting forest and wetlands to agriculture need to be carefully monitored and subject to environmental and fire risk assessment. Degraded forest and grasslands have been especially prone to fire; their rehabilitation is an important priority and effective land use planning should be implemented to prevent the clearing of marginal land. Further regulation of the forest industries is needed to reduce waste and the use of fire for clearing. In particular land tenure boundaries need to be defined in order to identify responsibilities for fire prevention and control, and customary land and resource use rights should be recognised and reinforced.

The development of environmental awareness throughout society was a significant feature of the fires and haze of 1997 and 1998. Here the mass media played an important role although, as McLellan shows in his chapter on Malaysian press coverage (Chapter 13), this was not always consistent and often subject to government influence. Government agencies were often unwilling to release information on air pollution, but were forced to do so by public concern and the need to take protective action. NGOs were important in providing information and also in encouraging legal action. The inadequacy of government responses often provided a stimulus for civil society activities.

Environment policy development is closely dependent on scientific knowledge. There is a need for dissemination of this knowledge and for it to be communicated to the policy makers. It is hoped that this book has made a contribution to our knowledge of the fire and haze phenomena in Southeast Asia and that it will lead to a closer interplay between science and policy.

LIST OF CONTRIBUTORS

I. Anderson

Natural Resources Institute, University of Greenwich, Central Avenue, Chatham Martime, Kent ME4 4TB, UK

E-mail: ffpcp@mdp.co.id

R. Balasubramanian

Department of Chemical & Environmental Engineering, National University of Singapore

E-mail: cherbala@nus.edu.sg

J. M. Bompard

c/o CIRAD-Forêt, Campus International de Baillarguet, B.P. 5035, 34032 Montpellier cedex 1, France

E-mail: bompard@wanadoo.fr

R. Bowen

Natural Resources Institute, University of Greenwich, Central Avenue, Chatham Martime, Kent ME4 4TB, UK

E-mail: ffpcp@mdp.co.id

M. Brauer

School of Occupational and Environmental Hygiene, The University of British Columbia, 3rd. Floor 2206 East Mall, Vancouver, BC V6T 1Z3, Canada

E-mail: brauer@unixg.ubc.ca

P. Brimblecombe

Department of Environmental Sciences, University of East Anglia, Norwich NR4 7TJ, UK

E-mail: P.Brimblecombe@uea.ac.uk

G. R. Carmichael

Center for Global and Regional Environmental Research, The University of Iowa, Iowa City, IA 52242, USA

E-mail: gcarmich@engineering.uiowa.edu

S. J. Davies

Institute of Biodiversity and Environmental Conservation, Universiti Malaysia, Sarawak, 94300, Kota Samarahan, Sarawak, Malaysia

and

Center for International Development, Harvard University, Cambridge, Massachusetts 02138, USA

E-mail: sjdavies@mailhost.unimas.my

P. Eaton

Department of Public Policy and Administration, University of Brunei Darussalam, Tungku Link BE 1410, Brunei Darussalam

E-mail: ppeaton@fbeps.ubd.edu.bn

D. Glover

Economy and Environment Programme for SE Asia (EEPSEA), Tanglin, PO Box 101, 912404 Singapore

E-mail: dglover@idrc.org.sg

A. Gouyon

Idé-Force, Paris, 8 rue de l'Isly, 75008 Paris, France

E-mail: ideforce@softel.fr

P. Guizol

CIRAD-Forêt attached to CIFOR, P.O. Box 5696, JKPWB, Jakarta 10065, Indonesia

E-mail: p.guizol@cgiar.org

J.R.E. Harger

Highspan International Environmental Consultants, 38 Vanessa Crescent, Glendowie, Auckland, New Zealand

E-mail: robinharger@hotmail.com

L.K. Kwoh

Centre for Remote Imaging, Sensing and Processing (CRISP), Faculty of Science, National University of Singapore, Lower Kent Ridge Road, 119260, Singapore

E-mail: lkkwoh@nus.edu.sg

J. S. Levine

Atmospheric Sciences Division, NASA Langley Research Center, Hampton, Virginia 23681-0001, USA

E-mail: j.s.levine@larc.nasa.gov

S.C. Liew

Centre for Remote Imaging, Sensing and Processing (CRISP), Faculty of Science, National University of Singapore, Lower Kent Ridge Road, 119260, Singapore

E-mail: phyliew@leonis.nus.edu.sg

H. Lim

Centre for Remote Imaging, Sensing and Processing (CRISP), Faculty of Science, National University of Singapore, Lower Kent Ridge Road, 119260, Singapore

E-mail: phylimh@leonis.nus.edu.sg

O.K. Lim

Centre for Remote Imaging, Sensing and Processing (CRISP), Faculty of Science, National University of Singapore, Lower Kent Ridge Road, 119260, Singapore

S.F. Lim

Malaysian Meteorological Service, Jalan Sultan, 46667 Petaling Jaya, Selangor D.E., Malaysia

E-mail: lim999@tm.net.my

J. McLellan

Department of English Language and Applied Linguistics, University of Brunei Darussalam, Tungku Link BE 1410, Brunei Darussalam

E-mail: jamesmcl@ubd.edu.bn

K. Ostermann

Atmospheric Science Program, The University of British Columbia, Room 217-1984 West Mall, Vancouver, BC V6T 1Z2, Canada

E-mail: osterman@interchange.ubc.ca

M.J. Phadnis

Woodrow Wilson School of Public and International Affairs, Princeton University, Princeton, NJ 08540, USA

E-mail: mphadnis@Princeton.edu

L. Potter

Department of Geographical and Environmental Studies, University of Adelaide, Adelaide, South Australia 5005, Australia

E-mail: lpotter@arts.adelaide.edu.au

M. Radojević

Department of Chemistry, University of Brunei Darussalam, Tungku Link BE 1410, Brunei Darussalam

E-mail: miro@fos.ubd.edu.bn

INDEX

D

E

F

G

H

I

S

T

U

V

W